Advanced Technologies in Optical Wireless Communications

Advanced Technologies in Optical Wireless Communications

Guest Editors

Cuiwei He
Chen Chen

MDPI

Basel • Beijing • Wuhan • Barcelona • Belgrade • Novi Sad • Cluj • Manchester

Guest Editors
Cuiwei He
School of Information Science
Japan Advanced Institute of
Science and Technology
(JAIST)
Nomi
Japan

Chen Chen
School of Microelectronics
and Communication
Engineering
Chongqing University
Chongqing
China

Editorial Office
MDPI AG
Grosspeteranlage 5
4052 Basel, Switzerland

This is a reprint of the Special Issue, published open access by the journal *Photonics* (ISSN 2304-6732), freely accessible at: https://www.mdpi.com/journal/photonics/special_issues/6D0IZP4237.

For citation purposes, cite each article independently as indicated on the article page online and as indicated below:

Lastname, A.A.; Lastname, B.B. Article Title. *Journal Name* **Year**, *Volume Number*, Page Range.

ISBN 978-3-7258-4981-9 (Hbk)
ISBN 978-3-7258-4982-6 (PDF)
https://doi.org/10.3390/books978-3-7258-4982-6

Contents

About the Editors

Cuiwei He

Cuiwei He received his Ph.D. degree in the topic of optical wireless communication from Monash University, Melbourne, Australia, in 2017. From 2017 to 2020, he worked as a Research Fellow at Monash University. From 2020 to 2021, he was a Postdoctoral Research Assistant at the Department of Engineering Science, University of Oxford, U.K. He is currently a Senior Lecturer at the School of Information Science, Japan Advanced Institute of Science and Technology (JAIST). His research interests include visible light communications and optical wireless communications.

Chen Chen

Chen Chen (Professor) received B.S. and M.Eng. degrees from the University of Electronic Science and Technology of China, Chengdu, China, in 2010 and 2013, respectively, and a Ph.D. from Nanyang Technological University, Singapore, in 2017. He was a Postdoctoral Researcher at the School of Electrical and Electronic Engineering, Nanyang Technological University, from 2017 to 2019. He is currently a professor at the School of Microelectronics and Communication Engineering, Chongqing University, Chongqing, China. His research interests include optical wireless communication, optical access networks, Internet of Things, and machine learning. He currently serves as Section Board Member for "Optical Communication and Network" of the MDPI journal *Photonics*.

photonics

MDPI

Editorial

Advanced Technologies in Optical Wireless Communications

Cuiwei He [1,*] and Chen Chen [2,*]

1 School of Information Science, Japan Advanced Institute of Science and Technology, Nomi 923-1211, Japan
2 School of Microelectronics and Communication Engineering, Chongqing University, Chongqing 400044, China
* Correspondence: cuiweihe@jaist.ac.jp (C.H.); c.chen@cqu.edu.cn (C.C.)

1. Introduction

Optical wireless communication (OWC) is expected to be a key component of future wireless communication networks, with a wide range of applications such as indoor visible-light communication (VLC) [1,2], visible-light positioning (VLP) [3], underwater OWC [4], and satellite communications [5]. These diverse application scenarios highlight the potential of OWC technologies to complement or even surpass traditional radio-frequency (RF)-based wireless systems in terms of bandwidth availability, security, and energy efficiency. However, they also introduce new technical demands related to channel modeling, optical beam steering, device integration, and system optimization across varying environments. While significant progress has been made in developing related technologies over the past decade, many critical challenges remain unsolved or require further research.

This Special Issue aims to address some of the ongoing challenges in OWC by presenting a collection of 16 peer-reviewed articles, which are listed in the "List of Contributions" section. These contributions come from research institutions and universities across nine countries, including Australia, Belgium, China, Japan, Germany, Greece, the Netherlands, Spain, and Ecuador, reflecting the global collaborative effort undertaken to advance research in this field. This Special Issue concludes by summarizing these articles and categorizing them into five topics: VLC, VLP, space optical communication, modulation and multiplexing techniques, and OWC with photonic integrated circuits (PICs).

2. Visible-Light Communication (VLC)

Indoor VLC is an important form of OWC that uses white-light-emitting diodes (LEDs) for indoor illumination and wireless data transmission simultaneously. The performance of a VLC system depends on various factors [6], including the characteristics of the LED transmitters, the sensitivity of the photodetector (PD) receivers, the optical components integrated into the system, and the signal modulation schemes used. Moreover, in a transmission system with multiple LED transmitters, the spatial configuration and alignment of the transmitters and receivers also play an important role in determining overall system performance.

This Special Issue presents five articles related to VLC. Like other communication systems, the performance of a VLC system is influenced by the bandwidths of both the LED transmitter and the electronic circuits used to drive the LEDs. To enhance the bandwidths of LED transmitters, Contribution 1 investigates how different crystal orientations of the GaN structure impact the light emission lifetime of LEDs and, consequently, their modulation bandwidth. Contribution 2 focuses on improvements to the LED driving circuit, particularly enhancements to boost/buck DC-DC converters.

Another notable challenge in VLC is mitigating interference from adjacent LED transmitters. Contribution 3 explores the use of a liquid lens to adjust the receiver's field of view (FOV), therefore reducing interference from nearby transmitters. In transmission scenarios involving multiple LED transmitters, the layout of these LEDs is crucial and must also satisfy illumination requirements. Contribution 4 investigates the optimization of LED layout while considering practical illumination distribution constraints. Finally, Contribution 5 presents a study on the channel characterization of visible-light bands in an industrial scenario and compares it with that of RF bands.

3. Visible-Light Positioning (VLP)

VLP is another important form of OWC, often considered a subset of VLC. In VLP, light signals emitted from LEDs are used to determine the position of objects or users within an indoor environment. By analyzing the properties of the received light signals, such as the received signal strength (RSS), angle of arrival (AoA), or time of flight (ToF), VLP systems can achieve higher accuracy in indoor localization compared to their RF counterparts, such as the global positioning system (GPS) [3]. Another advantage of VLP is that it can be seamlessly integrated with VLC, using the same LED infrastructure for both indoor positioning and wireless data transmission.

In this Special Issue, we present three articles on the topic of VLP. Since VLP systems typically require multiple transmitters and photodetectors for triangulation calculations, Contribution 6 adapts this configuration to support multiple-input single-output (MISO) and multiple-input multiple-output (MIMO) transmissions. Like MIMO used in RF wireless systems, this approach aims to enhance either data transmission rates or the reliability of received signals. Contribution 7 also investigates the integration of VLC and VLP. A key contribution of this work is the use of a solar cell array, rather than conventional photodetectors, for signal detection. An experimental setup is implemented to demonstrate the system's performance in terms of both data transmission and positioning accuracy. To achieve a more accurate positioning system using AoA, Contribution 8 studies the performance of an angular diversity receiver with aperture structures. This design was originally proposed in [7] for VLC MIMO communications and later adapted for a VLP system in [8]. As an extension of [8,9], this work focuses on optimizing the layout of the LEDs to improve the positioning accuracy. These improvements are demonstrated through simulations, considering the Cramér–Rao bound (CRB) as the performance metric.

4. Space Optical Communication

Space optical communication is another emerging and important form of OWC that enables high-speed data transmission between various types of satellites, as well as between satellites and ground stations. These systems typically operate across different orbits, including geostationary Earth orbit (GEO), low Earth orbit (LEO), and high-altitude platform stations (HAPS), each offering distinct advantages in terms of coverage and latency. As a complementary technology to terrestrial wireless and optical fiber networks, space optical communication is expected to play an important role in the next generation of global connectivity [10]. However, it also faces several technical challenges, such as solar interference, atmospheric disturbances, and the complexities of accurate tracking and alignment in dynamic space environments.

This Special Issue presents three articles that explore different aspects of space optical communication. Like other OWC systems, the performance of the transmission link in space optical communication highly depends on the sensitivity of the photodetector used. Avalanche photodiodes (APDs) are well known for their high gain and stable performance and are therefore promising candidates for space applications. Contribution 9 investigates the performance of APDs by simulating their behavior under various channel impairments, including turbulence-induced beam scintillation, pointing errors, and AoA fluctuations, using a detailed channel model and by analyzing the theoretical symbol error rate (SER). Moreover, unlike RF wireless and fiber optics, the signal received in space optical communication can be highly distorted. In recent years, there has been a trend toward exploring the use of AI-related algorithms for signal equalization [11]. Contribution 10 studies a neural network (NN)-based approach for signal equalization and compensation of distortions caused by free-space channels. Finally, Contribution 11 presents a model that examines the optical power requirements for satellite constellations throughout their orbital periods, taking multiple system parameters into account.

5. Modulation and Multiplexing Techniques

Similarly to strategies used in fiber optics and RF wireless communications, one of the most effective approaches to increasing the transmission data rate of OWC is the use of multiplexing techniques. In OWC, since light carries the transmitted data, many multiplexing methods commonly used in fiber optics can potentially be adapted for OWC. These methods include wavelength-division multiplexing (WDM) [12], polarization-division multiplexing (PDM) [13], and orbital angular momentum (OAM) multiplexing [14]. Moreover, because the OWC channel is wireless, multiplexing techniques from RF wireless systems can also be applied. Examples include multiple-input multiple-output (MIMO) [15,16], non-orthogonal multiple access (NOMA) [17], and orthogonal frequency-division multiple access (OFDMA) [18]. These multiplexing techniques are summarized in Figure 1. Despite the many possible ways to implement transmission multiplexing in OWC, a key challenge lies in managing crosstalk and interference between channels, which increase as the number of multiplexed channels increases.

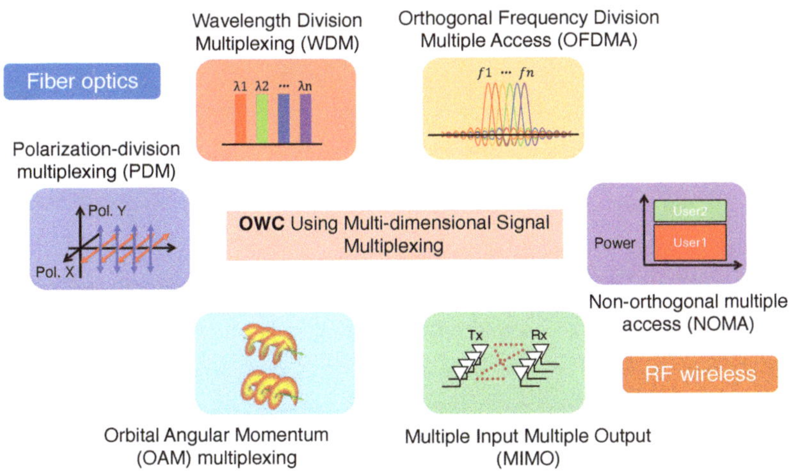

Figure 1. A wide range of transmission multiplexing techniques used in fiber optics and RF wireless communications can potentially be applied to OWC to achieve high-dimensional multiplexing.

This Special Issue includes four articles in the category of modulation and multiplexing techniques. In Contribution 12, an OWC system that combines MIMO with Faster-than-Nyquist (FTN) signaling is studied. This approach aims not only to increase the transmission data rate, but also to improve spectral efficiency. However, in this case, crosstalk occurs between different MIMO channels, and FTN introduces strong interference. To overcome this issue, a machine learning-based approach combining a convolutional neural network (CNN) and a bidirectional long short-term memory (Bi-LSTM) network is proposed for interference mitigation. Contribution 13 focuses on the study of NOMA in OWC. In particular, NOMA is integrated with the concept of cognitive radio (CR) to achieve high spectral efficiency.

Regarding modulation methods, OFDM is often considered in OWC as it is already widely used in many existing wireless communication standards [18]. However, because of the high peak-to-average power ratio (PAPR) of OFDM signals, clipping noise often occurs [19]. Contribution 14 investigates a receiver-based signal processing method designed to mitigate clipping noise in optical OFDM systems. In Contribution 15, the use of code shift keying (CSK) in an intensity modulation/direct detection (IM/DD)-based OWC system is studied. Unlike conventional CSK, which is based on orthogonal codes, this work focuses on the use of non-orthogonal codes to improve the transmission efficiency of OWC systems.

6. Photonic Integrated Circuits (PICs) for OWC

In recent years, there has been a growing trend toward integrating PICs into optical communication systems, especially in optical fiber communications, to meet the increasing demand for compactness, energy efficiency, and high-speed signal processing. The use of PICs in OWC remains relatively underexplored. This Special Issue includes a review article, Contribution 16, which highlights some of the recent progress in PIC development for OWC, particularly focusing on the PICs developed by Fraunhofer HHI.

7. Conclusions

In summary, OWC remains a rapidly developing technique that requires continuous research. This Special Issue highlights a diverse collection of articles within the field, covering many of the most active and popular areas of OWC research. We sincerely thank all of the authors who submitted their work to this Special Issue and contributed to its success.

Author Contributions: Writing—original draft preparation, C.H.; writing—review and editing, C.H. and C.C. All authors have read and agreed to the published version of the manuscript.

Funding: The work of Cuiwei He was supported by the Japan Society for the Promotion of Science (JSPS) Kakenhi, under Grants JP 23K13332 and JP 23KK0257.

Institutional Review Board Statement: Not applicable.

Informed Consent Statement: Not applicable.

Data Availability Statement: Not applicable.

Acknowledgments: We thank all of the authors who submitted their work to this Special Issue, and to the *Photonics* editorial staff, especially Pedri Zheng, for their invaluable assistance.

Conflicts of Interest: The authors declare no conflicts of interest.

List of Contributions

1. Faruki, M.J.; Bera, K.; Karmakar, N. Impact of crystal orientation on modulation bandwidth: Towards GaN LED-based high-speed visible light communication. *Photonics* **2024**, *11*, 542. https://doi.org/10.3390/photonics11060542
2. Aller, D.G.; Lamar, D.G.; Garcia-Mere, J.R.; Hernando, M.M.; Rodriguez, J.; Sebastian, J. Series/Parallel Boost/Buck DC/DC Converter as a Visible Light Communication HB-LED Driver Based on Split Power. *Photonics* **2025**, *12*, 402. https://doi.org/10.3390/photonics12050402

3. Bera, K.; Karmakar, N. Interference Mitigation in VLC Systems using a Variable Focus Liquid Lens. *Photonics* **2024**, *11*, 506. https://doi.org/10.3390/photonics11060506
4. Zhao, Q.; Zhang, W.; Fan, J.; Deng, L. Design of the Light Source Layout Optimization Strategy Based on Region Partition and Pre-Bias Compensation for Indoor Visible Light Communication Systems. *Photonics* **2023**, *10*, 1344. https://doi.org/10.3390/photonics10121344
5. Yin, Y.; Tang, P.; Zhang, J.; Hu, Z.; Jiang, T.; Xia, L.; Liu, G. Channel Characterization and Comparison in Industrial Scenario from Sub-6 GHz to Visible Light Bands for 6G. *Photonics* **2025**, *12*, 257. https://doi.org/10.3390/photonics12030257
6. Wang, R.; Sun, Y.; Liu, Z.; Gao, M.; You, X. Integrated Mobile Visible Light Communication and Positioning Systems Based on Decision Feedback Channel Estimation. *Photonics* **2024**, *11*, 537. https://doi.org/10.3390/photonics11060537
7. Xie, D.; Liu, Z.; Yu, C. Single-source VLCP system based on solar cell array receiver and right-angled tetrahedron trilateration VLP (RATT-VLP) algorithm. *Photonics* **2024**, *11*, 536. https://doi.org/10.3390/photonics11060536
8. Menendez, J.M.; Steendam, H. Optimisation of the Transmitter Layout in a VLP System Using an Aperture-Based Receiver. *Photonics* **2024**, *11*, 517. https://doi.org/10.3390/photonics11060517
9. Guo, W.; Wu, X.; Yang, L. Avalanche Photodiode-Based Deep Space Optical Uplink Communication in the Presence of Channel Impairments. *Photonics* **2025**, *12*, 562. https://doi.org/10.3390/photonics12060562
10. Gao, Y.; Jing, Q.W.; Liu, M.F.; Zong, W.H.; Hong, Y.Q. Deep Learning-Assisted High-Pass-Filter-Based Fixed-Threshold Decision for Free-Space Optical Communications. *Photonics* **2024**, *11*, 599. https://doi.org/10.3390/photonics11070599
11. Gioulis, M.; Kamalakis, T.; Alexandropoulos, D. Comprehensive Optical Inter-Satellite Communication Model for Low Earth Orbit Constellations: Analyzing Transmission Power Requirements. *Photonics* **2025**, *12*, 392. https://doi.org/10.3390/photonics12040392
12. Cao, M.; Yang, Q.; Zhou, G.; Zhang, Y.; Zhang, X.; Wang, H. A Hybrid Network Integrating MHSA and 1D CNN–Bi-LSTM for Interference Mitigation in Faster-than-Nyquist MIMO Optical Wireless Communications. *Photonics* **2024**, *11*, 982. https://doi.org/10.3390/photonics11100982
13. Liu, R.; Wang, Z.; Wang, X.; Lu, J.; Wang, Y.; Zhuo, Y.; Wu, R.; Wei, Z.; Liu, H. Performance Analysis of Soft Switching FSO/THz-RF Dual-Hop AF-NOMA Link Based on Cognitive Radio. *Photonics* **2023**, *10*, 1086. https://doi.org/10.3390/photonics10101086
14. Gao, Y.; Lian, J. Tolerance-Aided Interference Degradation for Optical OFDM in Power-Constrained Systems. *Photonics* **2023**, *10*, 1206. https://doi.org/10.3390/photonics10111206
15. Komuro, N.; Habuchi, H. Design and Analysis of Enhanced IM/DD System with Nonorthogonal Code Shift Keying and Parallel Transmission. *Photonics* **2025**, *12*, 166. https://doi.org/10.3390/photonics12020166
16. Qian, T.; Schuler, B.; Gupta, Y.D.; Deumer, M.; Andrianopoulos, E.; Lyras, N.K.; Kresse, M.; Weigel, M.; Reck, J.; Mihov, K.; et al. Hybrid Photonic Integrated Circuits for Wireless Transceivers. *Photonics* **2025**, *12*, 371. https://doi.org/10.3390/photonics12040371

References

1. Chi, N.; Zhou, Y.; Wei, Y.; Hu, F. Visible light communication in 6G: Advances, challenges, and prospects. *IEEE Veh. Technol. Mag.* **2020**, *15*, 93–102. [CrossRef]
2. Haas, H.; Yin, L.; Wang, Y.; Chen, C. What is lifi? *J. Light. Technol.* **2015**, *34*, 1533–1544. [CrossRef]
3. Armstrong, J.; Sekercioglu, Y.A.; Neild, A. Visible light positioning: A roadmap for international standardization. *IEEE Commun. Mag.* **2013**, *51*, 68–73. [CrossRef]
4. Zeng, Z.; Fu, S.; Zhang, H.; Dong, Y.; Cheng, J. A survey of underwater optical wireless communications. *IEEE Commun. Surv. Tutorials* **2016**, *19*, 204–238. [CrossRef]
5. Kaushal, H.; Kaddoum, G. Optical communication in space: Challenges and mitigation techniques. *IEEE Commun. Surv. Tutorials* **2016**, *19*, 57–96. [CrossRef]
6. He, C.; Chen, C. A review of advanced transceiver technologies in visible light communications. *Photonics* **2023**, *10*, 648. [CrossRef]
7. Wang, T.Q.; He, C.; Armstrong, J. Performance analysis of aperture-based receivers for MIMO IM/DD visible light communications. *J. Light. Technol.* **2016**, *35*, 1513–1523. [CrossRef]

8. Steendam, H.; Wang, T.Q.; Armstrong, J. Theoretical lower bound for indoor visible light positioning using received signal strength measurements and an aperture-based receiver. *J. Light. Technol.* **2016**, *35*, 309–319. [CrossRef]
9. Menéndez, J.M.; Steendam, H. On the optimisation of illumination LEDs for VLP systems. *Photonics* **2022**, *9*, 750. [CrossRef]
10. Toyoshima, M. Trends in satellite communications and the role of optical free-space communications. *J. Opt. Netw.* **2005**, *4*, 300–311. [CrossRef]
11. Shi, J.; Niu, W.; Ha, Y.; Xu, Z.; Li, Z.; Yu, S.; Chi, N. AI-enabled intelligent visible light communications: Challenges, progress, and future. *Photonics* **2022**, *9*, 529. [CrossRef]
12. Wang, Y.; Tao, L.; Huang, X.; Shi, J.; Chi, N. 8-Gb/s RGBY LED-based WDM VLC system employing high-order CAP modulation and hybrid post equalizer. *IEEE Photonics J.* **2015**, *7*, 7904507.
13. Chvojka, P.; Burton, A.; Pesek, P.; Li, X.; Ghassemlooy, Z.; Zvanovec, S.; Anthony Haigh, P. Visible light communications: Increasing data rates with polarization division multiplexing. *Opt. Lett.* **2020**, *45*, 2977–2980. [CrossRef] [PubMed]
14. Trichili, A.; Park, K.H.; Zghal, M.; Ooi, B.S.; Alouini, M.S. Communicating using spatial mode multiplexing: Potentials, challenges, and perspectives. *IEEE Commun. Surv. Tutorials* **2019**, *21*, 3175–3203. [CrossRef]
15. Zeng, L.; O'Brien, D.C.; Le Minh, H.; Faulkner, G.E.; Lee, K.; Jung, D.; Oh, Y.; Won, E.T. High data rate multiple input multiple output (MIMO) optical wireless communications using white LED lighting. *IEEE J. Sel. Areas Commun.* **2009**, *27*, 1654–1662. [CrossRef]
16. He, C.; Wang, T.Q.; Armstrong, J. Performance of optical receivers using photodetectors with different fields of view in a MIMO ACO-OFDM system. *J. Light. Technol.* **2015**, *33*, 4957–4967. [CrossRef]
17. Chen, C.; Zhong, W.D.; Yang, H.; Du, P. On the performance of MIMO-NOMA-based visible light communication systems. *IEEE Photonics Technol. Lett.* **2017**, *30*, 307–310. [CrossRef]
18. Armstrong, J. OFDM for optical communications. *J. Light. Technol.* **2009**, *27*, 189–204. [CrossRef]
19. He, C.; Armstrong, J. Clipping noise mitigation in optical OFDM systems. *IEEE Commun. Lett.* **2016**, *21*, 548–551. [CrossRef]

photonics

MDPI

Article

Impact of Crystal Orientation on Modulation Bandwidth: Towards GaN LED-Based High-Speed Visible Light Communication

Md Jahid Faruki *, Krishnendu Bera and Nemai Karmakar

Department of Electrical and Computer Systems Engineering, Monash University, Melbourne, VIC 3800, Australia
* Correspondence: mdjahid.faruki@monash.edu

Abstract: Light-emitting diodes (LEDs) with high modulation bandwidth are required for high-speed visible light communication applications. Crystal orientation in the GaN LED structure plays a key factor in its modulation bandwidth as the recombination lifetime is highly dependent on crystal orientation owing to the Quantum-Confined Stark Effect (QCSE). In this study, six different crystal orientation multi-quantum well (MQW) GaN LEDs are simulated to understand the impact of heterostructure orientation on modulation bandwidth, radiative recombination rates, and emission intensity. The results of this study demonstrate that semi-polar $10\bar{1}3$ MQW LEDs provide the highest bandwidth in the current density range of 9–20 kA/cm^2 compared to the other five orientations. For instance, the semi-polar $10\bar{1}3$-based LED offers a modulation bandwidth of 912.7 MHz at 20 kA/cm^2 current density. These results suggest that the semi-polar $10\bar{1}3$ orientation-based LED has the potential to support a high-speed visible light communication system.

Keywords: GaN LED; modulation bandwidth; visible light communications; crystal orientation

1. Introduction

Visible light communication (VLC) attracts significant research interest due to many of its alluring features, such as a freely available unlicensed spectrum, ability to co-exist with the WiFi spectrum without interference, assured security to keep the signal private and safe, dual usage of light for illumination and communication, and no association with any potential health hazard. VLC can bring a technological revolution if implemented successfully. The last 20 years have seen enormous development in solid-state optoelectronics, and group III-Nitride LEDs can now emit the whole visible spectrum. Owing to high external quantum efficiency (EQE), over 56% and efficiency higher than 100 lumen/W, GaN LEDs are perceived as potential transmitters for the VLC system [1]. All primary colors can be produced using (In/Al)GaN LEDs, allowing for LEDs ranging from red to ultraviolet. Therefore, the GaN-based LED [2] is a strong candidate for VLC application.

A major challenge of the high-speed VLC is the low modulation bandwidth of the light transmitter. For high-speed data communication, an optical transmitter (e.g., LED) is required which can deliver a high modulation bandwidth. The modulation bandwidth of the commercially available LEDs is quite low. One of the potential reasons behind la ow modulation bandwidth is that the LEDs are primarily designed for lighting, and the communication aspect was not considered. The modulation bandwidth of the off-the-shelf LEDs is within 20 MHz [3], which is the main constraint to establishing a LED-based high-speed VLC system. One of the limiting factors of the modulation bandwidth is the RC time constant [4,5]; however, the RC time constant can be minimized by using a micro-LED [1,6].

The concept of the micro-LED has revolutionized the field of high-speed VLC systems. The micro-LED structure is like the LED structure, while the size is in the micro-meter dimension (lateral dimension of less than 100 µm). Therefore, the RC time constant of the micro-LED is much less [7]. A high current density can be induced in the micro-LED, thus a high modulation bandwidth can be achieved. It can offer a modulation bandwidth in the range of a few hundred MHz, making it suitable for VLC applications. Thus, the micro-LED is perceived to be the best candidate for the next generation of VLC systems, providing tens of Gbps speed communication owing to its promise of a high modulation bandwidth [8–11]. The current study assumes a micro-LED structure, while the size is in the micro-meter dimension (lateral dimension of less than 100 μm). Due to the reduced capacitance of the small active region [6], the RC time constant of the proposed micro-LED is much less and insignificant in the modulation bandwidth performance of the proposed LED.

It is well understood that the modulation bandwidth of the LED is the major bottleneck to attaining high-speed data transmission (tens of Gbps). The modulation bandwidth of the wurtzite GaN LED (the most thermodynamically stable crystal) hugely depends on the Quantum-Confined Stark Effect (QCSE) as the crystal orientation significantly affects the LED's differential carrier lifetime, impacting the modulation bandwidth. The electric field within the quantum well causes a tilting of the conduction and valence bands in the well, separating the electrons and holes. This phenomenon is known as QCSE. Spontaneous polarization is caused by internal electric fields within the quantum wells which exist even without an externally applied bias. The electric field across the quantum well separates the electrons and holes from each other spatially (reducing the probability of radiative recombination) [12,13]. When there is reduced overlap between electron and hole wave functions due to QCSE, radiative recombination rates decrease [1], resulting in a longer radiative lifetime with a low modulation bandwidth. On the other hand, if the QCSE is minimized, a shorter recombination lifetime and a high modulation bandwidth can be achieved.

For high-speed VLC communications, the LED should ideally have a high modulation bandwidth in the range of hundreds of MHz. The modulation bandwidth and recombination rates of GaN LEDs are significantly influenced by the crystal orientation of the GaN structure due to QCSE, as explained in the previous section. Considerable progress has been made in improving the modulation bandwidth performance of GaN devices across various crystal orientations [9–11,14–20], including polar [9,10,17], semi-polar [11,14,15,18], and nonpolar [14,16,20] orientations. Despite numerous studies conducted on various crystal orientations, such as polar, semi-polar, and nonpolar, a comparative study is yet to be conducted to understand the impact of different orientations on the modulation bandwidth, the radiative recombination rate, and emission intensity. Crystal orientations significantly affect the recombination lifetime, thus significantly influencing the LED's modulation bandwidth. Also, increasing the current density decreases the recombination lifetime and, therefore, increases the LED's modulation bandwidth. As LEDs are current-driving devices, in this study, when modulation bandwidth performances are compared for various crystal orientations, we fixed the current density and calculated the modulation bandwidth for each crystal orientation LED to understand the impact of crystal orientations on the LED's modulation bandwidth performance at a particular current density. This study maintains the same LED heterostructures but varies the crystal orientations to investigate how crystal orientation affects the modulation bandwidth performance. The goal is to identify which crystal orientation-based LEDs provide the highest bandwidth to be used as a transmitter for Gbps speed visible light data communications.

This study theoretically analyzes the impact of different crystal orientations on the behavior of the modulation bandwidth, radiative recombination rates, and emission intensity. Six different crystal orientations, namely (0001) Ga-polar-, $(000\bar{1})$ N-polar-, nonpolar-, semi-polar $10\bar{1}\bar{1}$-, semi-polar $10\bar{1}\bar{3}$-, and semi-polar $11\bar{2}2$-based MQW LEDs, are simulated to understand the impact of heterostructure orientation on the modulation bandwidth. This study will further our understanding of high-modulation-bandwidth LEDs to meet high-speed visible light communication industry requirements.

2. Device Structure and Parameters

The Simulator of Light Emitters based on Nitride Semiconductors (SiLENSe) Version 6.6 software tool has been chosen to simulate the structures. SiLENSe software enables one-dimensional (1D) simulation of heterostructures used in LEDs and laser diodes. The software simulates a band diagram as a function of the p-n junction bias, distribution of electrons and holes, electron and hole current density, radiative and non-radiative recombination rates, carrier energy levels and wave functions in the quantum wells, emission and gain spectra, internal quantum efficiency (IQE), injection efficiency, etc. The micro-LED structure has been described from the n-side to the p-side as SiLENSe software simulates the active region of the micro-LED. Only the central part of the heterostructure, which begins from the n-type current spreading layer to the p-type current spreading layer, is described in the simulation (one-dimensional), and the current is assumed to flow through the whole structure. In this study, the LED is simulated with planar current spreading. The first layer in the simulation is a thick n-type contact layer which is responsible for lateral current spreading.

The design schematic is given in Figure 1. A planar chip structure is simulated, with p and n contact electrodes formed on the top side of the LED heterostructure. The layer sequence is described starting from the substrate. A 2000 nm thick n-type GaN layer with a donor dopant concentration of 3×10^{18} cm^{-3} (Si-doped) is grown. On the top of the n-GaN layer, a 12 nm n-GaN barrier is defined with a donor concentration of 2×10^{18} cm^{-3}. Above the n-GaN barrier, a 4 nm InGaN quantum well (InGaN-QW) has been defined with ($In_{0.13}Ga_{0.87}N$) as the active area where the radiative recombination would happen (known as an active region or active layer). A total of three sets of quantum wells and barriers have been specified. Above the QWs, a 12 nm unintentionally doped GaN (uid-GaN) barrier with acceptor concentration 0 (at bottom) to 2×10^{19} cm^{-3} (at top) with 5×10^{18} cm^{-3} at the middle of the uid-GaN-barrier. The active region is capped with a 10 nm p-AlGaN electron blocking layer (AlN:GaN = 0.15:0.85) with an acceptor concentration of 2×10^{19} cm^{-3} placed to reduce the leakage of the electron from the active layer, as suggested by the literature [21]. On the top of the AlGaN layer, a p-type GaN layer with an acceptor doping concentration of 2×10^{19} cm^{-3} has been placed which has a thickness of 100 nm (Mg-doped). Finally, a heavily doped (acceptor concentration of 1.6×10^{20} cm^{-3}) p-type GaN layer (p++GaN) is placed with a thickness of 10 nm. The electron and hole mobility values for this structure are set as 100 cm^2/V.s and 10 cm^2/V.s, respectively [22]. The dislocation density is specified as 1×10^9 cm^{-2} [23–25].

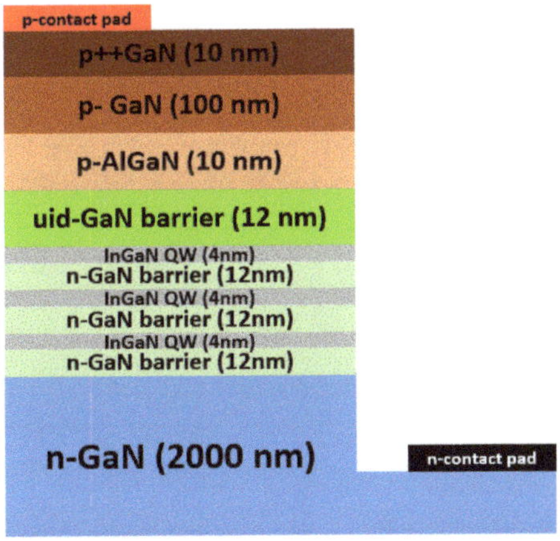

Figure 1. Cross-sectional view of the simulated MQW LED device.

The structure is simulated from 0 to 10 V bias with a step size of 0.025 V. The same structure is simulated for six different crystal orientations, namely (0001) Ga-polar, $(000\bar{1})$

N-polar, nonpolar, semi-polar $10\bar{1}\bar{1}$, semi-polar $10\bar{1}\bar{3}$, and semi-polar $11\bar{2}2$, to understand the impact of crystal orientation on the LED's modulation bandwidth.

3. Results and Discussion

The photon generation depends on the Quantum-Confined Stark Effect (QCSE), which is hugely dependent on the crystal orientation. The band diagram of the proposed LED structure is shown in Figure 2, depicting the energy bands and the behavior of electrons and holes within the device structure. It can be seen from Figures 2 and 3 that the band bending is larger for polar orientation ((0001) Ga-polar and (000$\bar{1}$) N-polar) compared to the nonpolar orientation. In polar orientation, the electron and hole wave functions (Figure 3) lead to separation/displacement (and reduced overlap). Therefore, the recombination rate or efficiency is lower than nonpolar orientation. In semi-polar orientations, this effect of QCSE is in between polar and nonpolar orientations, as shown in Figures 2 and 3. QCSE plays an influential role in determining the radiative recombination rates [1] and it is observed in this simulation study.

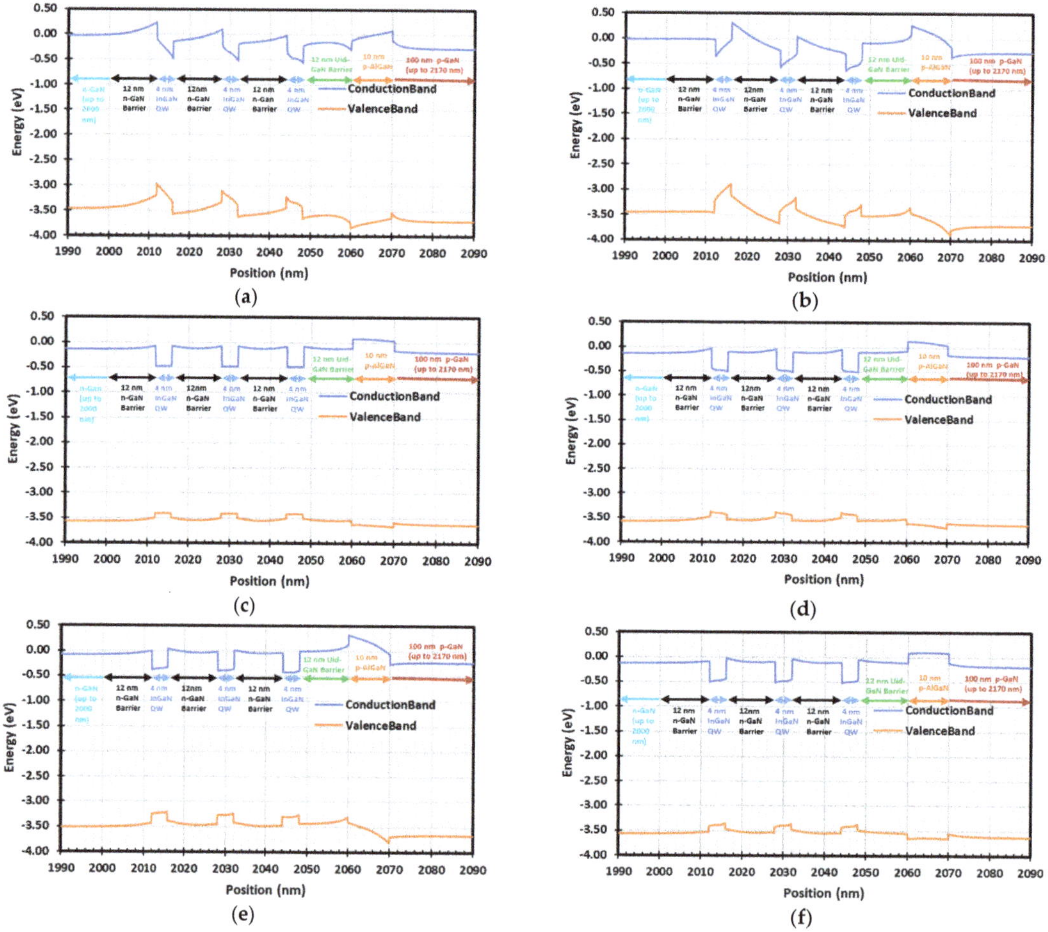

Figure 2. Band diagram of different crystal orientations at 3.6 V bias. (**a**) (0001) Ga-polar. (**b**) (000$\bar{1}$) N-polar. (**c**) Nonpolar. (**d**) Semi-polar $10\bar{1}\bar{1}$. (**e**) Semi-polar $10\bar{1}\bar{3}$. (**f**) Semi-polar $11\bar{2}2$.

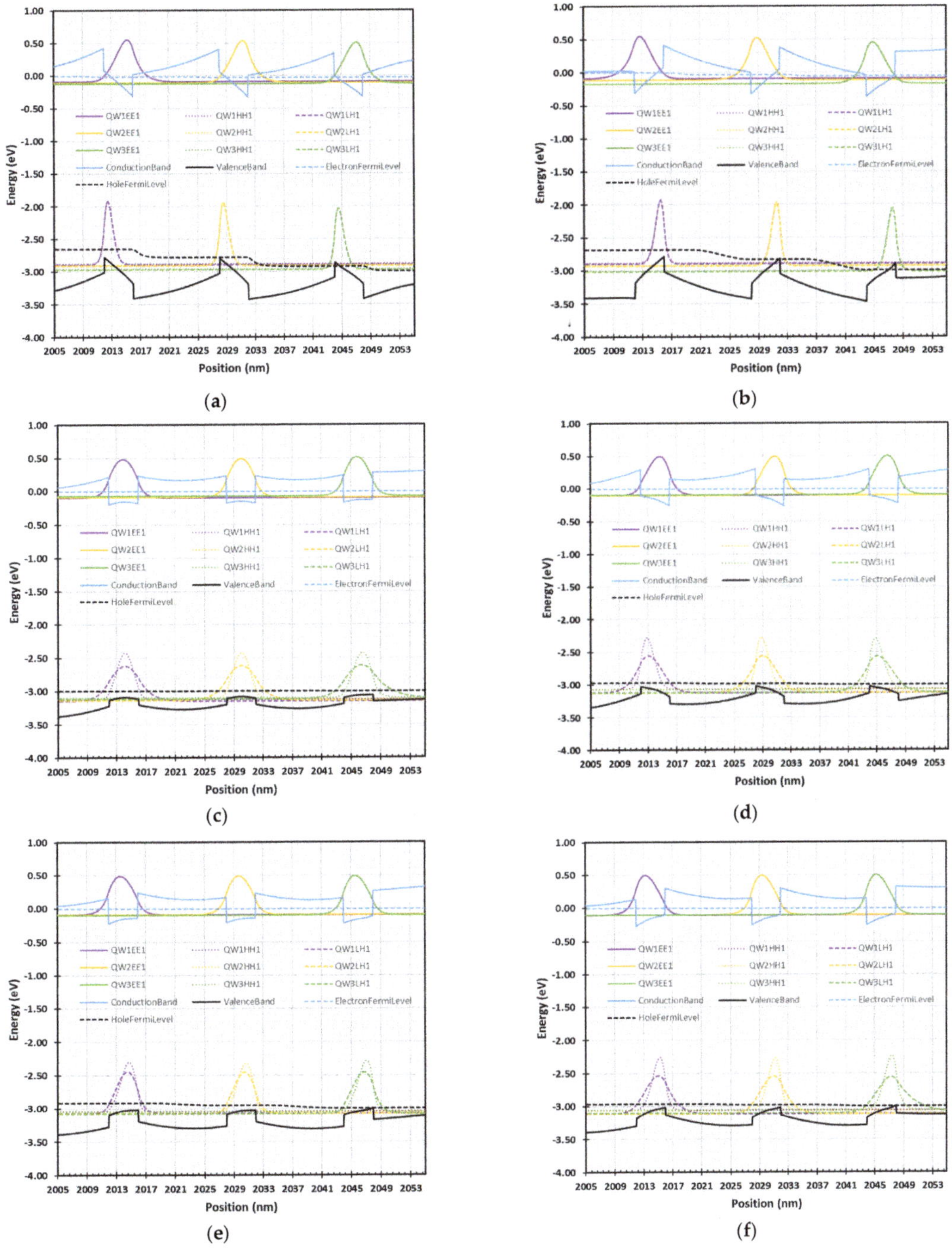

Figure 3. Carrier wave functions of different crystal orientations at 3.6 V bias (EE, HH, and LH denote electrons, heavy holes, and light holes). (**a**) (0001) Ga-polar. (**b**) (000$\bar{1}$) N-polar. (**c**) Nonpolar. (**d**) Semi-polar 10$\bar{1}\bar{1}$. (**e**) Semi-polar 10$\bar{1}\bar{3}$. (**f**) Semi-polar 11$\bar{2}$2.

Total Recombination rate R can be described as,

$$R = R^{rad} + R^{Non-rad}$$

$$R = R^{rad} + R^{SRH} + R^{Auger} \quad (1)$$

R^{rad} is the radiative recombination rate and $R^{Non-rad}$ is the non-radiative recombination rate. $R^{Non-rad}$ is the summation of R^{SRH} and R^{Auger}. R^{SRH} is the non-radiative Shockley–Read–Hall (SRH) recombination that occurs due to crystal defects and dislocations in the heterostructure. R^{Auger} is non-radiative Auger recombination which occurs at high carrier density where one electron and one hole recombine and transfer resulting energy to a third electron or hole in the high-energy band which then loses extra energy via thermal vibration.

The radiative recombination rate is expressed as,

$$R^{rad} = B.\, np\ [1 - \exp(-\frac{Fn - Fp}{kT})] \quad (2)$$

The SRH radiative coefficient $B = 2.4 \times 10^{-11}$ cm^{3}s^{-1} is considered as determined by previous study [24,26]. n and p refer to carrier concentration of electrons and holes, respectively. F_n and F_p represent the fermi levels of the electrons and holes, respectively. k denotes the Boltzmann constant, and T is the temperature.

The non-radiative Shockley–Read–Hall recombination rate ($R^{Non-rad}$) is calculated based on the Shockley–Read approach model [27] which relates the carrier lifetimes with the dislocation density. Further details of the estimation can be found in [24] and Silense Physics summary [28].

Figure 4 depicts the radiative and non-radiative recombination rates for all six simulated crystal orientations at 2 kA/cm^2 current density. Non-radiative recombination rates are the summation of non-radiative Shockley–Read–Hall (SRH) recombination and non-radiative Auger recombination rates. The recombination rates are much higher in the quantum wells than in the QW barriers, so the peaks occur at the QWs. The proposed LED structures have three QWs, and the recombination peaks occur at approximately 2046 nm, 2030 nm, and 2014 nm positions (at the first QW (the one closer to p-GaN), the second QW, and the third QW). The radiative recombination is highest in N-Polar at 2 kA/cm^2 current density. Most recombination occurs at first QW (the one closer to the p-GaN) at this low current density. For N-Polar and Ga-Polar orientations, approximately 90% radiative recombination occurs at the first QW, which means almost all the current is converted into recombination in the first QW. As there are higher concentrations of carriers in a particular QW that emits light, N-Polar and Ga-Polar orientations are expected to have a faster recombination rate and a higher modulation bandwidth. In the first QW, the peak radiative recombination rate for N-Polar is observed as 2.83×10^{28} cm^{-3}s^{-1}, whereas this value is 2.75×10^{28} cm^{-3}s^{-1} in Ga-Polar and 1.98×10^{28} cm^{-3}s^{-1} in semi-polar $10\overline{1}\overline{3}$. At this current density range, N-polar orientations are expected to outperform other orientations in terms of the modulation bandwidth.

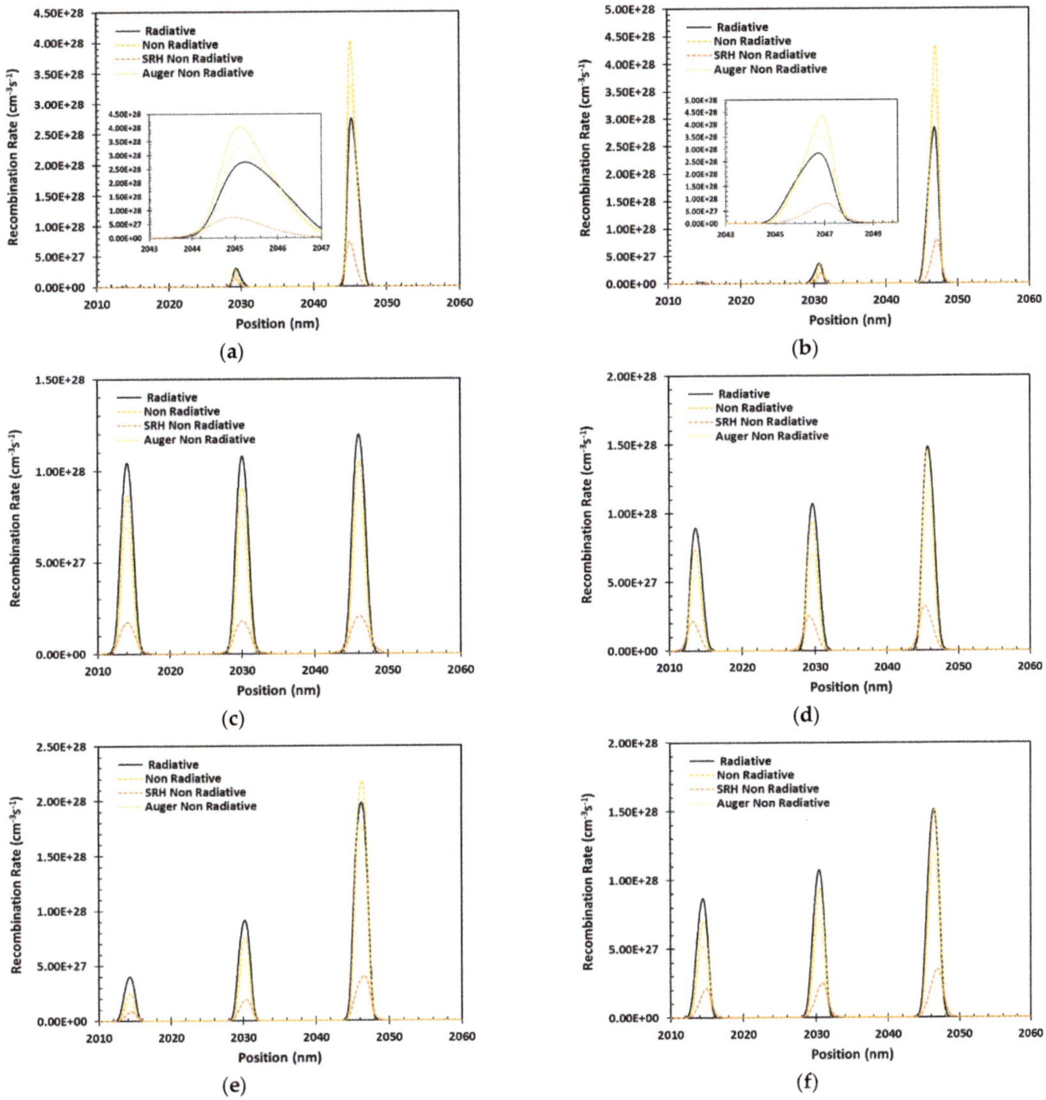

Figure 4. *Cont.*

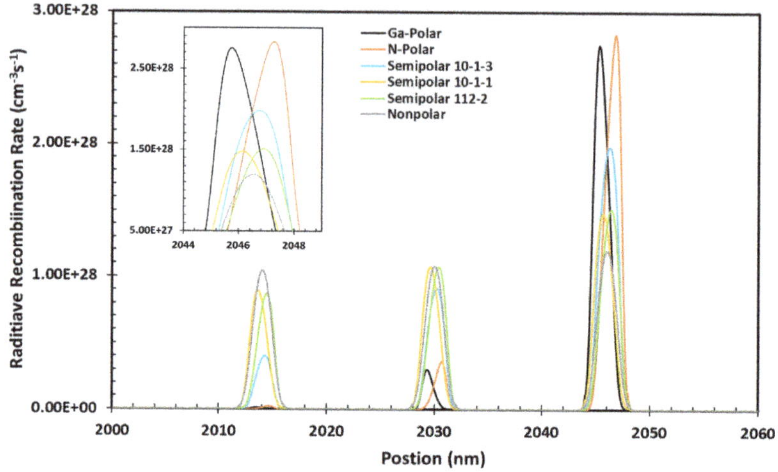

(g)

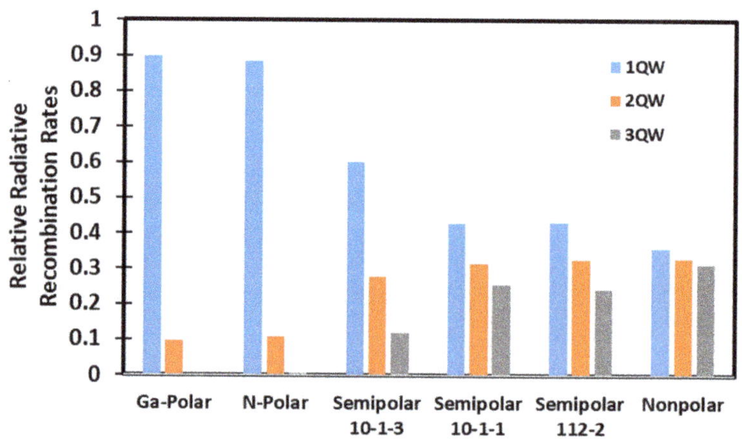

(h)

Figure 4. Six different crystal orientation-based LEDs at 2 kA/cm^2 current density: (**a**) (0001) Ga-polar. (**b**) $(000\bar{1})$ N-polar. (**c**) Nonpolar. (**d**) Semi-polar $10\bar{1}\bar{1}$. (**e**) Semi-polar $10\bar{1}\bar{3}$. (**f**) Semi-polar $11\bar{2}2$. (**a–f**): Radiative and non-radiative recombination rates; (**g**) radiative recombination rate at each of three QWs; (**h**) relative radiative recombination rates at each QWs. Note: E+n refers to 10^n.

The effective differential carrier recombination lifetime is defined [29] as follows,

$$\tau_{rec} = \frac{1}{A + 2BN + 3CN^2} \tag{3}$$

N refers to the carrier concentration in the active region, A is the Shockley–Read–Hall (SRH) non-radiative coefficient, B is the SRH radiative coefficient, and C is the higher-order non-radiative coefficient (Auger coefficient).

For 1D device simulation, Equation (3) can be calculated as,

$$\tau_{rec} = \left(\frac{dR}{dn}\right)^{-1} = \frac{dn}{dR} \tag{4}$$

where R and n refers to the radiative recombination rate and carrier concentration, respectively.

To integrate the recombination rate over the active region, R can be expressed as,

$$R = \frac{j_{rec}}{q} = \frac{j \times \eta_{inj}}{q} \tag{5}$$

j_{rec} refers to the recombination current density in the QWs. q refers the electron charge (1.60×10^{-19} Coulombs).

η_{inj} refers to the Injection efficiency given by,

$$\eta_{inj} = \frac{q}{j} \int_{active\ region} dz.\ R(z) \tag{6}$$

j refers to the current density and is expressed [28] as,

$$j = j_n + j_p = n_{2D}\mu_n \frac{dF_n}{dZ} + p_{2D}\mu_p \frac{dF_p}{dZ} \tag{7}$$

j_n and j_p are the electron's current density and the hole's current density, respectively. μ_n and μ_p are the electron mobility and the hole mobility, respectively. F_n and F_p are the fermi levels of the electrons and holes, respectively. The electron current density (j_n) and the hole current density (j_p) are assumed to be proportional to the gradient of the fermi level of the electrons ($\frac{dF_n}{dZ}$) and holes ($\frac{dF_p}{dZ}$), respectively. n_{2D} and p_{2D} refers to the sheet electron concentrations of the carriers in the active region and the sheet hole concentrations of the carriers in the active region, respectively.

$$n_{2D} = \int_{active\ region} dz.\ n(z) \tag{8}$$

$$p_{2D} = \int_{active\ region} dz.\ p(z) \tag{9}$$

The average carrier concentration is estimated as

$$n = \sqrt{n_{2D} \times p_{2D}} \tag{10}$$

Based on the above estimation, Equation (4) can be expressed as,

$$\tau_{rec} = \frac{\Delta(\sqrt{n_{2D} \times p_{2D}})}{\Delta\left(\frac{j \times \eta_{inj}}{q}\right)} = \frac{(\sqrt{n_{2D} \times p_{2D}})_2 - (\sqrt{n_{2D} \times p_{2D}})_1}{\left(\frac{j \times \eta_{inj}}{q}\right)_2 - \left(\frac{j \times \eta_{inj}}{q}\right)_1} \tag{11}$$

Equation (11) shows that the differential lifetime τ_{rec} relies on current density (j), Injection efficiency (η_{inj}) and the sheet electron and hole concentrations of the carriers in the active region (n_{2D} and p_{2D}). When the bias condition changes, it impacts the value of j, η_{inj}, n_{2D}, p_{2D}, which in turn affects τ_{rec}.

The modulation bandwidth of the micro-LED is dependent on the effective differential recombination lifetime (τ_{rec}), as expressed as follows [30],

$$f_{-3dB} = \frac{1}{2\pi\ \tau_{rec}} \tag{12}$$

The higher the bias voltage, the higher the injected carrier concentration in the active region, resulting in a lower carrier lifetime (higher modulation bandwidth) as per Equations (3) and (12).

Higher bias voltage increases the active region's carrier density, lowering the effective carrier lifetimes. It is found that the turn-on voltage is approximately 3 V. After 3 V, the structure starts to draw a high current, generating a high current density. The characteristics of current density with respect to bias voltage are presented in Figure 5a, for (0001) Ga-polar orientation as a demonstration. It can be seen the current density keeps increasing as the bias voltage increases. In this study, approximately 20 kA/cm^2 current density is considered the maximum limit (this is achieved at 3.9 V biasing for this Ga-Polar orientation). Higher than this limit might cause heating issues. Figure 5b shows that the differential lifetime is reduced to 0.36 ns, resulting in a theoretical modulation bandwidth of 445 MHz (at 4 V biasing). The current densities at different bias voltages for various heterostructure orientations are plotted in Figure 5c. When the bias voltages increase, the current density increases for all crystal orientations. The current density of the Ga-polar orientation sharply increases beyond 3.6 V due to the injection of more electrons and holes into the active region compared to other crystal orientations. This behavior is caused by changes in the concentration of electrons and holes, due to polarization. The current density increases steeply after 3.6 V bias due to changes in the electron and hole quasi-Fermi levels, conduction and valence bands caused by polarization. Also, it can be observed that the current density of the Ga-polar at any bias voltage (such as 3.6 V) is not the lowest, but it exhibits the lowest performance in terms of the modulation bandwidth. This behavior of the Ga-polar is attributed to low injection efficiency (η_{inj}) and low sheet electron and hole concentrations of the carriers in the active region (n_{2D} and p_{2D}).

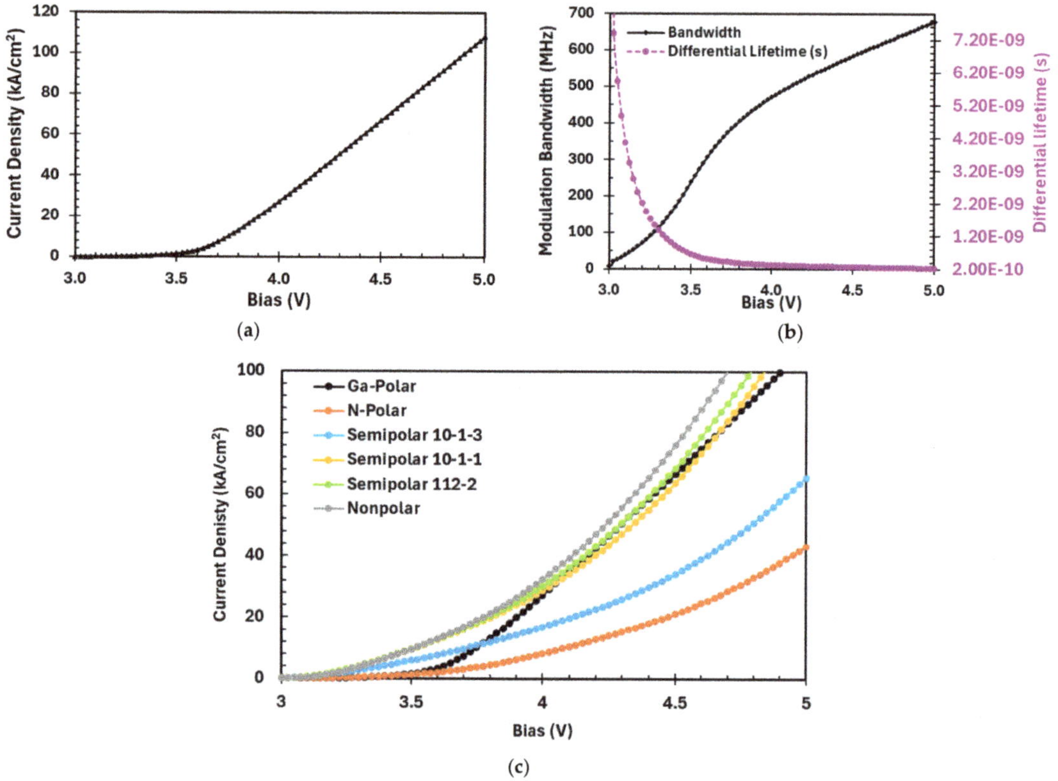

Figure 5. (**a**) (0001) Ga-polar: Current density induced with respect to the bias voltage; (**b**) (0001) Ga-polar: Theoretical modulation bandwidth and differential carrier lifetime at different bias voltage; (**c**) current densities at different biases for all heterostructure orientations. Note: E-n refers to 10^{-n}.

The maximum attainable modulation bandwidth for all six simulated orientations is shown in Figure 6. In the 0–9 kA/cm^2 current density, N-polar demonstrates the highest modulation bandwidth, whereas in the range of 9–20 kA/cm^2 current density, semi-polar $10\overline{1}3$ orientation exhibits the highest modulation bandwidth. As can be seen from Table 1, with a low current density, N-Polar offers the highest modulation bandwidth compared to the semi-polar and nonpolar crystal orientations. For instance, at 2 kA/cm^2, N-polar offers a modulation bandwidth of 304.6 MHz, Ga-polar offers a modulation bandwidth of 270 MHz, whereas semi-polar $10\bar{1}3$, semi-polar $10\bar{1}\bar{1}$, and semi-polar $11\bar{2}2$, and nonpolar orientations offer a modulation bandwidth of 245.5 MHz, 237.5 MHz, 231.1 MHz, and 224.5 MHz, respectively (Figure 6b). This high offering of N-polar and Ga-polar is attributed to high carrier concentration in one particular single QW which causes a shorter carrier lifetime than the simulated orientations. In the 0–9 kA/cm^2 current density range, most of the light is emitted from the first QW (the one closer to the p-GaN). For instance, at 2 kA/cm^2 current density, more than 88% of radiative recombination occurs at the first QW of the N-Polar structure, and the injection efficiency is observed as high as 0.99. Therefore, almost all the injected electron (current) is converted into recombination in the first QW. As there are higher concentrations of carriers in the first QW (which is emitting light), N-Polar orientation demonstrates the highest radiative recombination rate of 2.83×10^{28} cm^{-3}s^{-1} and the highest modulation bandwidth of 304.6 MHz compared to the other five orientations. Also, N-polar orientation is performing the best, attributing to a higher ratio of radiative to auger recombination coefficients. For instance, in the first quantum well (the one closer to the p-type), the radiative recombination rate of N-polar orientation is found as 2.83×10^{28} cm^{-3}s^{-1}, whereas in the Ga-polar, semi-polar $10\bar{1}3$ and nonpolar the radiative recombination rates are recorded as 2.75×10^{28} cm^{-3}s^{-1}, 1.98×10^{28} cm^{-3}s^{-1}, and 1.20×10^{28} cm^{-3}s^{-1}, assisting N-polar to perform better.

Table 1. Comparison of radiative recombination rates occurring at each quantum well (QW) under various applied current densities. Injection efficiency and modulation bandwidth are also provided.

Orientations	At 2 kA/cm^2 Current Density					At 20 kA/cm^2 Current Density				
	Relative Radiative Recombination (%)			Injection Efficiency	Mod. Bandwidth (MHz)	Relative Radiative Recombination (%)			Injection Efficiency	Mod. Bandwidth (MHz)
	1 QW	2 QW	3 QW			1 QW	2 QW	3 QW		
Ga-Polar	90.02	9.82	0.16	0.91	270.0	72.96	23.90	3.14	0.30	445.7
N-Polar	88.37	10.85	0.78	0.99	304.6	58.25	30.58	11.17	0.92	819.7
Semi-polar $10\bar{1}3$	60.15	27.82	12.03	1.00	245.5	39.30	32.62	28.07	0.97	912.7
Semi-polar $10\bar{1}\bar{1}$	42.86	31.43	25.71	0.99	237.5	35.52	32.79	31.69	0.92	906.4
Semi-polar $11\bar{2}2$	43.06	32.64	24.31	0.99	231.1	35.81	32.78	31.40	0.92	903.5
Nonpolar	35.82	32.84	31.34	0.98	224.5	34.51	33.04	32.45	0.85	839.6

In the 0–3 kA/cm^2 current density, Ga-Polar offers the second-highest modulation bandwidth as it demonstrates the second-lowest carrier lifetime due to the second-highest recombination rate among these six simulated structures. For instance, at 2 kA/cm^2 current density, Ga-Polar exhibits a radiative recombination rate of 2.75×10^{28} cm^{-3}s^{-1} and a modulation bandwidth of 270 MHz (second highest). This behavior is attributed to having 90% of total electron injection (due to current) being converted into recombination in the first QW and also having 0.91 injection efficiency at 2 kA/cm^2 current density.

In the 0–9 kA/cm^2 current density range, for four other crystal orientations simulated in this study (except N-Polar and Ga-Polar), the injected current is distributed well among all three QWs. For instance, at 2 kA/cm^2 current density, the radiative recombination ratio in 1st QW:2nd QW:3rd QW are as follows: semi-polar $10\bar{1}3$ (60.15:27.82:12.03), semi-polar $10\bar{1}\bar{1}$ (42.86:31.43:25.71), semi-polar $11\bar{2}2$ (43.06:32.64:24.31) and nonpolar (35.82:32.84:31.34). This implies that the recombinations happening per QW for these four crystal orientations are approximately 3-fold lower than N-Polar and Ga-Polar. Due to lower carrier concentration per QW at a low current density of 2 kA/cm^2, these four crystal orientations show slower recombination rates and smaller modulation bandwidths. Note that, in the 0–3 kA/cm^2 current density range, the non-polar depicts the lowest modulation bandwidth due to a lower injection efficiency than other semi-polar orientations as shown in Table 1.

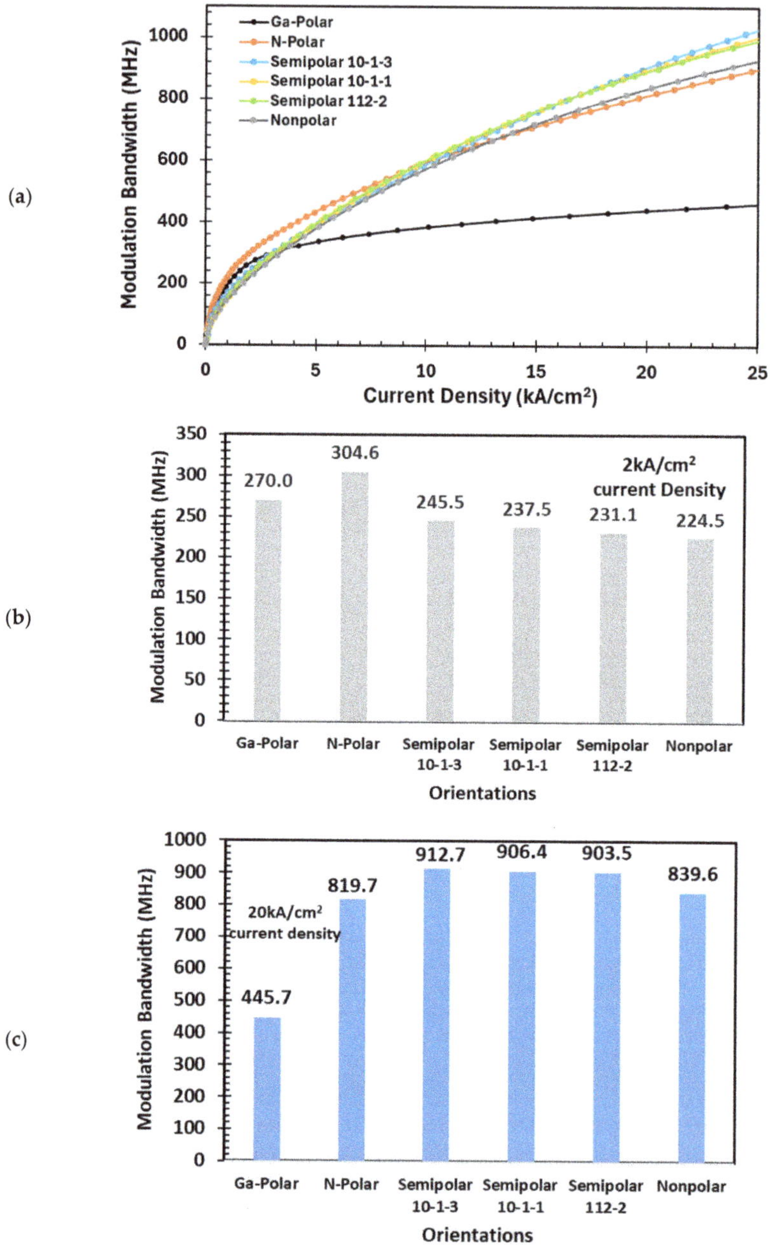

Figure 6. Modulation bandwidth of different crystal orientation-based LEDs: (**a**) in 0–25 kA/cm^2 current density range, (**b**) at 2 kA/cm^2 current density, and (**c**) at 20 kA/cm^2 current density.

When the current density is in the range of 3–20 kA/cm^2, Ga-polar exhibits the lowest modulation bandwidth owing to decreased injection efficiency. Above 3 kA/cm^2 current density, the injection efficiency decreased dramatically (drops from 0.9 to 0.3 at 3 kA/cm^2 to 20 kA/cm^2) compared to other crystal structures, so the injected electron (current) converted into the recombination is lower. Thus, the recombination lifetime is longer, resulting in a lower bandwidth. Owing to the low injection efficiency, the Ga-Polar orientation LED offers a low modulation bandwidth of 445.7 MHz at a high current density of 20 kA/cm^2. It indicates that the ratio of the current density converted to the recombination rates from the QW layers to the total current density is much lower, which may limit the running of the Ga-polar at such a high current.

As shown in Table 1 and Figure 6, when subjected to high current density (9–20 kA/cm^2), semi-polar $10\bar{1}\bar{3}$ offers the highest modulation bandwidth compared to the Polar and nonpolar crystal orientations. For instance, at 20 kA/cm^2, semi-polar $10\bar{1}\bar{3}$ offers a modulation bandwidth of 912.7 MHz, which is higher than semi-polar $10\bar{1}\bar{1}$, semi-polar $11\bar{2}2$, nonpolar, N-polar, and Ga-Polar orientations that offer a modulation bandwidth of 906.4 MHz, 903.5 MHz, 839.6 MHz, 819.7 MHz, and 445.7 MHz, respectively (see Figure 6c). This high offering of semi-polar and nonpolar orientation is attributed to the larger wave-function overlap and shorter carrier lifetime than the polar structure. Semi-polar $10\bar{1}\bar{3}$ orientation performs better owing to a higher ratio of radiative to auger recombination coefficients [10]. For instance, in the first quantum well (the one closer to the p-type), the radiative recombination rates of semi-polar $10\bar{1}\bar{3}$ is found as 7.33×10^{28} cm^{-3}s^{-1} while for semi-polar $10\bar{1}\bar{1}$, semi-polar $11\bar{2}2$, nonpolar, N-polar and Ga-Polar orientations nonpolar orientations it is observed as 6.5×10^{28} cm^{-3}s^{-1}, 6.5×10^{28} cm^{-3}s^{-1}, 5.85×10^{28} cm^{-3}s^{-1}, 1.02×10^{29} cm^{-3}s^{-1}, and 5.76×10^{28} cm^{-3}s^{-1}, respectively, which allows semi-polar $11\bar{2}2$ to perform better. At high current density, the injected electrons (current) are distributed to all three QWs as the number of carriers and carrier concentrations are very high. The current is converted into recombination in all three QWs. The carrier concentration is almost homogeneously distributed for all three semi-polar and nonpolar orientations at 20 kA/cm^2 current density. In the first, second and third QWs, the radiative recombination ratios are as follows: semi-polar $10\bar{1}\bar{3}$ (39.30:32.62:28.07), semi-polar $10\bar{1}\bar{1}$ (35.52:32.79:31.69), semi-polar $11\bar{2}2$ (35.81:32.78:31.40) and nonpolar (34.51:33.04:32.45), as depicted in Figure 7. As the carrier concentration is already high at 20 kA/cm^2 current density and the carriers are distributed almost equally, and the semi-polar and nonpolar orientations demonstrate a better modulation bandwidth. It is worth noting that in the Ga-polar orientation, approximately 72% of total recombination occurs in the first quantum well. One might assume that this would result in a high modulation bandwidth. However, in reality, the modulation bandwidth for Ga-Polar is the lowest (445.7 MHz) at 20 kA/cm^2 current density due to a very low injection efficiency of only 0.3 and low sheet electron and hole concentrations of the carriers in the active region. Table 2 shows the percentage of radiative recombination rates at a particular QW under two different current densities (2 kA/cm^2 and 20 kA/cm^2). In 20 kA/cm^2 current density, N-polar offers as high as 819 MHz owing to carrier concentrations distributed into each of the three QWs and higher injection efficiency. Note that at high current density, there are large no. of carriers available to be distributed to each of three QWs, and all three QWs are contributing to a lower radiative lifetime, resulting in a high modulation bandwidth. Also, the injection efficiency of N-polar is as high as 0.92, enabling it to offer a high modulation bandwidth of 819.7 at 20 kA/cm^2 current density. The efficiency droop is calculated for all six simulated orientation LED structures by comparing the result attained at low current density (i.e., 2 kA/cm^2) and high current density (i.e., 20 kA/cm^2). The efficiency droop for Ga-polar, N-polar, semi-polar $10\bar{1}\bar{3}$, semi-polar $10\bar{1}\bar{1}$, semi-polar $11\bar{2}2$, and nonpolar LEDs are observed as 70%, 26%, 28%, 32%, 32%, and 37%, respectively. Based on this result, it is obvious that semi-polar $10\bar{1}\bar{3}$ orientation LED heterostructure is the optimum choice as it offers the highest modulation bandwidth of 912.7 MHz and a high injection efficiency of 97%, a low efficiency droop of 28%.

(a)

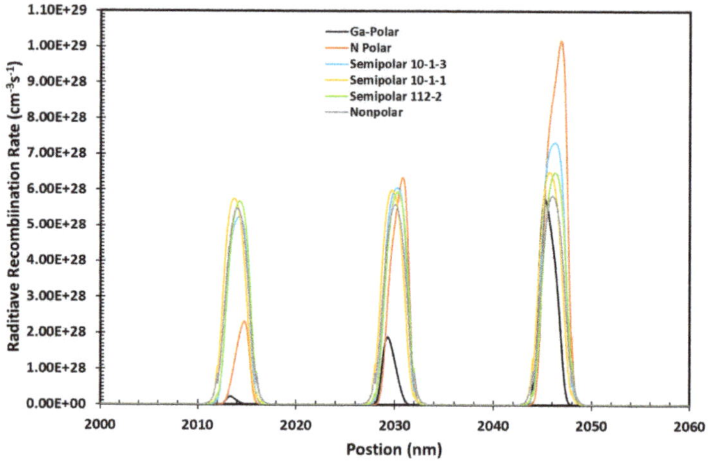

(b)

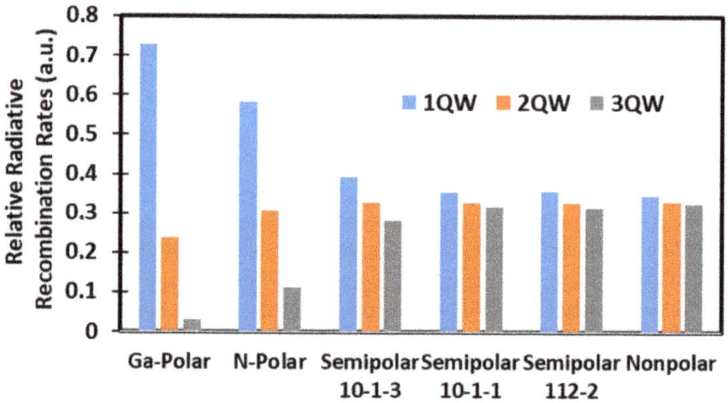

Figure 7. At 20 kA/cm^2 current density: (**a**) radiative recombination rate at each of three QWs; (**b**) relative radiative recombination rates at each QWs. Note: E+n refers to 10^n.

Table 2. Percentage of the radiative recombination rates at a particular QW at two different current density.

Orientations	1 QW at 2 kA/cm^2	1 QW at 20 kA/cm^2	2 QW at 2 kA/cm^2	2 QW at 20 kA/cm^2	3 QW at 2 kA/cm^2	3 QW at 20 kA/cm^2
Ga-Polar	90.02	72.96	9.82	23.90	0.16	3.14
N-Polar	88.37	58.25	10.85	30.58	0.78	11.17
Semi-polar $10\overline{1}\overline{3}$	60.15	39.30	27.82	32.62	12.03	28.07
Semi-polar $10\overline{1}\overline{1}$	42.86	35.52	31.43	32.79	25.71	31.69
Semi-polar $11\overline{2}2$	43.06	35.81	32.64	32.78	24.31	31.40
Nonpolar	35.82	34.51	32.84	33.04	31.34	32.45

A comparison plot is given in Figure 8, where total light emissions for all crystal orientations are compared at 20 kA/cm^2 current density. In all simulated orientations, the emission is in the blue region with the following values—semi-polar $10\overline{1}\overline{3}$ occurs at 409 nm wavelength with a peak emission intensity of 1.54×10^{21} cm^{-2}s^{-1}nm^{-1}, semi-polar $10\overline{1}\overline{1}$ occurs at 403 nm wavelength with a peak emission intensity of 1.5×10^{21} cm^{-2}s^{-1}nm^{-1}, semi-polar $11\overline{2}2$ occurs at 404 nm wavelength with a peak emission intensity of 1.48×10^{21} cm^{-2}s^{-1}nm^{-1}, nonpolar occurs at 400 nm wavelength with a peak emission intensity of 1.47×10^{21} cm^{-2}s^{-1}nm^{-1}, N-polar occurs at 413 nm wavelength with a peak emission intensity of 1.2×10^{21} cm^{-2}s^{-1}nm^{-1}, and Ga-polar occurs at 416 nm wavelength with a peak emission intensity of 5.02×10^{20} cm^{-2}s^{-1}nm^{-1}. The semi-polar $10\overline{1}\overline{3}$ orientation outperforms the other orientations owing to a higher radiative combination and less QCSE effect. Figure 9 depicts the peak emission wavelength of different crystal orientation-based LEDs in the 0–25 kA/cm^2 current density range. The peak emission wavelength remained stable for all six simulated crystal orientations, indicating the color stability of the LEDs. The peak emission wavelength for each orientation is between 400 nm and 430 nm, with a slight blue shift observed as the current density increased.

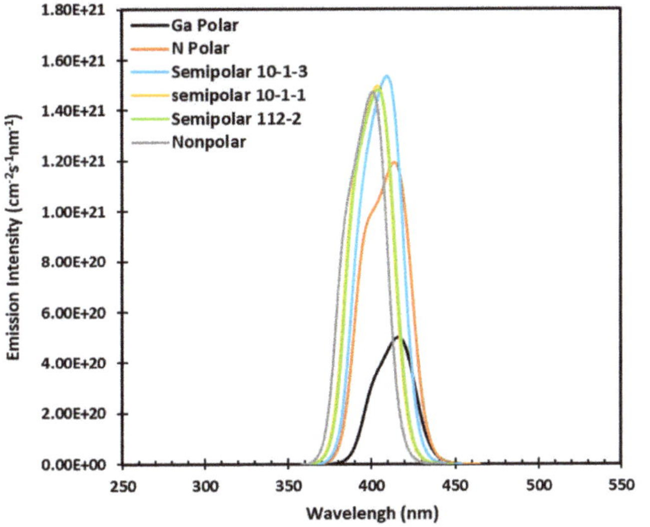

Figure 8. Total emission intensity from each heterostructure at 20 kA/cm^2 current density. Note: E+n, refers to 10^n.

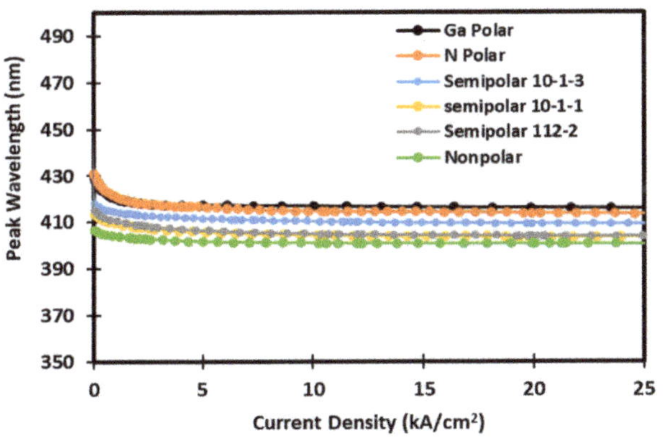

Figure 9. Peak emission wavelength of different crystal orientation-based LEDs in 0–25 kA/cm^2 current density range.

According to the results obtained in this study, the semi-polar $10\overline{1}\overline{3}$ is observed as the best orientation to attaining a high modulation bandwidth and has the potential to support a high-speed visible light communication system. It demonstrated a modulation bandwidth of over 912 MHz for semi-polar $10\bar{1}3$ orientation, at 20 kA/cm^2 current density, whereas semi-polar $10\bar{1}1$ and semi-polar $11\bar{2}2$ orientation LEDs have demonstrated a modulation bandwidth of approximately 905 MHz, indicating significant potential. For the semi-polar case, this study obtained a modulation bandwidth of more than 912 MHz at 20 kA/cm^2, whereas the study in [10] observed 325 MHz at 1 kA/cm^2 current density. The nonpolar orientation LED demonstrated an approximately 840 MHz modulation bandwidth at 20 kA/cm^2 current density, whereas previous work [16] reported 525 MHz. This simulation demonstrated a 400 MHz modulation bandwidth for Ga-polar cases at 20 kA/cm^2 current density. Previous works have shown 230 MHz [31] and 123 MHz at 1 kA/cm^2 [10] for polar samples. Significant advancements have been made in developing high-performance nonpolar, semi-polar, and polar GaN devices [14]. Researchers demonstrated some artificial semi-polar planes such as $20\bar{2}1$ and $20\bar{2}\bar{1}$, by employing precise growth conditions and specific patterning techniques on sapphire substrates, showing great promise [32]. Researchers have successfully grown single-phase nonpolar $1\bar{1}00$ m-plane GaN [33] on planar bulk m-plane GaN substrate and nonpolar $11\bar{2}0$ a-plane [34] GaN thin films on $1\bar{1}02$ r-plane sapphire. Furthermore, different orientations, including semi-polar ($11\bar{2}2$) were achieved on m-plane [35] and a-plane [36] GaN SiC substrates. Semi-polar ($10\bar{1}1$) [37] and nonpolar m-plane were also obtained in Si substrates [38], by implementing patterned v-grooves on Si (001) and (112) [39]. It is believed that affordable and large area nonpolar and semi-polar substrates production would be possible in the near future to meet the visible light communication industry demands. Our study will guide the epitaxy experts towards optimizing LED structure growth on nonpolar and semi-polar substrate orientations. The proposed semipolar and nonpolar-based LEDs have the potential to be used as transmitters for high-speed visible light communication system applications like high-speed wireless communication in the office and home, where 4 K video streaming and larger file transfers are crucial. Moreover, these LEDs can be used for Augmented Reality (AR), Virtual Reality (VR), and industrial automation for 5G-and-beyond connectivity towards Gbps-speed visible light communication.

4. Conclusions

In this study, planar chip structure has been simulated to demonstrate how different crystal orientations impact an LED's recombination rates and the modulation bandwidth. Among six different simulated orientations, semi-polar $10\bar{1}3$ LED demonstrated the highest modulation bandwidth owing to the higher radiative combination and reduced QCSE effect. At a current density of 20 kA/cm^2, $10\bar{1}3$ semi-polar offers a modulation bandwidth of 912.7 MHz. This finding demonstrates that semi-polar $10\bar{1}3$ MQW LED shows the potential for high-speed visible light communications. We believe this study will develop an understanding of and encourage further studies on the different crystal orientations of GaN to enable high-speed LED transmitters towards high-speed visible light communications.

Author Contributions: M.J.F., conceptualization, methodology, data curation, and manuscript writing; K.B., methodology and manuscript revision. N.K., manuscript revision, project resources and administration, and supervision. All authors have read and agreed to the published version of the manuscript.

Funding: This research received no external funding.

Institutional Review Board Statement: Not applicable.

Informed Consent Statement: Not applicable.

Data Availability Statement: The data supporting this study's findings are available from the corresponding author upon reasonable request. The request should be made via email.

Acknowledgments: The authors wish to acknowledge the Department of Electrical & Computer System Engineering, Monash University, for funding this study and Kirill Bulashevich from the STR group for technical support and discussion.

Conflicts of Interest: The authors declare that they have no known competing financial interests or personal relationships that could have appeared to influence the work reported in this paper.

References

1. Rajbhandari, S.; McKendry, J.J.; Herrnsdorf, J.; Chun, H.; Faulkner, G.; Haas, H.; Watson, I.M.; O'Brien, D.; Dawson, M.D. A review of gallium nitride LEDs for multi-gigabit-per-second visible light data communications. *Semicond. Sci. Technol.* **2017**, *32*, 023001. [CrossRef]
2. Muhammad, U.; Nawaz, N.; Saba, K.; Karimov, K.; Muhammad, N. Experimental and numerical analysis of the indium-content on the internal electromechanical field in GaN-based light-emitting diodes. *Optik* **2018**, *172*, 1193–1198. [CrossRef]
3. Zafar, F.; Bakaul, M.; Parthiban, R. Laser-diode-based visible light communication: Toward gigabit class communication. *IEEE Commun. Mag.* **2017**, *55*, 144–151. [CrossRef]
4. Xiao, H.; Xiao, X.; Wu, D.; Wang, R.; Wang, K.; Chiang, K.S. Effects of injection current on the modulation bandwidths of quantum-dot light-emitting diodes. *IEEE Trans. Electron Devices* **2019**, *66*, 4805–4810. [CrossRef]
5. Shi, J.-W.; Huang, H.-Y.; Sheu, J.-K.; Chen, C.-H.; Wu, Y.-S.; Lai, W.-C. The improvement in modulation speed of GaN-based Green light-emitting diode (LED) by use of n-type barrier doping for plastic optical fiber (POF) communication. *IEEE Photonics Technol. Lett.* **2006**, *18*, 1636–1638. [CrossRef]
6. James Singh, K.; Huang, Y.-M.; Ahmed, T.; Liu, A.-C.; Huang Chen, S.-W.; Liou, F.-J.; Wu, T.; Lin, C.-C.; Chow, C.-W.; Lin, G.-R. Micro-LED as a promising candidate for high-speed visible light communication. *Appl. Sci.* **2020**, *10*, 7384. [CrossRef]
7. Xie, E.; He, X.; Islim, M.S.; Purwita, A.A.; McKendry, J.J.; Gu, E.; Haas, H.; Dawson, M.D. High-speed visible light communication based on a III-nitride series-biased micro-LED array. *J. Light. Technol.* **2019**, *37*, 1180–1186. [CrossRef]
8. McKendry, J.J.; Massoubre, D.; Zhang, S.; Rae, B.R.; Green, R.P.; Gu, E.; Henderson, R.K.; Kelly, A.; Dawson, M.D. Visible-light communications using a CMOS-controlled micro-light-emitting-diode array. *J. Light. Technol.* **2011**, *30*, 61–67. [CrossRef]
9. Maaskant, P.P.; Shams, H.; Akhter, M.; Henry, W.; Kappers, M.J.; Zhu, D.; Humphreys, C.J.; Corbett, B. High-speed substrate-emitting micro-light-emitting diodes for applications requiring high radiance. *Appl. Phys. Express* **2013**, *6*, 022102. [CrossRef]
10. Monavarian, M.; Rashidi, A.; Aragon, A.; Oh, S.; Rishinaramangalam, A.; DenBaars, S.; Feezell, D. Impact of crystal orientation on the modulation bandwidth of InGaN/GaN light-emitting diodes. *Appl. Phys. Lett.* **2018**, *112*, 041104. [CrossRef]
11. Monavarian, M.; Rashidi, A.; Aragon, A.; Nami, M.; Oh, S.; DenBaars, S.; Feezell, D. Trade-off between bandwidth and efficiency in semipolar (20 2¯ 1¯) InGaN/GaN single-and multiple-quantum-well light-emitting diodes. *Appl. Phys. Lett.* **2018**, *112*, 191102. [CrossRef]
12. Zhu, D.; Humphreys, C.J. Solid-state lighting based on light emitting diode technology. *Opt. Our Time* **2016**, 87–118. [CrossRef]
13. Tabbakh, T.A.; Anandan, D.; Sheldon, M.J.; Tyagi, P.; Alfaifi, A. Recent Advancements in GaN LED Technology. In *Light-Emitting Diodes-New Perspectives*; IntechOpen: London, UK, 2022.
14. Monavarian, M.; Rashidi, A.; Feezell, D. A Decade of Nonpolar and Semipolar III-Nitrides: A Review of Successes and Challenges. *Phys. Status Solidi (a)* **2019**, *216*, 1800628. [CrossRef]
15. Romanov, A.; Baker, T.; Nakamura, S.; Speck, J. Strain-induced polarization in wurtzite III-nitride semipolar layers. *J. Appl. Phys.* **2006**, *100*, 023522. [CrossRef]
16. Rashidi, A.; Monavarian, M.; Aragon, A.; Okur, S.; Nami, M.; Rishinaramangalam, A.; Mishkat-Ul-Masabih, S.; Feezell, D. High-speed nonpolar InGaN/GaN LEDs for visible-light communication. *IEEE Photonics Technol. Lett.* **2017**, *29*, 381–384. [CrossRef]
17. Xie, E.; Stonehouse, M.; Ferreira, R.; McKendry, J.J.; Herrnsdorf, J.; He, X.; Rajbhandari, S.; Chun, H.; Jalajakumari, A.V.; Almer, O. Design, fabrication, and application of GaN-based micro-LED arrays with individual addressing by N-electrodes. *IEEE Photonics J.* **2017**, *9*, 1–11. [CrossRef]
18. Xu, F.; Qiu, P.; Tao, T.; Tian, P.; Liu, X.; Zhi, T.; Xie, Z.; Liu, B.; Zhang, R. High Bandwidth Semi-Polar InGaN/GaN Micro-LEDs With Low Current Injection for Visible Light Communication. *IEEE Photonics J.* **2023**, *15*, 1–4. [CrossRef]
19. Chai, H.; Yao, S.; Lei, L.; Zhu, Z.; Li, G.; Wang, W. High-speed parallel micro-LED arrays on Si substrates based on via-holes structure for visible light communication. *IEEE Electron. Device Lett.* **2022**, *43*, 1279–1282. [CrossRef]
20. Zhu, S.; Shan, X.; Qiu, P.; Wang, Z.; Yuan, Z.; Cui, X.; Zhang, G.; Tian, P. Low-power high-bandwidth non-polar InGaN micro-LEDs at low current densities for energy-efficient visible light communication. *IEEE Photonics J.* **2022**, *14*, 1–5. [CrossRef]
21. Schubert, E.F. *Light-Emitting Diodes*, 2nd ed.; Cambridge University Press: Cambridge, UK, 2003; p. 88.
22. Konoplev, S.S.; Bulashevich, K.A.; Karpov, S.Y. From large-size to micro-LEDs: Scaling trends revealed by modeling. *Phys. Status Solidi (a)* **2018**, *215*, 1700508. [CrossRef]
23. Bulashevich, K.; Karpov, S.Y. Is Auger recombination responsible for the efficiency rollover in III-nitride light-emitting diodes? *Phys. Status Solidi c* **2008**, *5*, 2066–2069. [CrossRef]
24. Karpov, S.Y.; Makarov, Y.N. Dislocation effect on light emission efficiency in gallium nitride. *Appl. Phys. Lett.* **2002**, *81*, 4721–4723. [CrossRef]
25. Bandić, Z.; Bridger, P.; Piquette, E.; McGill, T. The values of minority carrier diffusion lengths and lifetimes in GaN and their implications for bipolar devices. *Solid-State Electron.* **2000**, *44*, 221–228. [CrossRef]
26. Im, J.; Moritz, A.; Steuber, F.; Härle, V.; Scholz, F.; Hangleiter, A. Radiative carrier lifetime, momentum matrix element, and hole effective mass in GaN. *Appl. Phys. Lett.* **1997**, *70*, 631–633. [CrossRef]

27. Shockley, W.; Read Jr, W. Statistics of the recombinations of holes and electrons. *Phys. Rev.* **1952**, *87*, 835. [CrossRef]
28. STR-Soft. Silense Physics Summary. Available online: https://str-soft.com/devices/silense/ (accessed on 1 November 2023).
29. Green, R.P.; McKendry, J.J.; Massoubre, D.; Gu, E.; Dawson, M.D.; Kelly, A.E. Modulation bandwidth studies of recombination processes in blue and green InGaN quantum well micro-light-emitting diodes. *Appl. Phys. Lett.* **2013**, *102*, 091103. [CrossRef]
30. Lan, H.-Y.; Tseng, I.-C.; Kao, H.-Y.; Lin, Y.-H.; Lin, G.-R.; Wu, C.-H. 752-MHz Modulation Bandwidth of High-Speed Blue Micro Light-Emitting Diodes. *IEEE J. Quantum Electron.* **2018**, *54*, 1–6. [CrossRef]
31. Liu, X.; Tian, P.; Wei, Z.; Yi, S.; Huang, Y.; Zhou, X.; Qiu, Z.-J.; Hu, L.; Fang, Z.; Cong, C. Gbps long-distance real-time visible light communications using a high-bandwidth GaN-based micro-LED. *IEEE Photonics J.* **2017**, *9*, 1–9. [CrossRef]
32. Leung, B.; Wang, D.; Kuo, Y.S.; Han, J. Complete orientational access for semipolar GaN devices on sapphire. *Phys. Status Solidi (b)* **2016**, *253*, 23–35. [CrossRef]
33. Kim, K.C.; Schmidt, M.C.; Sato, H.; Wu, F.; Fellows, N.; Saito, M.; Fujito, K.; Speck, J.S.; Nakamura, S.; DenBaars, S.P. Improved electroluminescence on nonpolar m-plane InGaN/GaN quantum wells LEDs. *Phys. Status Solidi (RRL)–Rapid Res. Lett.* **2007**, *1*, 125–127. [CrossRef]
34. Craven, M.; Lim, S.; Wu, F.; Speck, J.; DenBaars, S. Structural characterization of nonpolar (1120) a-plane GaN thin films grown on (1102) r-plane sapphire. *Appl. Phys. Lett.* **2002**, *81*, 469–471. [CrossRef]
35. Cho, Y.; Sun, Q.; Lee, I.-H.; Ko, T.-S.; Yerino, C.; Han, J.; Kong, B.; Cho, H.; Wang, S. Reduction of stacking fault density in m-plane GaN grown on SiC. *Appl. Phys. Lett.* **2008**, *93*, 111904. [CrossRef]
36. Craven, M.; Wu, F.; Chakraborty, A.; Imer, B.; Mishra, U.; DenBaars, S.; Speck, J. Microstructural evolution of a-plane GaN grown on a-plane SiC by metalorganic chemical vapor deposition. *Appl. Phys. Lett.* **2004**, *84*, 1281–1283. [CrossRef]
37. Izyumskaya, N.; Zhang, F.; Okur, S.; Selden, T.; Avrutin, V.; Özgür, Ü.; Metzner, S.; Karbaum, C.; Bertram, F.; Christen, J. Optical studies of strain and defect distribution in semipolar (11 01) GaN on patterned Si substrates. *J. Appl. Phys.* **2013**, *114*, 113502. [CrossRef]
38. Ding, K.; Avrutin, V.; Izyumskaya, N.; Metzner, S.; Bertram, F.; Christen, J.; Ozgur, U.; Morkoc, H. Recent progress in nonpolar and semi-polar GaN light emitters on patterned Si substrates. *Gallium Nitride Mater. Devices XIII* **2018**, *10532*, 6–21.
39. Izyumskaya, N.; Okur, S.; Zhang, F.; Monavarian, M.; Avrutin, V.; Özgür, Ü.; Metzner, S.; Karbaum, C.; Bertram, F.; Christen, J. Optical properties of m-plane GaN grown on patterned Si (112) substrates by MOCVD using a two-step approach. In *Gallium Nitride Materials and Devices IX*; SPIE: San Francisco, CA, USA, 2014; pp. 257–263.

photonics

MDPI

Article

Series/Parallel Boost/Buck DC/DC Converter as a Visible Light Communication HB-LED Driver Based on Split Power †

Daniel G. Aller [1], Diego G. Lamar [2,*], Juan R. Garcia-Mere [2], Marta M. Hernando [2], Juan Rodriguez [2] and Javier Sebastian [2]

1 Airbus Crisa (an Airbus Defense and Space Company), 28760 Madrid, Spain; daniel.garcia-aller@airbus.com
2 Power Supplies Group, Electrical Engineering Department, The University of Oviedo, 33202 Gijon, Spain; garciamjuan@uniovi.es (J.R.G.-M.); mmhernando@uniovi.es (M.M.H.); rodriguezmjuan@uniovi.es (J.R.); sebas@uniovi.es (J.S.)
* Correspondence: gonzalezdiego@uniovi.es

† This paper is an extended version of our paper published in the 2020 IEEE 21st Workshop on Control and Modeling for Power Electronics, entitled "Design of a high performance VLC-LED driver for Visible Light Communication based on the split of the power".

Abstract: This paper presents a high-brightness LED (HB-LED) driver for visible light communication (VLC) based on two converters. The first is a high-frequency buck DC/DC converter and the second is a low-frequency boost DC/DC converter, connected in series with respect to the LED load and connected in parallel at the input, forming a series/parallel boost/buck DC/DC converter. It is well known that a VLC system needs to perform two different tasks: biasing the HB-LED load and transmitting the communication signal. These typically have different power requirements; the bias power is 75%, while the communication power is 25% of the total power. The requirements of each converter are also different; the communication signal requires a fast output response and, therefore, a high switching frequency, while the biasing control does not require a converter with a high output voltage response. The proposed architecture in this paper takes advantage of the differences between the two tasks and achieves high efficiency and high communication performance by means of splitting power between the two DC/DC converters. The high-frequency buck DC/DC converter tracks the communication signal, while the low-frequency boost DC/DC converter is responsible for lighting tasks. This technique enables high efficiency because most of the power is processed by the low-frequency converter, while a minor part of the power is processed by the high-frequency converter, achieving high communication performance. To provide experimental results, the proposed VLC HB-LED driver was built and validated by reproducing a 64-QAM with a bit rate up to 1.5 Mbps, reaching 91.5% overall efficiency.

Keywords: visible light communication (VLC); high-brightness LED; series/parallel boost/buck DC/DC converter; split power

1. Introduction

Wireless communication systems are essential for the present and future development of society. Furthermore, the speed required by each communication service is continuously growing due to the high transmission per second (i.e., bit rate) required by mainstream services. As a result, the radio frequency (RF) spectrum is already extremely congested, so enabling the data traffic predicted in the next few years requires further research into new wireless communication technologies [1].

https://doi.org/10.3390/photonics12050402

Visible light communication (VLC) [2] is one of the most promising solutions for alleviating the saturation of the RF spectrum. VLC uses the wide, unlicensed, visible light spectrum (430–750 THz range) to transmit information. The strength and potential of this approach become apparent when the communication task is merged with the lighting functionality of high-brightness LED (HB-LED) lamps [3,4].

Figure 1 shows the behavior of an HB-LED load and the intrinsic behavior in a VLC system. A VLC system needs to accomplish two tasks: ensure the desired lighting level by controlling the average current through the HB-LEDs, and reproduce the communication signal by applying a high-frequency communication waveform to the HB-LEDs. As Figure 1 shows, for a specific average current nI_{avg}, a light flux $n\phi_{avg}$ is emitted and a voltage is obtained. This voltage $nV_{avg}(T)$ depends on the temperature T of the HB-LEDs. If a temperature shift. ($T_2 > T_1$) takes place, then the HB-LEDs go from a lower temperature T_1 to a higher temperature T_2. Here, the necessary $nV_{avg}(T)$ decreases for the same current nI_{avg}. This undesired effect means a control system is needed for the HB-LEDs [5]. If this effect is not corrected, the quality and performance of the communication could be affected [6]. Therefore, the driver has to control the average current so that it is always working in the middle of the HB-LED's linear region. This point (i.e., the middle of the HB-LED's linear region) always maximizes the range for the communication signal, nV_{Ω}. Therefore, we define the nV_{Ω} as a region between the threshold voltage $nV_{th}(T)$ and the maximum voltage $nV_{max}(T)$. In other words, it represents the linear part of the voltage–current relationship.

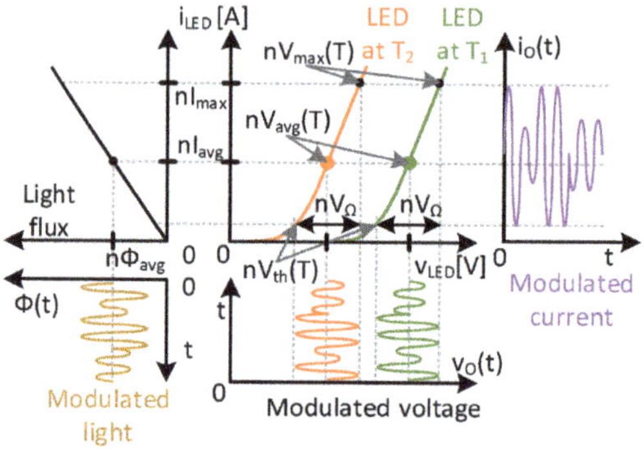

Figure 1. Temperature shift in HB-LED curves (string of HB-LEDs).

Turning our attention to nV_{Ω}, the communication controller must generate the communication signal within the linear part of the voltage–current relationship (i.e., $nV_{th}(T)$). If the temperature changes, the slope of the curve and the maximum HB-LED current do not change (Figure 1). This is particularly important because the working range (i.e., $nV_{th}(T)$) is also unchanged. Provided that the communication is always lower than the nV_{Ω}, and provided the bias task corrects the current properly (i.e., nI_{avg}), distortion of the emitted light is minimized [6].

The most popular approach for reproducing passband modulation schemes involves the use of a bias-T architecture. In this approach, a high-efficiency DC/DC converter provides a positive DC voltage to bias the HB-LEDs, while the signal to be transmitted is reproduced using a linear power amplifier (LPA) [7–9]. Both voltage components are then combined and injected into the HB-LED set with a bias-T circuit made up of a coupling capacitor and a choke inductor. Class A and Class A/B LPAs are typically used due to their high linearity in reproducing high-bandwidth signals. Nevertheless, their reported power efficiencies are very poor (10–40%), thus degrading the overall efficiency of the system and limiting the transmitted power, which in turn limits the communication range. This drawback is particularly significant in the design of VLC drivers for high-power HB-LED loads (office lighting, car headlights, outdoor lighting, etc.) where the extra power losses require the use of additional heatsinks and cooling systems [10].

Another way to implement a VLC transmitter is to modify the traditional HB-LED driving stage by integrating the communication task, producing a VLC HB-LED driver. To overcome the efficiency issue, the use of fast-response DC/DC converters has been proposed as an alternative for delivering both the bias voltage and the communication signal. Nevertheless, the high switching frequency required by the DC/DC converter to accurately reproduce the communication signal and the need to process all the bias power (much higher than the communication power) at that frequency leads to high switching losses and/or extremely complex designs from a hardware perspective (multilevel and multiphase topologies, 6th- to 12th-order filters, etc.) [11–13]. Therefore, one way to further improve efficiency is by splitting the power between two DC/DC converters: one fast-output current DC/DC converter in charge of communication and another, slow-output current DC/DC converter in charge of bias tasks (responsible for delivering the bias power and then maintaining illumination at the desired level in response to temperature changes) [14–16]. In [14], two buck DC/DC converters were placed with the output in series with respect to the HB-LED load: one low-frequency converter for the lighting tasks at low frequency and a high-frequency converter for the communication signal. This architecture achieved high efficiency and communication capability. However, the main drawback of this approach is the need for two different input voltages that are isolated from each other. In [15], the need for two different input voltages (not isolated from each other) was addressed by using two buck DC/DC converters in a two-input buck (TIBuck) configuration. The main drawback of this topology is the reduction in resolution for tracking the voltage across the HB-LED load versus temperature changes. In [16], a Class A/B LPA handles the communication tasks and a buck DC/DC converter controls the lighting. Because Class A/B processes very little power, the efficiency penalty is reduced. However, the signal to be transmitted does not take advantage of the full linear region of the HB-LED load.

This paper presents a VLC HB-LED driver based on a low-frequency boost DC/DC converter and a high-frequency buck DC/DC converter. It is called a series/parallel boost/buck DC/DC converter. The two DC/DC converters have a connection in parallel at its output and are connected in series with the HB-LED load. The proposed topology uses the technique of splitting the power to achieve the best performance. The low-frequency boost DC/DC converter is responsible for controlling the HB-LED lighting task, while the high-frequency buck DC/DC converter handles communication performance. This technique allows for most of the bias power to be processed by the low-frequency converter (achieving high efficiency). Meanwhile, the high-frequency converter delivers the communication power, taking advantage of the entire linear region of the HB-LED load (achieving high communication performance). This configuration follows the same principles as other split-power proposals for VLC but without the need for two input voltages (isolated from each other) [14] and without being unable to take advantage of the buck DC/DC converter useful duty cycle range, which always provides the maximum resolution to track the output voltage (taking advantage of the entire linear region of the HB-LED) [15].

This paper is organized as follows. Section 2 reviews the analysis of the series/parallel boost/buck DC/DC converter. Section 3 describes the power flow between the two converters and an efficiency analysis. The experimental results are given in Section 4.

Section 5 discusses the main characteristics of the converter and compares its performance with the state of the art [14–16], and finally, the conclusions are described in Section 6.

2. Analysis of the Series/Parallel Boost/Buck DC/DC Converter as a VLC HB-LED Driver

2.1. Principle of Operation

A block diagram of the proposed topology is shown in Figure 2a, and an example of the waveforms is shown in Figure 2b. Both converters are connected to the same input voltage V_1, and their outputs are connected in series with respect to the HB-LED string load. The HB-LED string load comprises HB-LEDs. The boost DC/DC converter is responsible for generating a DC voltage V_{bias}, while the buck DC/DC converter generates the communication signal $v_{sig}(t)$, as shown in the block diagram in Figure 2a.

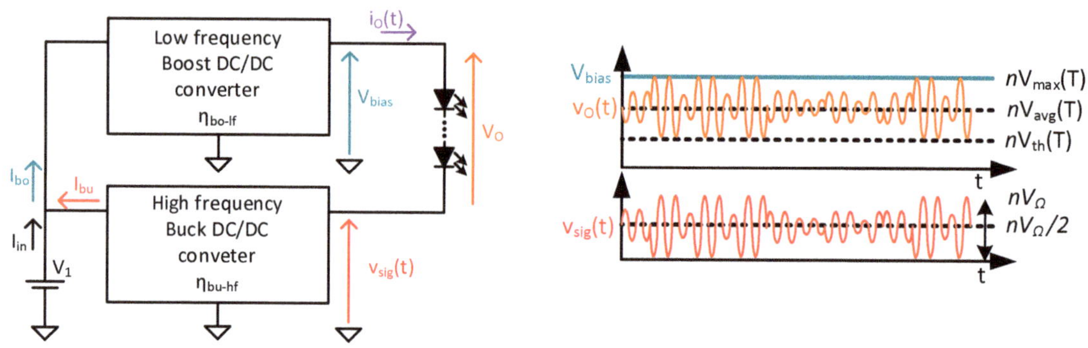

Figure 2. (**a**) Block diagram of the proposed series/parallel boost/buck DC/DC converter acting as a VLC HB-LED driver. (**b**) Main waveforms of the proposed series/parallel boost/buck DC/DC converter acting as a VLC HB-LED driver.

In order to make the n HB-LED string work within the linear region (Figure 1), the HB-LED voltage $v_o(t)$ has to be between the threshold voltage $nV_{th}(T)$ and the maximum voltage $nV_{max}(T)$, where Vth(T) and $V_{max}(T)$ are the threshold and maximum voltages of a single HB-LED, and $nV_{avg}(T)$ is the voltage in the middle of the linear region of the HB-LED string. For the sake of simplicity, the HB-LED string's working range is defined according to the middle point of the linear region $nV_{avg}(T)$. It is important to highlight that the dependency on temperature is noted as (T).

The useful voltage range of the HB-LED load for communication (i.e., the peak-to-peak maximum amplitude allowing for the communication signal due to its linearity) is nV_{Ω}, defined as follows:

$$nV_{\Omega} = nV_{\max}(T) - nV_{\text{th}}(T), \tag{1}$$

where nV_{Ω} has no dependency on temperature. For the sake of simplicity, the limits of the useful voltage range of the HB-LED can be obtained from the middle point of the linear region, $nV_{avg}(T)$.

$$nV_{max} = nV_{avg}(T) + \frac{nV_{\Omega}(T)}{2}, \tag{2}$$

$$nV_{th} = nV_{avg}(T) - \frac{nV_{\Omega}(T)}{2}. \tag{3}$$

Considering (2) and (3), we can conclude that controlling $nV_{avg}(T)$ when the temperature varies always allows the HB-LED string to work at the middle point of the linear region, which defines the useful voltage range of the HB-LED load for communication, nV_{Ω}. According to Figure 2a, the expression of the voltage across the HB-LED string is

$$v_o(t) = V_{bias} - v_{sig}(t). \tag{4}$$

V_{bias} (the voltage of the boost DC/DC converter) only has a DC component (neglecting its ripple), while $v_{sig}(t)$ (the voltage of the buck DC/DC converter) can be represented by its average value (V_{sig}) plus the amplitude of the signal ($\Delta v_{sig}(t)$):

$$v_{sig}(t) = V_{sig} - \Delta V_{sig}(t). \tag{5}$$

2.2. Resolution of the Buck DC/DC for Tracking the Output Voltage

The target is to keep the average voltage of the string of HB-LEDs, V_o = avg(v_o(t)), equal to $nV_{avg}(T)$ at any temperature. Moreover, it is important to note that V_o only depends on the average value of the output voltage of both converters, which is

$$V_o = nV_{avg}(T) = V_{bias} - V_{sig}. \tag{6}$$

Equation (6) shows that the biasing point can either be controlled by the average value of the boost or the buck DC/DC converters. If control of the temperature drift is implemented via the action of the boost DC/DC converter, we can conclude that it is only the boost DC/DC converter output voltage, V_{bias}, that compensates for temperature changes, making the average value of the buck DC/DC converter independent of the temperature. This is very important because the buck DC/DC converter always has the maximum resolution in its duty cycle (i.e., $d_{bu}(t)$) for tracking $v_o(t)$.

If $d_{bu}(t)$ is the duty cycle of the buck DC/DC converter, then $v_{sig}(t)$ can be expressed as a function of $d_{bu}(t)$ and V_1:

$$v_{sig}(t) = V_1 d_{bu}(t). \tag{7}$$

Substituting (7) into (4), the value of $v_o(t)$ becomes

$$v_O(t) = V_{bias} - V_1 d_{bu}(t), \tag{8}$$

where [0, 1] is the range of $d_{bu}(t)$; the highest output voltage occurs for $d_{bu}(t) = 0$, and the lowest output voltage occurs for $d_{bu}(t) = 1$. Considering previous definitions, the range of the output voltage is $[V_{bias} - V_1, V_{bias}]$. At this point, two design rules can be defined if we want the entire useful voltage range of the HB-LED load (i.e., nV_{Ω}) to be used for communication.

$$nV_{max}(T) = V_{bias}, \tag{9}$$

$$nV_{th}(T) = V_{bias} - V_1. \tag{10}$$

Figure 3a shows that the full range of $d_{bu}(t)$ defines useful voltages in the linear region of the HB-LED load, maximizing the resolution of the buck DC/DC converter for tracking the output voltage for communication tasks.

Considering (1), (9), and (10), the following expression is obtained:

$$V_1 = nV_{\Omega}. \tag{11}$$

Therefore, V_1 determines the value of the voltage that maximizes the trackable output voltage range of the buck DC/DC, which only depends on nV_{Ω}. As previously mentioned, nV_{Ω} does not depend on temperature. This is extremely important because maximization of the output voltage range of the buck DC/DC for tracking the communication signal $v_{sig}(t)$

is independent of the temperature. Figure 3 shows the representation of the output voltage range of the buck DC/DC converter (by means of the $d_{bu}(t)$ useful range) versus the useful voltage range of the HB-LED load and their impact on the resolution when the temperature changes from T_2 to T_1, with $T_1 > T_2$. If (9) and (10) are fulfilled by the boost DC/DC converter, two conclusions can be drawn. First, the boost DC/DC converter is responsible for temperature changes. Second, the boost DC/DC converter provides the maximum resolution to the buck DC/DC converter for tracking its output voltage (i.e., $v_{sig}(t)$), and therefore, the output voltage across the HB-LED load (i.e., $v_o(t)$) versus temperature changes. Thus, the series/parallel boost/buck DC/DC converter improves on the performance reported in [14,16] because it uses only a single-input voltage source (unlike [14], which uses two). Moreover, the proposed converter is better than that in [15] in terms of resolution versus temperature changes. However, both converters process all power, and the main switches are rated for the input voltage. This is the price for using only a single voltage source and the improvement in terms of resolution with temperature changes.

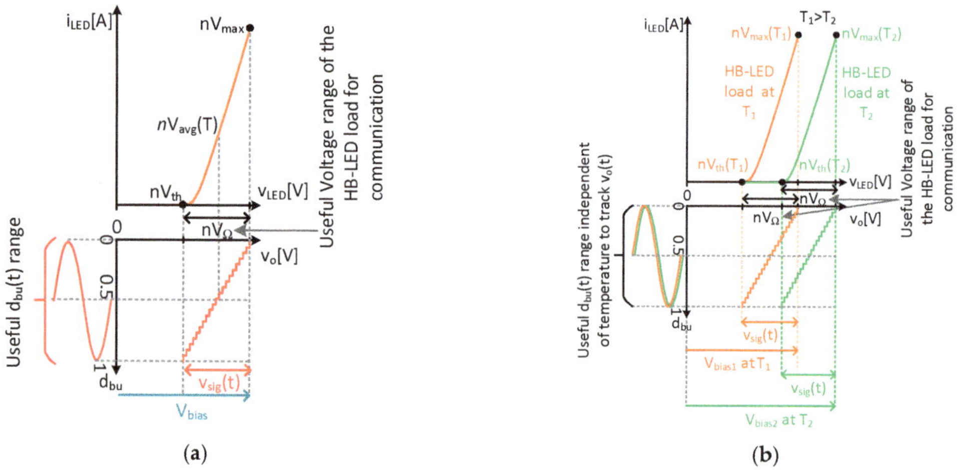

Figure 3. Representation of the output voltage range of the buck DC/DC converter versus the useful voltage range of the HB-LED load and their impact on resolution for tracking the output voltage. (**a**) Without temperature change. (**b**) With an increase in temperature.

2.3. Controlling the Average Value of the Voltage Across the HB-LED Load

As previously mentioned, the boost converter is responsible for controlling the average value of $v_o(t)$. With D_{bias} being the duty cycle of the boost DC/DC converter, V_{bias} can be expressed as

$$V_{bias} = \frac{V_1}{1 - V_{bias}}. \tag{12}$$

By substituting (12) into (6), and assuming that the duty cycle of the buck converter is 0.5 to maximize the useful $d_{bu}(t)$ range, the average value of output voltage can be rewritten as

$$V_o = nV_{avg}(T) = \frac{V_1}{1 - V_{bias}} - \frac{V_1}{2}. \tag{13}$$

It is important to note that all previous equations define the behavior and the design of both the buck DC/DC and boost DC/DC converters, ensuring maximum resolution at the output of the buck converter when the temperature shifts in the HB-LED load.

3. Power Flow and Efficiency Analysis

An important consideration to study is the power flow between the two converters to achieve power balance and efficiency of the series/parallel boost/buck DC/DC converter. Note that the boost DC/DC converter injects power into the HB_LED load, but the buck DC/DC converter drains power from the HB_LED load.

3.1. Power Flow Analisys

As Figure 2a shows, I_{bo} is required by the boost DC/DC converter from the voltage source V_1. In contrast, the buck DC/DC converter injects I_{bu} into V_1. Therefore, I_{in} can be written as

$$I_{\text{in}} = I_{\text{bo}} - I_{\text{bu}}. \tag{14}$$

The boost converter injects an average power to the HB-LED load, generating the voltage V_{bias} and the I_o current. In this case, the power of the boost DC/DC converter is

$$P_{\text{bo-out}} = I_{\text{o}} \cdot V_{\text{bias}}. \tag{15}$$

One part of the power is processed by the HB-LED load, P_{LED}. By using (4), P_{LED} can be expressed as

$$P_{\text{LED}} = I_{\text{o}} \cdot V_{\text{o}} = I_{\text{o}} \cdot (V_{\text{bias}} - V_{\text{sig}}), \tag{16}$$

where V_o is the average value of $v_o(t)$ and V_{sig} is the average value of $v_{sig}(t)$. The other part of the power is processed by the buck DC/DC and drained from the HB-LED load by injecting I_{bu} to V_1. Thus, the buck DC_/DC converter drains $P_{bu\text{-}in}$, which can be written as

$$P_{\text{bu-in}} = I_{\text{o}} \cdot V_{\text{sig}}. \tag{17}$$

3.2. Simplification of Series/Parallel Boost/Buck DC/DC

This converter's way of processing power allows it to be simplified. A buck DC/DC converter that drains rather than supplies current can be analyzed as a boost DC/DC converter. Due to this, one possible series/parallel boost/buck DC/DC Converter implementation is that shown in Figure 4.

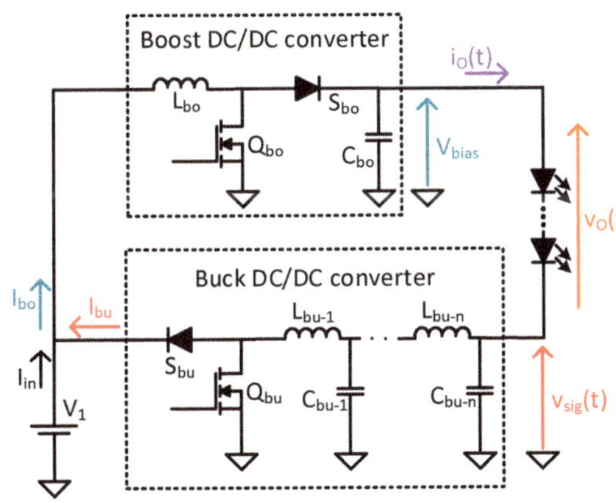

Figure 4. Representation of the simplification of the series/parallel boost/buck DC/DC converter, where the switching of the two converters is asynchronous. This allows the control signals of both MOSFETs to be linked to the ground.

If the switching of both converters is asynchronous, it is possible to connect both MOSFETs to the ground, making implementation easier because an isolator is not needed to control the floating MOSFET in the buck DC/DC converter.

3.3. Efficiency Analysis

Efficiency analysis was carried out by examining the efficiency of each converter in the series/parallel boost/buck DC/DC converter. Figure 2a defines the efficiency of the boost DC/DC converter as $\eta_{\text{bo-lf}}$, and the efficiency of the buck DC/DC converter as $\eta_{\text{bu-hf}}$. Moreover, Figure 2a defines the input power as

$$P_{\text{in}} = I_{\text{in}} \cdot V_1 = V_1 \cdot (I_{\text{bo}} - I_{\text{bu}}). \tag{18}$$

The output power is defined in (16). Therefore, the efficiency of the series/parallel boost/buck DC/DC converter (i.e., η_t) can be easily defined by the output power divided by the input power:

$$\eta_t = \frac{P_{\text{LED}}}{P_{\text{in}}} = \frac{I_o \cdot (V_{\text{bias}} - V_{\text{sig}})}{V_1 \cdot (I_{\text{bo}} - I_{\text{bu}})} = \frac{I_o \cdot V_{\text{bias}} - I_o \cdot V_{\text{sig}}}{V_1 \cdot I_{\text{bo}} - V_1 \cdot I_{\text{bu}}}. \tag{19}$$

The input power of the boost DC/DC converter can be expressed as a function of the input voltage and the input current:

$$P_{\text{bo-in}} = I_{\text{bo}} \cdot V_1. \tag{20}$$

The output power of the buck DC/DC converter can be expressed as a function of the input voltage and its output current:

$$P_{\text{bu-out}} = I_{\text{bu}} \cdot V_1. \tag{21}$$

By substituting (15), (17), (20), and (21) into (19), the efficiency η_t can be rewritten as

$$\eta_t = \frac{P_{\text{bo-out}} - P_{\text{bu-in}}}{P_{\text{bo-in}} - P_{\text{bu-out}}}. \tag{22}$$

The input power of the boost DC/DC converter (i.e., $P_{\text{bo-in}}$) can be rewritten as a function of its output power and its efficiency (i.e., $\eta_{\text{bo-lf}}$):

$$P_{\text{bo-in}} = \frac{P_{\text{bo-out}}}{\eta_{\text{bo-lf}}}. \tag{23}$$

Following the same procedure, the output power of the buck DC/DC converter can be expressed as a function of its input power (i.e., $P_{\text{bu-in}}$) and its efficiency (i.e., $\eta_{\text{bu-hf}}$):

$$P_{\text{bu-out}} = P_{\text{bu-in}} \cdot \eta_{\text{bu-hf}}. \tag{24}$$

Taking into account (23) and (24), (22) can be rewritten as

$$\eta_t = \eta_{\text{bo-lf}} \cdot \frac{1 - \frac{P_{\text{bu-in}}}{P_{\text{bo-out}}}}{1 - \eta_{\text{bo-lf}} \cdot \eta_{\text{bu-hf}} \cdot \frac{P_{\text{bu-in}}}{P_{\text{bo-out}}}}. \tag{25}$$

In order to simplify the analysis of (25), a new parameter will be defined. This parameter is α, defined as

$$\alpha = \frac{P_{\text{bu-in}}}{P_{\text{bo-out}}} = \frac{V_{\text{sig}} \cdot I_o}{V_{\text{bias}} \cdot I_o} = \frac{V_{\text{sig}}}{V_{\text{bias}}}, \tag{26}$$

where α depends on the amplitude of the communication signal (i.e., V_{sig}) and the output voltage of the boost DC/DC converter that defines the bias of the HB-LED load (i.e., V_{bias}).

Now, putting (26) into (25), the overall efficiency of the series/parallel boost/buck DC/DC converter is

$$\eta_t = \eta_{bo\text{-}lf} \cdot \frac{1-\alpha}{1-\eta_{bo\text{-}lf} \cdot \eta_{bu\text{-}hf} \cdot \alpha}. \tag{27}$$

Equation (27) shows the dependency of η_t on the efficiency of each converter in the series/parallel boost/buck DC/DC converter architecture. For example, in an edge case, when there is no signal, V_{sig} equals zero and, therefore, the overall efficiency is the efficiency of the boost DC/DC converter.

It is important to note that the series/parallel boost/buck DC/DC converter supplies an HB-LED load for communication. As explained in Section 2.2, the idea is to take advantage of all useful voltages in the linear region of the HB-LED load. This means that the value of V_{sig} must be 0.5 V_{Ω}. By using (7), (9), (11), and (26), the α parameter can be expressed as a function of the HB-LED characteristics:

$$\alpha = \frac{V_{\Omega}}{2V_{max}}. \tag{28}$$

Taking typical values of commercial HB-LEDs [17], V_{Ω} = 1 and V_{max} = 4, these values define an α parameter of 0.125. Taking advantage of this calculation, it is very interesting to evaluate how each converter's efficiency (i.e., buck DC/DC converter and boost DC/DC converter) influences the overall efficiency of the series/parallel boost/buck DC/DC converter. Figure 5 presents this analysis, evaluating the η_t in two cases. In the first case, the efficiency of the boost converter is kept constant at 0.95 (i.e., $\eta_{bo\text{-}lf}$ = 0.5) and the efficiency of the buck DC/DC converter varies between 0.8 and 1 (solid line in Figure 5). In the second case, the efficiency of the buck DC/DC converter is kept constant at 0.95 and the efficiency of the boost DC/DC converter varies between 0.8 and 1 (dotted line in Figure 5). As Figure 5 shows, the range of variation in the η_t is lower when the efficiency of the buck DC/DC converter varies and the efficiency of the boost DC/DC converter is kept constant. It is clear that the efficiency of the proposed series/parallel boost/buck DC/DC converter depends more on the efficiency of the boost DC/DC converter (i.e., $\eta_{bo\text{-}lf}$). This is because the boost DC/DC converter processes higher power levels, making its efficiency much more important. This is a notable benefit because the boost DC/DC converter operates at a lower switching frequency, and it is easier to achieve higher efficiency here than in the buck DC/DC converter, which operates at a higher frequency.

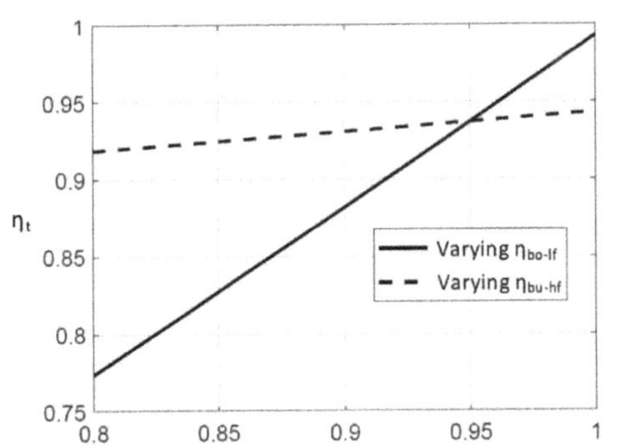

Figure 5. Effect of varying the efficiency of each individual DC/DC converter in η_t.

4. Experimental Results

The design and experimental results of a VLC driver prototype based on the parallel/series buck/boost converter proposed above are presented below. Figure 6a shows the prototype, and Figure 6b shows a schematic and block diagram of the control used.

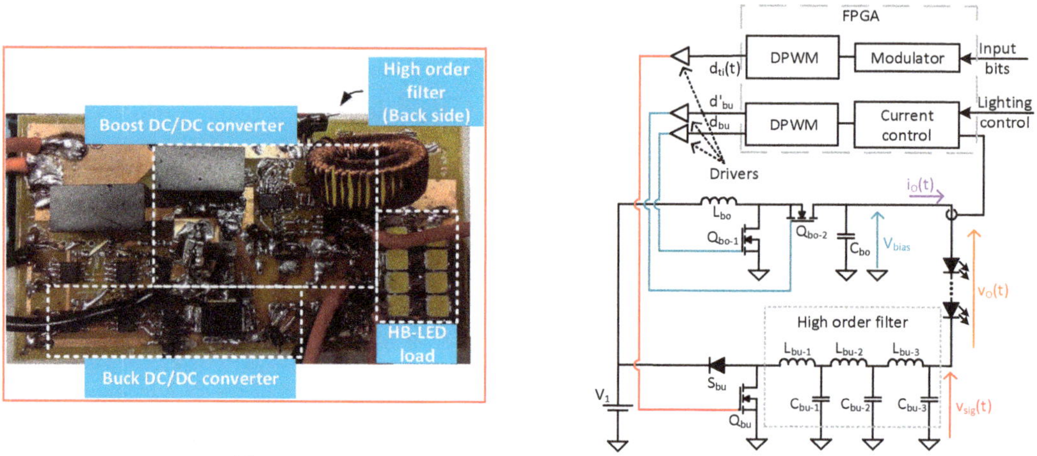

(**a**) (**b**)

Figure 6. (**a**) Photograph of the prototype. (**b**) Schematic and control diagram of the series/parallel boost/buck DC/DC converter.

As explained above, both converters share the same input voltage and are connected to the same set of HB-LEDs as a load. These HB-LEDs are differentially connected between the outputs of the boost DC/DC converter and the buck DC/DC converter. Control of both converters was implemented using a Nexys A7 FPGA [17]. The FPGA inputs were the bits to be transmitted, the lighting control, and the current flowing through the set of HB-LEDs. The FPGA implemented an average current control through the HB-LEDs and thus controlled the gate signals of the boost DC/DC converter. In this way, control of the average current flowing through the HB-LED load was implemented by controlling the value of the average output voltage. Moreover, the FPGA also controlled the buck DC/DC converter, which operated in an open loop, reproducing the communication signals generated by the modulator block.

The load used consisted of 8 XLamp MX-3 PC-LEDs [18] in series, which was the initial design parameter of the parallel/series buck/boost converter. The amplitude of the linear zone of each LED was $V\Omega = 1$ V. By applying (11), the input voltage was calculated, obtaining 8 V. As explained above, this voltage value is what allowed the maximum resolution to be obtained in the output voltage of the buck DC/DC converter. The average current through the set of HB-LEDs remained at 0.25 A, at the midpoint of the linear zone. The different parts of the specific design of the VLC driver architecture are outlined below.

4.1. Design of the Boost DC/DC Converter

The boost DC/DC converter is designed to operate as a traditional HB-LED driver with a switching frequency of 100 kHz, working in a closed loop. This converter controls the average voltage across the HB-LED load to control the average current flowing through them. Taking advantage of the slow dynamics of the temperature change in the HB-LEDs, neither a high switching frequency nor a fast response time in the converter is required. A trade-off regarding the choice of switching frequency must be highlighted: increasing the switching frequency allows the converter to be smaller (especially the reactive elements), increasing the switching losses and putting the converter's switching harmonic content closer in frequency to the signal bandwidth. The design of the converter elements does not differ from the traditional design of a boost DC/DC converter used as a traditional HB-LED driver, except for the considerations related to attenuating the switching noise. This noise is produced by the output voltage ripple of the boost DC/DC converter. The output capacitor must be suitable for decreasing the impact on the communication signal:

$$C_{bo} = \frac{\Delta Q}{\Delta 2 V_{\text{bias}}} = \frac{I_O D_{bo} T_{bo}}{\Delta 2 v_{\text{bias}}}. \quad (29)$$

where ΔQ is the total charge of the capacitor, D_{bo} and T_{bo} are the duty cycle and period of the boost DC/DC converter, respectively, and Δv_{bias} is its output voltage ripple.

The value of the output capacitor C_{bo} was calculated by applying (29) and limiting the ripple Δv_{bias} to 2%. No additional analysis was required for the input choke calculation, and traditional design rules were followed. The values of both reactive elements can be found in Table 1. Regarding component selection, the boost DC/DC converter was implemented in synchronous mode using the CSD88539 MOSFET integrated circuit, with Q_{bo-1} and Q_{bo-2} on the same chip. The driver circuitry was implemented with an ISL6700 half-bridge gate driver. A list of the main converter components is given in Table 1.

Table 1. List of main components of the boost DC/DC converter.

Q_{bo-1} and Q_{bo-2}	Boost Driver	Lbo	Cbo
CSD88539	ISL6700	71 µH	9 µF

4.2. Communication Scheme

A 64-QAM single-carrier modulation scheme was chosen. The carrier frequency was f_{sig} = 1 MHz and the symbol period was six signal periods. With these parameters, the modulation scheme had a theorical maximum bit rate of 1.5 Mbps because the error was not calculated. This modulation was chosen in order to test the ability of the series/parallel boost/buck DC/DC to reproduce and advance scheme modulation.

4.3. Design of the Buck DC/DC Converter

The buck DC/DC converter must be able to reproduce a 1 MHz modulated signal with low distortion and high efficiency. The switching frequency f_{bu}, cutoff frequency f_c, and filter order were designed based on the chosen modulation scheme [14]. In addition, a trade-off must be found between performance, resolution, and filter order. The higher the frequency, the higher the switching losses and the lower the duty cycle resolution. However, the filter design is simplified, allowing the required filter order to be reduced. On the other hand, reducing the frequency allows for reduced switching losses and increased duty-cycle resolution, but it complicates the filter design. The filter must be able to correctly pass the communication spectrum and sufficiently attenuate the switching frequency components and harmonics. This imposes some restrictions on how closely the converter switching frequency and the maximum communication frequency can be approximated depending on the filter order [14]. The switching frequency chosen was 10 MHz, a decade above the f_{sig} carrier frequency. The filter used is a 6th-order low-pass filter with a cut-off frequency of 2.5 MHz. The reactive values are shown in Table 2.

Table 2. Circuit components of the VLC HB-LED driver.

L_{bu1}	C_{bu1}	L_{bu2}	C_{bu2}	L_{bu3}	C_{bu3}
1.7 μH	9.9 nF	2.2 μH	9.9 nF	1.9 μH	5.72 nF

The remaining components were the PD84010S-E RF MOSFET as Q_{bu} and the UPS115UE3 diode as S_{bu}. Due to the high-switching frequency, the EL7155CSZ gate driver was used. A list of the converter components is given in Table 3.

Table 3. List of main components of the buck DC/DC converter.

S_{bu}	Buck Drivers	HB-LEDs
UPS115UE3	EL7155CSZ	XLamp MX-3

4.4. Experimental Waveforms

The designed prototype was experimentally evaluated by reproducing a communication signal (i.e., 64-QAM modulation scheme). During operation, the converter's performance was measured, and the most representative waveforms were obtained.

The input power was measured with two high-precision multimeters, measuring the average value of the input voltage and current. The output power, due to the high-frequency components of the signal, was calculated by extracting the signals using an oscilloscope and calculating the power through them. The architecture's performance was measured by transmitting a randomly generated 64-QAM sequence at a 0.5 m distance, achieving an efficiency of 91.5% with 7 W of output power.

The most representative waveforms of the prototype are shown in Figure 7. The voltages $v_{gs\text{-}bo}(t)$ and $v_{gs\text{-}bu}(t)$ are the gate signals that control the boost DC/DC and buck DC/DC converters, respectively. There was a significant difference in their operation frequency, with $v_{gs\text{-}bo}(t)$ being a 100 kHz signal and $v_{gs\text{-}bu}(t)$ being 10 MHz. The voltages $V_{bias}(T)$ and $v_{sig}(t)$ are the output voltages of the boost DC/DC and buck DC/DC converters, respectively. As can be seen, the voltage $V_{bias}(T)$ was constant and had negligible ripple. The variations in this voltage were produced by the temperature changes in the HB-LED load. However, the output current feedback loop kept the average current at I_o = 0.25 A. Without taking into account any temperature effects and according to the HB-LED manufacturer's datasheet, at a room temperature of 25 °C, the maximum voltage across the HB-LED for a current of 0.5 A should be V_{bias} = 32 V. In the operating example shown in Figure 7, after a few minutes of operation, during which the temperature was stable, the maximum voltage dropped to V_{bias} = 25 V, 21% lower, keeping the overall linear zone of the HB-LED load available to track the signal.

In addition, the buck DC/DC converter generated the communication signal $v_o(t)$ at its output, producing the variations in the HB-LED load necessary for VLC transmission. Figure 8 shows that the gate voltage $v_{gs\text{-}bu}(t)$ varied its duty cycle in order to generate the variations in the output voltage $v_o(t)$. The maximum amplitude of $v_o(t)$ voltage was 7.5 V, maintaining the usability of the linear zone of the HB-LED load. Voltage $v_{sig}(t)$ showed correct reproduction of the signal. This shows the correct design of the output filter and the correct attenuation of high-frequency noise, which comes from switching.

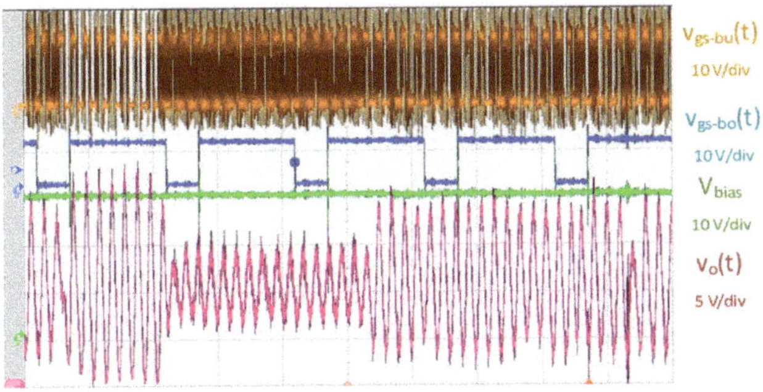

Figure 7. Gate signal of the buck and boost DC/DC converters $v_{gs\text{-}bu}(t)$ (orange) and $v_{gs\text{-}bo}(t)$ (blue), output voltage of the boost DC/DC converter (green) and output voltage ($v_o(t)$) of the signal (magenta).

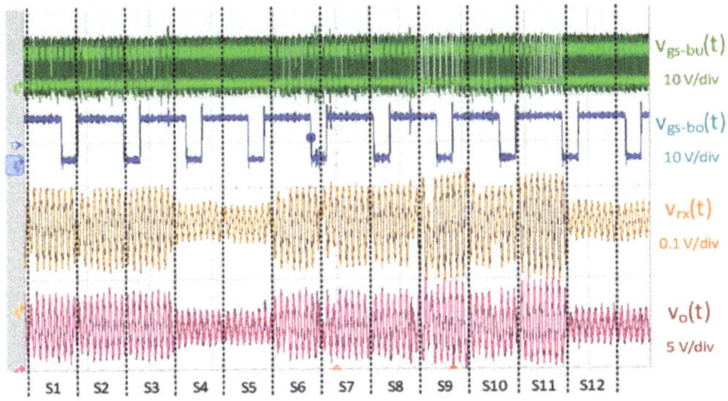

Figure 8. Gate signals of the buck and boost DC/DC converters $v_{gs\text{-}bu}$ (green) and $v_{gs\text{-}bo}$ (blue), output voltage of the buck converter $v_o(t)$ (magenta) and the voltage of the optical receiver v_{rx} (yellow) during a transmission.

In order to analyze the prototype's communication capability and correct transformation of the VLC signal into light, a longer transmission was analyzed in more detail, as shown in Figure 8, with 12 different symbols emitted. In this case, $v_o(t)$ and the light signal received from the HB-LED load $v_{rx}(t)$ were obtained and compared. The light signal (i.e., $v_{rx}(t)$) was obtained using a high-bandwidth Thorlabs PDA10A-EC optical receiver [19]. The receiver was placed in front of the HB-LED load at 0.5 m and provided a voltage output signal proportional to the light variations it received. It can be seen that $v_o(t)$ and $v_{rx}(t)$ were proportional, without apparent distortion. Incorrect voltage level design or incorrect polarization control could cause distortion in the emitted light signal, causing the HB-LED load to operate outside the linear range. Neither of these effects is observed in Figure 8.

It is important to note that some tests could be used to calculate the error, such as the bit error rate (BER), error–vector magnitude (EVM_{rms}), or the error according to the reception distance, but performing these tests depends on the demodulator as well as the optical receiver, which must be designed for this specific application. The design of the demodulator and the optical receiver are outside the scope of this paper. The main contribution of the present paper is the presentation of the series/parallel boost/buck DC/DC as a VLC transmitter for splitting power.

5. Comparison and Discussion with Other Approaches

Table 4 shows a comparison to the state of the art of power-efficient VLC drivers based on switching converters able to reproduce advanced modulation schemes—all using the power splitting technique.

Table 4. Comparison of power-efficient VLC drivers based on switching converters, which use the power splitting technique.

Ref.	Topology	Input Voltage Sources	Modulation Scheme	P_O (W)	$i_{omax}/2I_o$	f_S (MHz)	H (%)	Distance (cm)	Bit Rate (Mbps)	EVM_{rms} (%)
[14]	Two-phase buck DC/DC converter with high order filer + synchronous buck DC/DC converter	2 [1*]	OFDM	10	0.8	10 [2*]	93.6	20	7.5	15
[15]	Fly-buck DC/DC converter + Class B power amplifier	1	OFDM	20	0.42	0.1 [3*]	94	50	6	5
[16]	TIBuck DC/DC converter	1	64-QAM	7	1	10 [2*]	94	40	1.5	12
This work	Series/parallel boost/buck DC/DC Converter	1	64-QAM	7	1	10 [2*]	91.5	50	1.5 [4*]	-

[1*] One is isolated. [2*] This is the switching frequency of the DC/DC converter in charge of reproducing the communication signal. [3*] This is the switching frequency of the DC/DC in charge of both biasing and supplying Class B power amplifier. [4*] This is the theorical maximum bit rate because the error was not calculated.

The proposed series/parallel boost/buck DC/DC converter outperformed traditional buck DC/DC converter solutions in terms of efficiency, bit rate, and moderate EVM_{rms}, even operating at a higher switching frequency. Moreover, the hardware complexity of the proposed solution is similar to the buck DC/DC converter solutions presented for designing high-bandwidth VLC drivers (i.e., two-phase structures with high-order filters).

The proposed series/parallel boost/buck DC/DC converter's natural way of splitting power simplifies previous approaches using this technique because it does not need two input voltage sources (one of which is isolated) nor one isolated DC/DC converter. In addition to simplicity, strong communication capabilities and high efficiency are maintained with good boost DC/DC converter design.

Moreover, the numbers for the series/parallel boost/buck DC/DC converter are very similar to those of the TIBuck DC/DC converter. However, the new architecture overcomes the main drawback of the TIBuck solution, which is the reduced resolution for tracking the voltage across the HB-LED load. The series/parallel boost/buck DC/DC always works in the full linear region of the HB_LED load. The price is reduced efficiency (91.5% vs. 94%)

Table 4 highlights the ratio between the maximum amplitude and the average value of the output current through the HB-LED load (i.e., $i_{omax}/2I_o$), which provides an approach for assessing the ratio between the communication power and lighting power. The series/parallel boost/buck DC/DC converter maximizes the communication power compared to other approaches, which minimize the amplitude of communication to achieve high efficiency (e.g., [19]).

6. Conclusions

The proposed series/parallel boost/buck DC/DC converter can achieve high efficiency and high communication performance by means of splitting the power between two different DC/DC converters. The design is based on two DC/DC converters connected to the same input voltage, with their outputs connected in series in relation to the HB-LED load. A low-switching-frequency boost DC/DC converter works with an output current feedback loop, which is responsible for delivering most of the power and controlling the lighting (i.e., bias) of the HB-LEDs. In contrast, an open loop, fast-response buck DC/DC converter generates the communication signal. Advantages of this approach include high efficiency, high communication performance, input voltage flexibility, and improved duty cycle resolution for tracking the output voltage. The overall power efficiency is improved by splitting the power between these two converters. Most power is processed by the boost DC/DC converter (with very high efficiency), and only a small part is processed by the buck DC/DC converter (with the efficiency compromised by the high switching frequency).

Another advantage is the flexibility in terms of input voltage. In contrast to other topologies based only on buck converters, the input voltage can be lower than the threshold voltage of the LED string. An additional effect of lower input voltages is increased resolution of the duty cycle for tracking the output voltage in the high-frequency buck DC/DC converter versus temperature changes. The experimental results demonstrate an efficiency of 91.5% in reproducing a 64-QAM digital modulation, with a bit rate of 1.5 Mbps.

Author Contributions: D.G.A. and J.R. conceived of and performed the experiments. D.G.A. and J.S. developed the adaptation of solutions for VLC and the design procedure. D.G.L. and D.G.A. analyzed the data and wrote the paper. M.M.H., J.R.G.-M. and J.S. contributed to the definition, the design, and the evaluation of the experiments and the revision of the paper. All authors have read and agreed to the published version of the manuscript.

Funding: This work was funded by the Principality of Asturias under the project GRU-GIC-24-036, the "Severo Ochoa" Program, project grant no. BP20-181, and by the FEDER funding.

Data Availability Statement: No new data were created or analyzed in this study.

Conflicts of Interest: There is no conflict of interest.

References

1. Cisco Systems. Cisco Annual Internet Report—Cisco Annual Internet Report (2018–2023) White Paper. Available online: https://www.cisco.com/c/en/us/solutions/collateral/executive-perspectives/annual-internet-report/white-paper-c11-741490.html (accessed on 18 April 2025).
2. *802.15.7-2018*; IEEE Standard for Local and Metropolitan Area Net-Works–Part 15.7: Short-Range Optical Wireless Communications. IEEE: New York, NY, USA, 2019; Revision of IEEE Std 802.15.7-2011. pp. 1–407. [CrossRef]
3. Matheus, L.E.M.; Vieira, A.B.; Vieira, L.F.M.; Vieira, M.A.M.; Gnawali, O. Visible Light Communication: Concepts, Applications and Challenges. *IEEE Commun. Surv. Tutor.* **2019**, *21*, 3204–3237. [CrossRef]
4. Rehman, S.U.; Ullah, S.; Chong, P.H.J.; Yongchareon, S.; Komosny, D. Visible Light Communication: A System Per-spective—Overview and Challenges. *Sensors* **2019**, *19*, 1153. [CrossRef] [PubMed]
5. Karunatilaka, D.; Zafar, F.; Kalavally, V.; Parthiban, R. LED Based Indoor Visible Light Communications: State of the Art. *IEEE Commun. Surv. Tutor.* **2015**, *17*, 1649–1678. [CrossRef]
6. Mesleh, R.; Elgala, H.; Little, T.D.C. On the performance degradation of optical wireless OFDM communication systems due to changes in the LED junction temperature. In Proceedings of the 2013 20th International Conference on Telecommunications, Casablanca, Morocco, 6–8 May 2013; pp. 1–5.
7. Khalid, A.M.; Cossu, G.; Corsini, R.; Choudhury, P.; Ciaramella, E. 1-Gb/s Transmission Over a Phosphorescent White LED by Using Rate-Adaptive Discrete Multitone Modulation. *IEEE Photonics J.* **2012**, *4*, 1465–1473. [CrossRef]
8. Huang, X.; Shi, J.; Li, J.; Wang, Y.; Chi, N. A Gb/s VLC Transmission Using Hardware Preequalization Circuit. *IEEE Photonics Technol. Lett.* **2015**, *27*, 1915–1918. [CrossRef]
9. Chun, H.; Rajbhandari, S.; Faulkner, G.; Tsonev, D.; Xie, E.; McKendry, J.J.D.; Gu, E.; Dawson, M.D.; O'Brien, D.C.; Haas, H. LED Based Wavelength Division Multiplexed 10 Gb/s Visible Light Communications. *J. Light. Technol.* **2016**, *34*, 3047–3052. [CrossRef]
10. Kazimierczuk, M.K. *RF Power Amplifiers*; John Wiley & Sons, Ltd.: Hoboken, NJ, USA, 2014; ISBN 978-1-118-84437-3.
11. Teixeira, L.; Loose, F.; Alonso, J.M.; Barriquello, C.H.; Reguera, V.A.; Costa, M.A.D. A Review of Visible Light Communication LED Drivers. *IEEE J. Emerg. Sel. Top. Power Electron.* **2022**, *10*, 919–933. [CrossRef]
12. Loose, F.; Teixeira, L.; Duarte, R.R.; Costa, M.A.D.; Barriquello, C.H. On the Use of the Intrinsic Ripple of a Buck Converter for Visible Light Communication in LED Drivers. *IEEE J. Emerg. Sel. Top. Power Electron.* **2018**, *6*, 1235–1245. [CrossRef]
13. Teixeira, L.; Loose, F.; Barriquello, C.H.; Reguera, V.A.; Costa, M.A.D.; Alonso, J.M. On Energy Efficiency of Visible Light Communication Systems. *IEEE J. Emerg. Sel. Top. Power Electron.* **2021**, *9*, 6396–6407. [CrossRef]
14. Rodriguez, J.; Lamar, D.G.; Miaja, P.F.; Aller, D.G.; Sebastian, J. Power-Efficient VLC Transmitter Based on Pulse-Width Modulated DC–DC Converters and the Split of the Power. *IEEE Trans. Power Electron.* **2019**, *34*, 1726–1743. [CrossRef]
15. Aller, D.G.; Lamar, D.G.; García-Mere, J.R.; Arias, M.; Rodriguez, J.; Sebastian, J. Performance Evaluation of the Two-Input Buck Converter as a Visible Light Communication High-Brightness LED Driver Based on Split Power. *Sensors* **2024**, *24*, 6392. [CrossRef] [PubMed]
16. Loose, F.; Teixeira, L.; Brum, J.P.; Barriquello, C.H.; Reguera, V.A.; Alonso, J.M.; Costa, M.A.D. Efficient Hybrid Buck Converter for Visible Light Communication in LED Drivers. *IEEE Trans. Ind. Electron.* **2022**, *69*, 1877–1887. [CrossRef]

17. Nexys A7: FPGA Trainer Board Recommended for ECE Curriculum. Available online: https://digilent.com/shop/nexys-a7-fpga-trainer-board-recommended-for-ece-curriculum/?srsltid=AfmBOooqaTmTTIZnRTuzBPsKMtRNslH92JbC72icYmMg-Nhh1N71M-nI (accessed on 18 April 2025).
18. Cree Components. XLamp MX-3 LEDs—Cree Com-Ponents. Available online: https://www.cree-led.com/products/leds/xlamp/ (accessed on 7 November 2024).
19. Thorlabs. PDA10A-EC Si Fixed Gain Detector, 200–100 nm, 150 MHz BW. Available online: https://www.thorlabs.com/thorproduct.cfm?partnumber=PDA10A-EC (accessed on 18 April 2025).

photonics MDPI

Article

Interference Mitigation in VLC Systems using a Variable Focus Liquid Lens

Krishnendu Bera * and Nemai Karmakar

Department of Electrical and Computer Systems Engineering, Monash University, Melbourne, VIC 3800, Australia; nemai.karmakar@monash.edu
* Correspondence: krishnendu.bera@monash.edu

Abstract: The field of view (FOV) is an important parameter of a visible light communication (VLC) receiver. A variable FOV can be useful to mitigate interference from neighboring cells in a multi-cell VLC network. The existing works on dynamic FOV VLC receivers have used a mechanical iris to control the receiver's FOV, making the VLC receiver bulky, slow, and power-consuming. In this article, an electronically controlled variable focus liquid lens is used to vary the FOV of the receiver dynamically. A low-cost microcontroller-based feedback control system controls the effective FOV of the receiver to reject signals from unwanted transmitters, thus maximizing the signal-to-interference plus noise ratio (SINR) of the received signal. Experimental results show that the proposed technique effectively improves SINR performance in a multi-cell VLC network. To the best of the authors' knowledge, this is the first reported work to utilize an electronically controlled compact liquid lens for designing a dynamic FOV VLC receiver.

Keywords: VLC network; optical wireless; Li-Fi; optical interference; interference mitigation; receiver FOV; liquid lens

1. Introduction

Visible light communication (VLC) is an emerging wireless communication technology that utilizes visible light as a data carrier. VLC can complement future wireless communication systems to address the spectrum crunch caused by the exponential growth of data traffic [1–3]. Researchers have already achieved speed in the order of gigabits/s in VLC systems [4–8]. However, one of the major drawbacks of VLC systems is their small coverage and short range [8]. Multiple VLC transmitters are generally used to cover an entire area due to the small communication coverage and line-of-sight nature of VLC systems. In fact, this brings the advantage of high spatial density in VLC systems compared to radio frequency communication systems. One of the key factors affecting multi-cell VLC systems' performance is the receiver's field of view (FOV) [9]. FOV refers to the angle of visibility between the transmitter and receiver from the receiver's point of view. FOV is a crucial parameter in VLC systems as it determines the number of transmitters 'seen' by the receiver and the amount of light received by the receiver. A smaller FOV leads to a higher outage probability. A larger FOV, on the other hand, increases the likelihood of interference from nearby unwanted light sources, including neighboring VLC transmitters, which can degrade the signal quality and reduce the data transmission rate. Therefore, there is a trade-off between the outage probability and the data transmission rate. The issue becomes more serious in dense VLC transmitter deployment scenarios. Therefore, the FOV must be carefully considered when designing VLC systems to ensure optimal performance. The importance of receiver FOV in VLC is depicted in Figure 1.

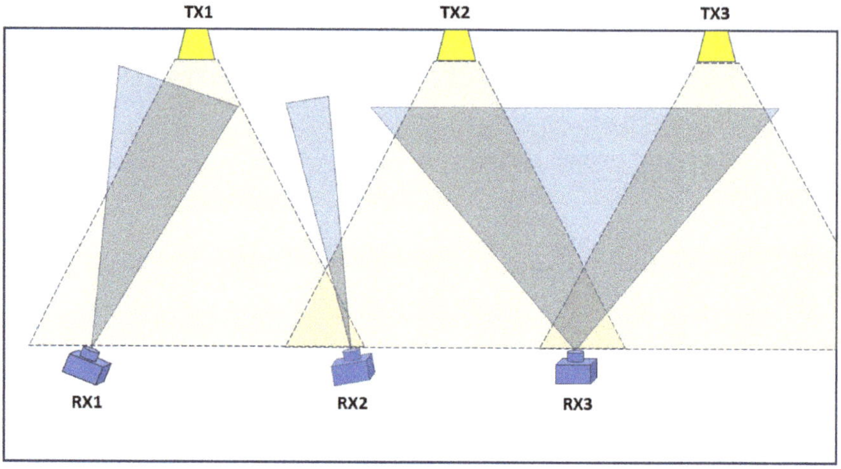

Figure 1. Importance of receiver's FOV in a VLC system: FOV of Receiver 1 (RX1) is aligned with transmitter 1 (TX1) only. RX2 is not aligned with any transmitter due to a small FOV. RX3 receives interfering signals from both TX2 and TX3.

To reduce unintended interference from neighboring transmitters in a multi-cell VLC architecture, different optimization techniques are used at the transmitter end, including optimization of individual transmitter's time-scheduling, power allocation, and zero-forcing precoding [10,11]. These techniques are sensitive to users' density and locations [11]. At the receiver end, angular diversity receivers, proposed to mitigate interference are non-planar and bulky [12–16]. The polarization of the optical signal has been exploited to mitigate interference of unwanted signals [17]. A liquid crystal device (LCD) has been used in front of the photo-diode (PD) to block signals from unwanted directions and improve interference performance [18]. The optimization of the receiver's field of view has also been proposed [11]. Using simulation results, it has been shown that communication performance can be improved significantly using a dynamic field-of-view receiver [9,19,20]. For hardware-level implementation, a mechanical iris has been used to dynamically change the receiver's field of view to improve communication performance in dense multi-cell VLC networks [21]. A mechanical iris commonly found in digital cameras is a mechanical system where the FOV of the sensor can be changed mechanically according to the need. In VLC systems, the optimum FOV is mainly determined by the position of the receiver with respect to different transmitters. A larger iris opening leading to a larger FOV allows more signal to be detected by the receiver and a smaller iris opening leads to minimum interference from neighboring cells. A particle swarm optimization of the receiver's FOV has been proposed using a similar mechanical setup [11]. However, a motor-controlled mechanical iris is bulky, slow, and power-consuming, thus impractical for communication systems, especially for compact handheld electronic devices. A non-mechanical, compact system is necessary for the VLC receiver's front end. In this paper, a variable focus liquid lens is utilized to adapt the FOV of the high-speed VLC receiver to minimize inter-cell interference of a VLC network and to maximize the signal-to-interference plus noise ratio (SINR) of the received signal.

The rest of the paper is organized as follows: Section 2 presents the simulation results, Section 3 explains the working principle of the proposed receiver, Section 4 describes the experiment process and discusses the experimental results, Section 5 discusses the advantages and potential application scenarios of the proposed system, and finally, in Section 6, conclusions are drawn.

2. Simulation Results

To study the effect of the receiver's FOV on communication performance, a multi-cell VLC system is simulated in MATLAB®. An LED or a diffused laser diode (LD) can be considered a Lambertian source, with an optical beam distribution as expressed below:

$$R(\phi) = \begin{cases} \frac{(m_l+1)}{2\pi} cos^{m_l}(\phi), & \text{if } -\pi/2 \le \phi \le \pi/2 \\ 0, & \text{elsewhere}, \end{cases} \quad (1)$$

where ϕ and ψ are irradiance and incidence angles, respectively. $m_l = -\frac{ln2}{ln(cos\Phi_{1/2})}$ is the Lambertian emission order, which is a function of the half-power semi-angle of the transmitter, $\Phi_{1/2}$.

The line-of-sight channel gain can be expressed as:

$$H_{los} = \begin{cases} \frac{A_r cos(\psi)(m_l+1)}{2\pi d^2} cos^{m_l}(\phi) G_{lens}, & \text{if } 0 \le \psi \le \psi_{max} \\ 0, & \text{elsewhere}, \end{cases} \quad (2)$$

where G_{lens} is the gain of the receiver lens; A_r is the area of PD; and d is the Euclidean distance between the transmitter and PD. In a room with multiple transmitters, the SINR can be expressed as:

$$SINR = \frac{(R_{pd} H_{los} P_{t,\alpha})^2}{(Bw \times INC)^2 + \sum_{i \neq \alpha} (R_{pd} H_i P_{t,i})^2}, \quad (3)$$

where $R_{pd}(A/W)$ is the detector responsivity; $Bw(Hz)$ is the bandwidth of the VLC channel; $INC(A/Hz)$ denotes the input noise current of the PD.

A standard room size (5 m × 5 m × 3 m) has been chosen. The transmitter power, detector diameter, responsivity of PD, bandwidth of the system, and input noise current are 10 W, 0.4 mm, 0.2 A/W, 20 MHz and 16 pA/$\sqrt{\text{Hz}}$ respectively. The half-power beam width of each transmitter has been taken as 60° to simulate the lighting-grade LEDs. The receiver is assumed to always face the roof perpendicularly, and the receiver plane is at a height of 0.85 m above the floor. The signal from each transmitter is degraded by the interference caused by the other three transmitters. The simulation result for the receiver's FOV of 45° is shown in Figure 2. The maximum SINR is achieved when the receiver is just under the individual transmitters. The maximum simulated SINR for this case is 27 dB. The simulation result for the receiver's FOV of 30° is shown in Figure 3. It is found that the peak SINR has improved by 6 dB compared to the previous case. However, due to reduced FOV, the performance degrades when the receiver moves away from the transmitter. As a result, the minimum SINR has dropped from −5 dB to −12 dB. The negative SINR indicates that the interference from neighboring transmitters is greater than the desired received signal. The drop in the minimum SINR means that the outage probability of the communication system increases with a lower FOV of the receiver. The trade-off between communication performance and FOV is very critical for VLC systems.

Next, the multi-cell VLC system is simulated with a variable FOV. All the other simulation parameters remain the same except the FOV of the receiver. It is assumed that the receiver's FOV is dynamic and it can be adjusted with a resolution of 1° to maximize the SINR performance. At each receiver position, the FOV is varied 0° to 90° with a resolution of 1°, and the FOV corresponding to the maximum attainable SINR at that point is considered as ideal. The simulation result for this case is shown in Figure 4. The maximum and the average SINR have improved to 39.3 dB and 17.6 dB, respectively.

Therefore, theoretically, it is established that a dynamic FOV of the VLC receiver can reduce the inter-cell interference, thus improving the overall throughput of the multi-cell VLC network. In this paper, a variable-focus liquid lens is utilized to implement the dynamic FOV of the high-speed VLC receiver to minimize the inter-cell interference of a VLC network and to maximize the SINR of the received signal. The experimental details and results are presented in the next sections.

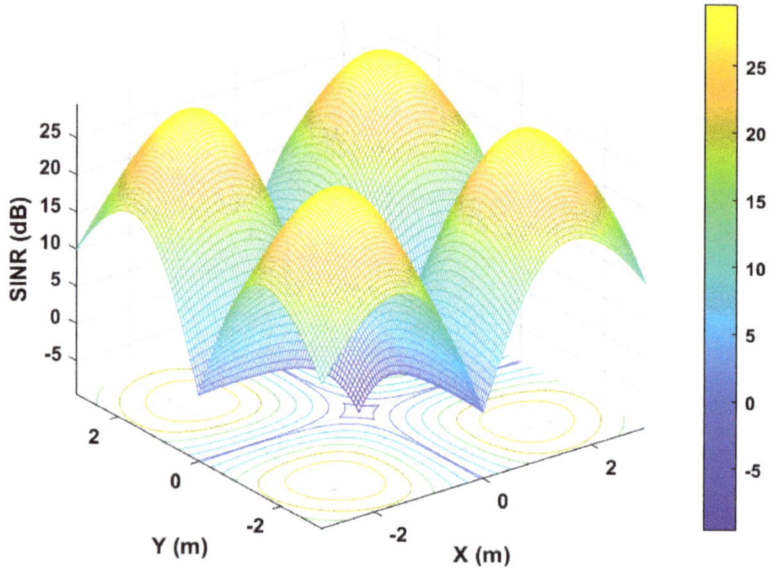

Figure 2. Simulated SINR distribution of a multi-cell VLC network for fixed receiver's FOV of 45°; maximum SINR is 27 dB and average SINR is 13.6.

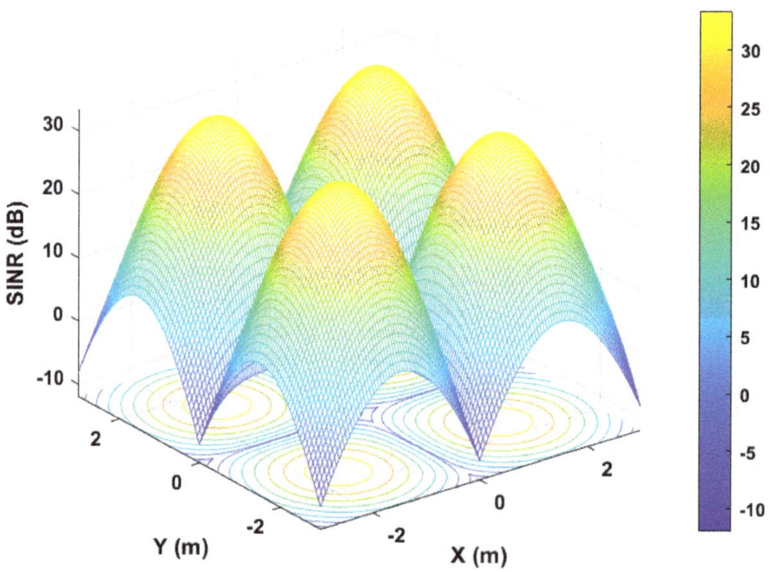

Figure 3. Simulated SINR distribution of a multi-cell VLC network for fixed receiver's FOV of 30°; maximum SINR is 33 dB and average SINR is 13.3.

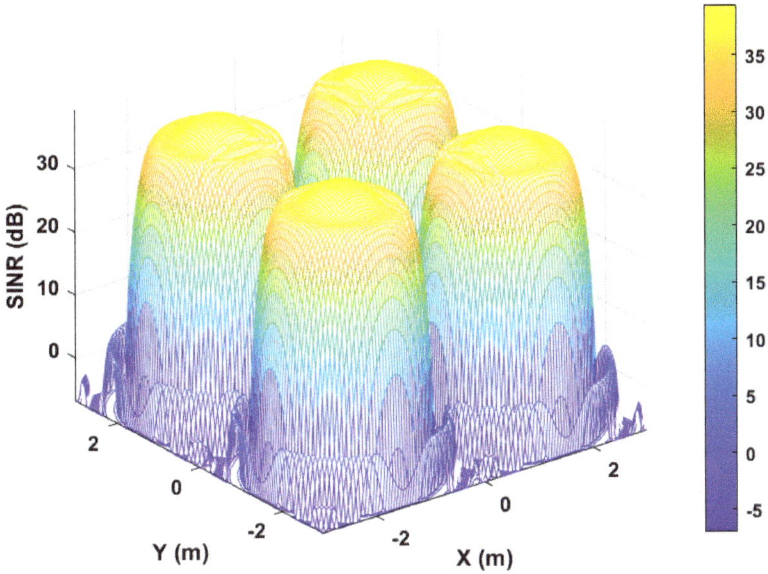

Figure 4. Simulated SINR distribution of a multi-cell VLC network with dynamic receiver's FOV; maximum SINR is 39.3 dB and average SINR is 17.6.

3. System Description

The goal of this work is to optimize the receiver FOV at the hardware level at the receiver front end. The intention is to receive the transmitted signal from one transmitter at the minimum FOV of the receiver. A mechanical iris-based dynamic FOV receiver has been proposed in the literature [19,21], as shown in Figure 5. In this work, it is implemented using a dynamic FOV lens assembly. The lens assembly consists of two lenses: one variable-focus liquid lens and the other, a fixed focus lens, as shown in Figure 6. The liquid lens (Corning variable-focus lens A-25H0-D0-P10) has an electronically variable focal length starting from −28 mm to +28 mm. The basic structure of a liquid lens consists of a small chamber filled with a conductive liquid, typically water or oil, and separated from the surrounding environment by a hydrophobic (water-repellent) coating. The liquid is typically contained within a transparent glass or plastic membrane. The key principle behind the operation of a liquid lens is electrowetting. Electrowetting refers to the phenomenon where the wetting properties of a liquid on a solid surface can be modified by applying an electric field. In the case of a liquid lens, an electric voltage is applied across electrodes placed on either side of the liquid chamber. By varying the voltage across the electrodes, the curvature of the liquid droplet can be controlled, thereby changing the focal length of the lens. The fixed-focus aspheric lens (Thorlabs ACL1210U) has a focal length of 10.5 mm. The optical signal enters the liquid lens from the free space channel and then passes through the fixed focus lens. Finally, the optical signal is received by the high-speed photo-detector (New Focus 1601FS-AC). The focal length of the liquid lens is varied electronically using a control signal. Thus, the effective focal length of the lens assembly can be expressed [22] according to Equation (4).

$$\frac{1}{f_{eff}} = \frac{1}{f_{liquid}} + \frac{1}{f_{fixed}} + \frac{d}{f_{liquid} \cdot f_{fixed}} \tag{4}$$

where f_{eff} is the effective focal length of the lens assembly, f_{liquid} is the focal length of the liquid lens, f_{fixed} is the focal length of the fixed aspheric lens and d is the distance between the liquid lens and the aspheric lens, as shown in Figure 6. The distance (d) between the liquid lens and the fixed focus lens is kept at 3 mm. It is kept at the minimum possible distance in the setup to minimize non-linearity in the control signal. The relationship between a lens' focal length and its field of view is inversely proportional. In general, as the focal length of a lens increases, the field of view decreases, and vice versa. It can be varied according to Equation (5). Thus, the FOV of the lens assembly can be electronically varied.

$$FOV_{eff} = 2\tan^{-1}\left(\frac{S}{2f_{eff}}\right) \tag{5}$$

where S is the size of the sensor, in this case, the diameter of the photo-diode. By varying the control voltage of the liquid lens from 31 V to 64 V, the effective FOV of the lens assembly can be varied from 6.1° to 16.5°, as shown in Figure 7, and it can be seen that the relationship is practically linear.

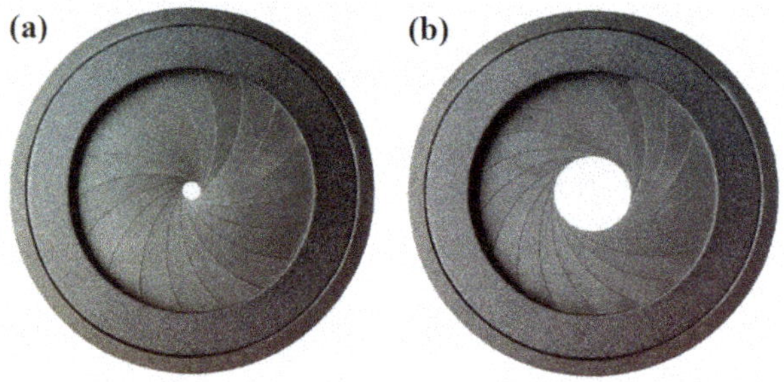

Figure 5. A mechanical iris for changing the FOV dynamically, as proposed in the literature [19,21]: (**a**) a smaller iris opening leading to smaller FOV and (**b**) a larger iris opening leading to larger FOV.

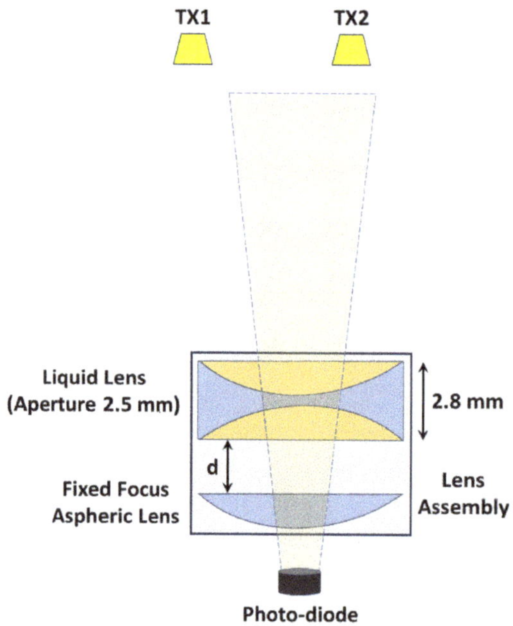

Figure 6. A schematic diagram of the proposed dynamic FOV lensing setup for selecting the desired signal and avoiding unwanted interfering signals.

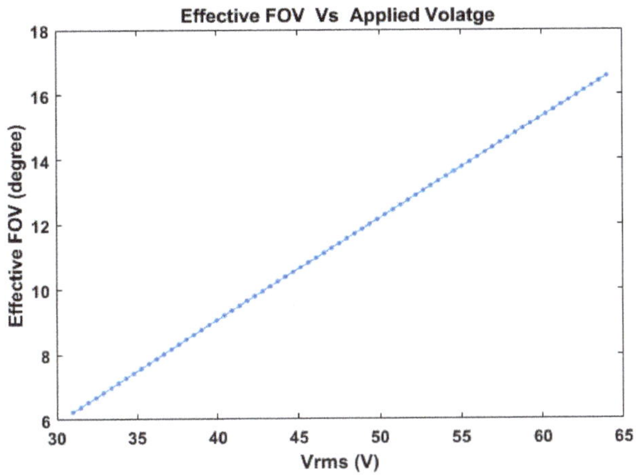

Figure 7. The effective FOV of the lens assembly versus the applied control voltage (the linear relationship between the effective FOV and the control voltage is beneficial for easy control of the FOV).

The block diagram of the proposed variable FOV VLC receiver is shown in Figure 8. The transmitted signals enter the receiver from the free-space channel through the liquid lens. After passing through the liquid lens, the light passes through the fixed focus aspheric lens and is collected by the photo-detector. The combination of these two lenses develops an overall lens assembly with a controllable FOV. The photo-detector converts the optical signal into an electrical signal. The output of the photo-detector module is passed through a broadband power divider designed and fabricated in-house. One output of the power divider goes to a digital oscilloscope for monitoring and the other output is connected to a control unit. The control unit processes the signal and provides a control signal to the liquid lens so that the FOV of the lens assembly can be optimized for minimum interference from neighboring transmitters. In this system, two laser diodes have been used as two transmitters. Both transmitters are blue laser diodes transmitting the same wavelength (450 nm) of light and the same optical power. Both transmitters send different on–off-keying (OOK) modulated signals sampled at a maximum of 1700 M Samples/s. One transmitter acts as a noise source for the other transmitter. The control unit is implemented using a low-cost Arduino Atmega 2560 microcontroller together with an IRF520N power MOSFET module. The control unit detects both the OOK and the noise signal. If the interference from the noise signal is high then the control unit sends a pulse-width-modulated signal to the MOSFET module. The output of the power MOSFET module adjusts the focal length of the liquid lens so that only the desired signal is detected by the photo-diode and the interfering light goes out of the FOV of the photo-diode receiver.

Each frame of the transmitted signals from both transmitters consists of two parts: (a) a unique identifier and (b) an actual modulated signal, as shown in Figure 9. The unique identifier is a 10-bit sequence code and is different for two transmitters. The transmitters are synchronized and the two identifier bit sequences from the two corresponding transmitters are transmitted in time division multiplexed form. At the receiver side, the control unit detects the identifier bit sequence to understand which signal is being received. The algorithm running in the control unit is presented through a flow chart in Figure 10. If the control unit cannot detect any identifier bit sequence, it sends a control signal to the liquid lens to increase the FOV of the receiver. If the control unit detects identifier bit sequences from both transmitters, it indicates that the receiver FOV is larger than necessary. It then sends a lower control voltage to the IRF520N power MOSFET module to reduce the FOV. Conversely, if the control unit identifies only one identifier bit sequence, then it indicates that the receiver is receiving a signal from only one transmitter. In that case, there is no interference from neighboring transmitters. Therefore, no FOV modification is required. For more than two transmitters, the number of identifier sequences needs to be increased. As the length of the identifier bit sequence is significantly smaller compared to the actual OOK signal, this does not degrade the communication speed.

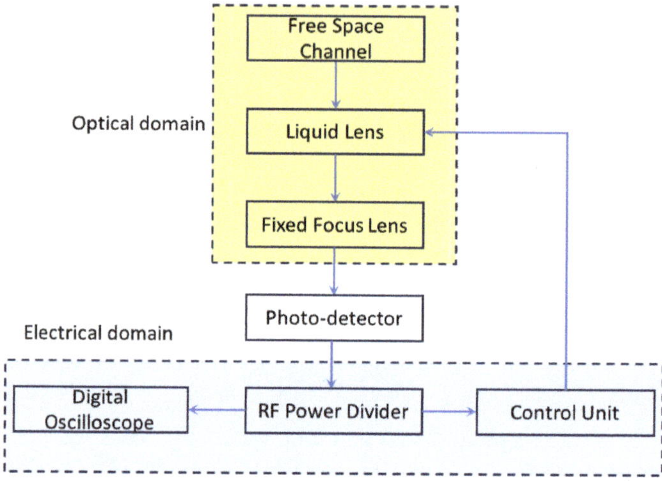

Figure 8. The block diagram of the proposed dynamic FOV VLC receiver design.

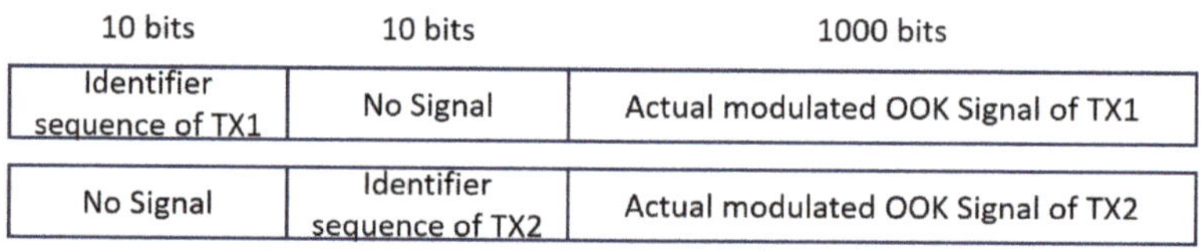

Figure 9. Contents of the signal frames from two transmitters (**above**) transmitter 1 (**below**) transmitter 2.

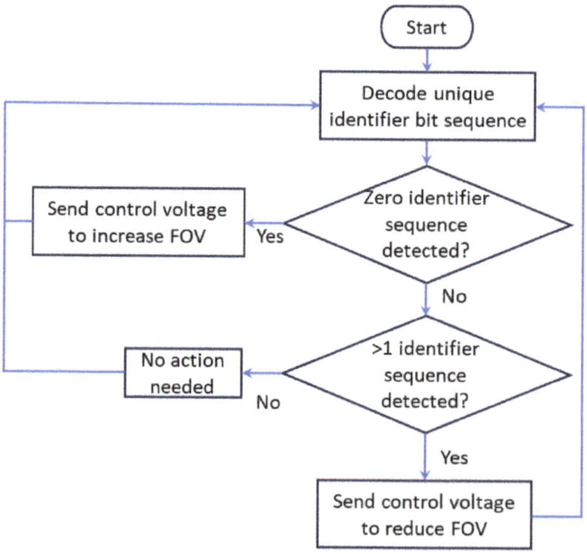

Figure 10. The flow chart of the control unit algorithm used to optimize the FOV of the receiver.

4. Experiments and Results

The experiments were performed in the laboratory with the optical components placed on an optical table for proper alignment. Two 450 nm blue laser diodes (Thorlabs PL450B) are used as transmitters: one acting as a desired transmitter and the other as an interfering signal source. The transmitter laser outputs are diffused and then directed towards the receiver, as shown in Figure 11.

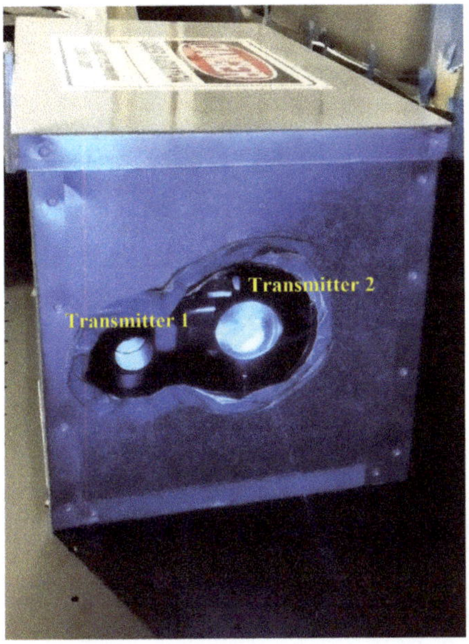

Figure 11. The transmitter side of the experimental platform on the optical table: two laser diode transmitters, one acting as a desired transmitter and the other as an interfering source.

The receiver side of the setup is shown in Figure 12 with the main blocks marked. The lens assembly is shown in detail in Figure 13. The distance between the transmitter and the receiver is 1.25 m. The two transmitters were placed with varying distances between them, starting from 5 cm up to 50 cm. For each position, the system performance was evaluated in terms of received SINR. It was found that the receiver can mitigate the interference with two transmitters at as close proximity as 10 cm. In practical scenarios, however, indoor lights are not installed so closely. Therefore, the performance evaluation of the proposed system was performed for the two transmitters placed at a 50 cm distance while keeping practical scenarios in mind. The SINR performance of a fixed FOV lens and our proposed dynamic FOV lens assembly are compared.

The SINR is calculated by measuring the received desired signal power, the interfering signal power, and the system noise power by turning off each transmitter at a time at each position of taking the reading. On the optical table, the receiver's position was kept fixed because the manual alignment of the lens assembly and the photo-detector is difficult and time-consuming, and it is easier to change the positions of the transmitters. One at a time, two LD transmitters were placed at a distance of 50 cm along the perpendicular axis of the communication direction. The two LD transmitter's positions were shifted, with one transmitter's position starting from 0 cm up to 50 cm at 5 cm intervals without changing the distance between them. The receiver considers the nearer transmitter as the desired one and the further one as interference. This method was adopted to replicate the situation as if the transmitters are at fixed positions and the receiver is moving, as happens in practical scenarios. The distance between the two transmitters was kept at

50 cm as practical indoor lights are generally not installed at any closer intervals than 50 cm. Effectively, the measured SINR at different receiver positions is plotted in Figure 14 for both the dynamic FOV receiver and fixed FOV receiver. It can be seen that the dynamic FOV receiver performs better at all positions, and the improvement is more significant when the receiver is nearer the middle of both transmitters. On average, the SINR improvement is more than 8 dB. The improvements are less significant when the receiver is either at the edge or in the middle of the two transmitters. At the edges, the interference is already insignificant, thus the improvement is minimal. In the middle, the interference is too severe to be mitigated. Apart from these two extreme positions, the SINR improvements are significant. Thus, a dynamic FOV receiver can be useful for interference mitigation in multi-cell VLC networks, especially in dense deployments. As the lenses and the PD are all circular and thus symmetrical, the one-dimensional receiver position shifting is enough to validate the effectiveness of the proposed system.

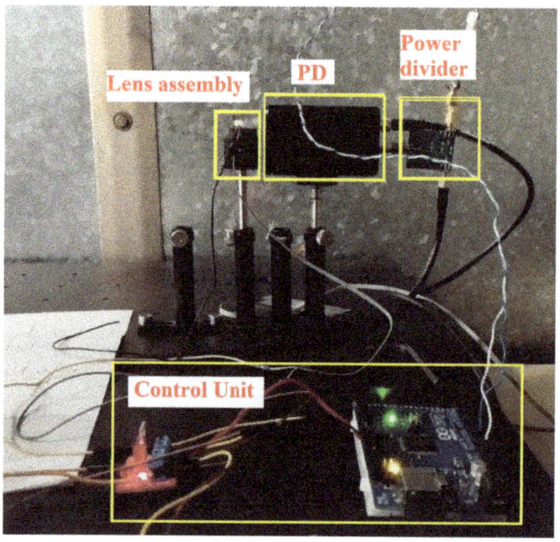

Figure 12. The receiver side of the experimental platform on the optical table.

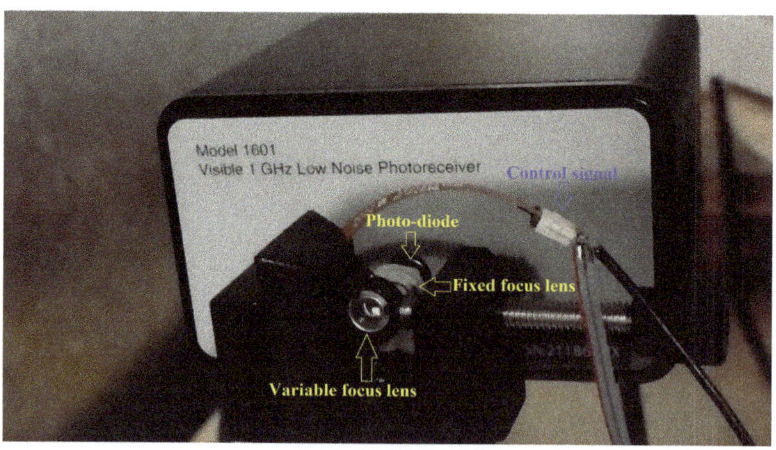

Figure 13. The lens assembly of the VLC receiver.

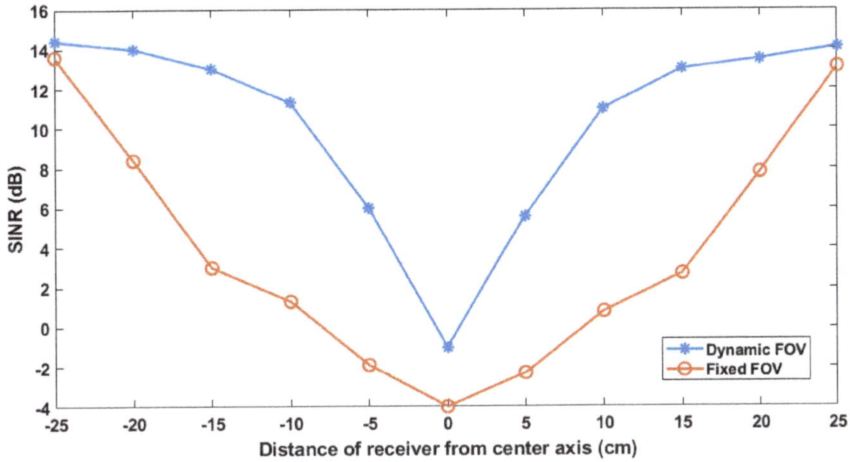

Figure 14. SINR plots at different receiver positions both for the proposed dynamic FOV receiver and the fixed FOV receiver.

Communication Performance

The communication performance at each point is evaluated and shown in Figure 15. For a fair evaluation, the maximum possible error-free data rate has been considered. The OOK modulation scheme was used for its simplicity. The sampling rate is varied and the maximum possible sample rate with no bit error is 1700 MHz, corresponding to a 1700 Mbps data rate, ignoring the identifier sequence. When the receiver was moved away from the transmitter, the interference increased. Thus, the data rate degrades gradually. However, the data rate fell sharply when the FOV of the receiver was fixed. On the other hand, for liquid lens-based dynamic FOV, the data rate was more stable. Except for exactly in the middle of the two transmitters, the data rate is higher than 400 Mbps. At all points, the dynamic FOV receiver performs better.

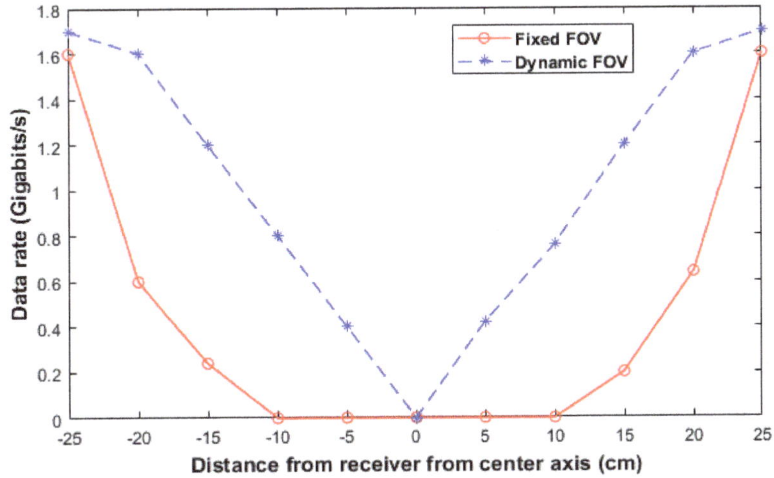

Figure 15. Error-free data rates at different receiver positions both for the proposed dynamic FOV receiver and the fixed FOV receiver.

5. Discussion and Potential Applications

In the literature, most works [12–16] have used angular diversity receivers to reduce interference in a VLC network. These receivers are non-planar and they consume more power due to multiple PDs involved. Mechanical iris-based dynamic FOV receivers [11,19,21] can mitigate interference using a single PD. However, they are bulky and more power-consuming (a few hundred mW) due to their mechanical operation and the involvement of an additional mechanical motor. Mechanical iris can also be noisy, which is not desired. The liquid lens-based receiver proposed in this paper is electronically controllable, compact (2.8mm thickness), silent, and less power-consuming (14 mW to be precise). As a result, the proposed receiver can be more suitable for hand-held devices. A comparison between this work and the related literature is presented in Table 1.

Table 1. Comparison with recent works on receiver design for interference mitigation in VLC systems

Ref.	Mechanism	Size/Shape	No of PDs	Power Consumption
[12–16]	Angular diversity receiver	Non-planar and bulky	4 or more	Higher due to multiple PDs (4 times compared to single PD systems)
[11,19,21]	Mechanical iris	Planar but large (few cm)	1	Higher due to mechanical iris and motor (few hundred mW)
This work	Liquid lens	Planar and compact (few mm)	1	Lower (14 mW)

The maximum power consumption of the liquid lens is 14 mW. There is no measurable power loss for the liquid lens compared to a solid glass-based lens. The power divider is a 10:90 divider. Therefore, only 10% of the received signal is used to control the liquid lens. The remaining 90% of the signal is processed for calculating the SINR. Therefore, the noise added by the passive power divider circuit is included in the measured SINR.

The method will work irrespective of LED or LD. The size of the light source should be smaller for better performance. The proposed method is more suitable for high-speed VLC systems where the transmitters are generally laser diodes (smaller size even after including the diffuser).

In indoor conditions, this proposed system can reduce interference without the risk of a signal outage. For PD-based receivers, photo-diode saturation is another issue that degrades the received signal quality in outdoor conditions. Within the photo-diode FOV, if there is any unwanted bright light source, such as the Sun or headlights from other cars, then the photo-diode becomes saturated. Optical filters, to some extent, can prevent saturation by blocking unwanted wavelengths of incident light but they degrade the signal power too. A variable FOV using a variable focus lens at the receiver can help avoid PD saturation by dynamically adjusting the FOV depending on the outdoor conditions.

Dynamic FOV receivers can also be very useful for visible light positioning systems. One of the issues of VLP systems is the small field of view VLC receivers and the requirement of dense lighting installation because of that. A wide FOV receiver, on the other hand, is affected by higher short noise, more interference, and distorted images in the case of image sensor receivers. A wide FOV lens for an image sensor may not be perfect for a front-faced camera in a smartphone with the primary purpose of photography. A variable FOV receiver can adjust the FOV depending on the channel conditions and application scenario.

6. Conclusions

The FOV of a receiver is a critical parameter in a multi-cell VLC system. An electronically controlled variable FOV receiver can be a solution for maintaining an LOS connection with the strongest signal intensity as well as reducing interference from unwanted sources.

In this work, an electronically controllable variable focus liquid lens is utilized to design a dynamic FOV VLC receiver for mitigating interference from neighboring cells. Experimental results show that the proposed dynamic FOV receiver significantly improves the SINR compared to a fixed FOV receiver. The proposed solution is compact, faster, and consumes less power compared to existing mechanical iris-based solutions. In addition to multi-cell VLC systems, the proposed solution has potential applications in indoor positioning and intelligent transport systems.

Author Contributions: Conceptualization, K.B.; methodology, K.B.; software, K.B.; validation, K.B.; investigation, K.B.; resources, N.K.; writing—original draft, K.B.; writing—review and editing, N.K.; supervision, N.K.; project administration, N.K. All authors have read and agreed to the published version of the manuscript.

Funding: This research received no external funding.

Institutional Review Board Statement: Not applicable.

Informed Consent Statement: Not applicable.

Data Availability Statement: Data presented in this paper are available upon request. The request should be made through email.

Conflicts of Interest: The authors declare no conflicts of interest.

References

1. Burchardt, H.; Serafimovski, N.; Tsonev, D.; Videv, S.; Haas, H. VLC: Beyond point-to-point communication. *IEEE Commun. Mag.* **2014**, *52*, 98–105. [CrossRef]
2. Ayyash, M.; Elgala, H.; Khreishah, A.; Jungnickel, V.; Little, T.; Shao, S.; Rahaim, M.; Schulz, D.; Hilt, J.; Freund, R. Coexistence of WiFi and LiFi toward 5G: Concepts, opportunities, and challenges. *IEEE Commun. Mag.* **2016**, *54*, 64–71. [CrossRef]
3. Ghassemlooy, Z.; Ebrahimi, F.; Rajbhandari, S.; Olyaee, S.; Tang, X.; Zvanovec, S. Investigation of a hybrid ofdm-pwm/ppm visible light communications system. *Opt. Commun.* **2018**, *429*, 65–71.
4. Chi, Y.C.; Hsieh, D.H.; Lin, C.Y.; Chen, H.Y.; Huang, C.Y.; He, J.H.; Ooi, B.; DenBaars, S.P.; Nakamura, S.; Kuo, H.C.; et al. Phosphorous diffuser diverged blue laser diode for indoor lighting and communication. *Sci. Rep.* **2015**, *5*, 18690. [CrossRef] [PubMed]
5. Lee, C.; Shen, C.; Cozzan, C.; Farrell, R.M.; Speck, J.S.; Nakamura, S.; Ooi, B.S.; DenBaars, S.P. Gigabit-per-second white light-based visible light communication using near-ultraviolet laser diode and red-, green-, and blue-emitting phosphors. *Opt. Express* **2017**, *25*, 17480–17487. [CrossRef] [PubMed]
6. Wu, T.C.; Chi, Y.C.; Wang, H.Y.; Tsai, C.T.; Cheng, C.H.; Chang, J.K.; Chen, L.Y.; Cheng, W.H.; Lin, G.R. White-lighting communication with a $Lu_3Al_5O_{12}$: Ce^{3+}/$CaAlSiN_3$: Eu^{2+} glass covered 450-nm InGaN laser diode. *J. Light. Technol.* **2018**, *36*, 1634–1643. [CrossRef]
7. Yeh, C.H.; Chow, C.W.; Wei, L.Y. 1250 Mbit/s OOK wireless white-light VLC transmission based on phosphor laser diode. *IEEE Photonics J.* **2019**, *11*, 7903205. [CrossRef]
8. Bera, K.; Parthiban, R.; Karmakar, N. Coverage Improvement of Laser Diode-Based Visible Light Communication Systems Using an Engineered Diffuser. *IEEE Access* **2023**, *11*, 122833–122841. [CrossRef]
9. Abdalla, I.; Rahaim, M.B.; Little, T.D. Interference mitigation through user association and receiver field of view optimization in a multi-user indoor hybrid RF/VLC illuminance-constrained network. *IEEE Access* **2020**, *8*, 228779–228797. [CrossRef]
10. Eltokhey, M.W.; Khalighi, M.A.; Ghassemlooy, Z. Dimming-aware interference mitigation for NOMA-based multi-cell VLC networks. *IEEE Commun. Lett.* **2020**, *24*, 2541–2545. [CrossRef]
11. Eltokhey, M.W.; Khalighi, M.A.; Ghassemlooy, Z. Optimization of Receivers' Field of Views in Multi-User VLC Networks: A Bio-Inspired Approach. *IEEE Wirel. Commun.* **2022**, *29*, 132–139. [CrossRef]
12. Eldeeb, H.B.; Selmy, H.A.; Elsayed, H.M.; Badr, R.I. Interference mitigation and capacity enhancement using constraint field of view ADR in downlink VLC channel. *IET Commun.* **2018**, *12*, 1968–1978. [CrossRef]
13. Palacios Játiva, P.; Román Canizares, M.; Azurdia-Meza, C.A.; Zabala-Blanco, D.; Dehghan Firoozabadi, A.; Seguel, F.; Montejo-Sánchez, S.; Soto, I. Interference mitigation for visible light communications in underground mines using angle diversity receivers. *Sensors* **2020**, *20*, 367. [CrossRef]
14. Hosney, M.; Selmy, H.A.; Srivastava, A.; Elsayed, K.M. Interference mitigation using angular diversity receiver with efficient channel estimation in MIMO VLC. *IEEE Access* **2020**, *8*, 54060–54073. [CrossRef]
15. Younus, S.H. Interference mitigation in multiuser WDM VLC systems using differential receiver. *Trans. Emerg. Telecommun. Technol.* **2022**, *33*, e4512. [CrossRef]
16. Al-Sakkaf, A.G.A.; Morales-Céspedes, M. Interference Management for VLC Indoor Systems based on Overlapping Field-of-View Angle Diversity Receivers. *IEEE Access* **2024**, *12*, 51431–51449. [CrossRef]

17. Atta, M.A.; Bermak, A. A polarization-based interference-tolerant VLC link for low data rate applications. *IEEE Photonics J.* **2018**, *10*, 7901911. [CrossRef]
18. Krohn, A.; Harlakin, A.; Arms, S.; Pachnicke, S.; Hoeher, P.A. Impact of liquid crystal based interference mitigation and precoding on the multiuser performance of VLC massive MIMO arrays. *IEEE Photonics J.* **2022**, *14*, 7348112. [CrossRef]
19. Abdalla, I.; Rahaim, M.B.; Little, T.D. Dynamic FOV tracking receiver for dense optical wireless networks. In Proceedings of the 2019 IEEE Global Communications Conference (GLOBECOM), Waikoloa, HI, USA, 9–13 December 2019; pp. 1–6.
20. Dixit, V.; Kumar, A. BER analysis of dynamic FOV based MIMO-NOMA-VLC system. *AEU-Int. J. Electron. Commun.* **2021**, *142*, 153989. [CrossRef]
21. Abdalla, I.; Rahaim, M.B.; Little, T.D. Dynamic FOV visible light communications receiver for dense optical networks. *IET Commun.* **2019**, *13*, 822–830. [CrossRef]
22. Ahmad, F.; Raghunathan, V. Design of variable link length indoor visible light communication system using focus tunable liquid lens. In Proceedings of the Broadband Access Communication Technologies XV, Online, CA, USA, 6–12 March 2021; Volume 11711, pp. 96–102.

photonics

Article

Design of the Light Source Layout Optimization Strategy Based on Region Partition and Pre-Bias Compensation for Indoor Visible Light Communication Systems

Qiong Zhao [1,*], Weilin Zhang [1], Jiacheng Fan [1] and Lijun Deng [2]

[1] School of Communications and Information Engineering, Xi'an University of Posts & Telecommunications, Xi'an 710121, China; zhangwlasshole@stu.xupt.edu.cn (W.Z.); fanjiacheng@stu.xupt.edu.cn (J.F.)
[2] School of Physics and Electrical Engineering, Weinan Normal University, Weinan 714099, China; denglijun@mail.nwpu.edu.cn
* Correspondence: zhaoqiong@xupt.edu.cn

Abstract: Optimization of the light source layout is an important issue for indoor visible light communication systems, as it affects the received optical power distribution and user perception. In this paper, we propose a local optimization strategy for the light source layout that balances optimization effectiveness with optimization efficiency. First, we divide the optimization region into multiple sub-regions with different sizes and optimization priorities, where the sizes and optimization priorities of the individual sub-regions are determined based on the effect minimization principle among the sub-regions. We then calculate the pre-bias factor based on the equivalent mapping, which can compensate for the effect of the light sources in the latter optimized sub-region on the source layout optimization in the current sub-region. Finally, we search for the coordinate of a single light source for each sub-region using the variance of the squared distance between the projection of the light source on the receiving plane and the received point as a fitness function. Simulation results show that the proposed optimization strategy performs well when the vertical distance between the ceiling and the receiving plane is not less than 2.85 m.

Keywords: visible light communication; light source layout; optical power; region partition; pre-bias compensation

1. Introduction

In recent years, with the rapid development of wireless communication services such as big data, smart home, and shared bicycle, there is a growing demand for high-speed communication. However, the traditional mobile communication system, its capacity, and spectrum resource limitations have been unable to meet the requirements of increasing information capacity and transmission rate. In order to achieve high-speed and high-quality communication and solve the problem of the increasingly exhausted electromagnetic spectrum, utilizing new spectrum resources has become one of the trends of mobile communication. As a promising short-distance transmission technology, indoor visible light communication (VLC) has attracted a growing amount of attention [1,2]. VLC is a communication method that utilizes light-emitting diode (LED) as the signal transmitting node, free space as the transmission medium, and receives the signal through photo-diode (PD). Compared with conventional radio frequency (RF) communication technology, VLC has the characteristics of large modulation bandwidth, high security, high signal-to-noise ratio, and no electromagnetic interference. To date, VLC technology has been considered as a complement to other wireless communication technologies. In addition, indoor VLC must satisfy both illumination and data communications. Due to these unique dual characteristics, it can be applied in communication scenarios such as airports, train stations, offices, homes, and underwater. And in order to realize the application of indoor VLC, it is essential to arrange the light source.

In indoor VLC systems, light signals are transmitted and received by light-emitting diodes (LEDs) and photo-diodes (PDs) [3]. Because of the propagation characteristics of the visible light, the line-of-sight (LOS) component is dominant compared to the reflection components. In this way, the optical power distribution in the receiving plane is determined by the light source layout. In indoor VLC, the light sources is placed on the ceiling of the room, and the user would appear at any random point. It is obvious that users always want better unobstructed light and a good communication experience. In a typical indoor environment, multiple LEDs are usually used to satisfy the illumination requirements. These LEDs are often placed at the center of the room's ceiling, which can lead to uneven illumination and optical power reception on the receiving plane and even blind spots. In this situation, both the illumination and optical power decay rapidly from the center to the edge of the receiving plane. Moreover, the non-uniform received optical power will affect the quality of service (QOS) of indoor VLC systems [4–7]. Therefore, the optimization of the light source layout is one of the key issues for indoor VLC systems.

Several light source layout optimization schemes have been reported in the literature so far. Komine and Nakagawa analyzed the illumination distribution under indoor VLC array light sources by numerical simulation, and studied the influence of the multi-path effect and signal-to-noise ratio (SNR) of a room of typical size (5 m $\times$ 5 m $\times$ 3 m) [8]. On the basis of this theory, some scholars have proposed different methods to optimize the light source layout. Wang and Yu proposed a circular layout of 16 LEDs, and introduced a quality factor Q to reduce the fluctuation in the SNR [9]. Pei and Jing proposed a new polygon layout method in three-dimensional space, which further optimizes the circular layout [10]. Several researchers have incorporated optimization algorithms into the optimization procedure of light source layout. Liu and Tang proposed a square plus ellipse layout, optimized by using the simulated annealing and particle swarm algorithms. The uniformity of the illumination and SNR is further improved [11]. Che and Wang proposed a hybrid immune clonal bat algorithm to optimize the layout of the light source, which enhances the fair distribution of optical power among indoor users [12]. Wang and Xu proposed an optimization scheme to improve the uniformity of the received optical power by optimizing the layout of the light source based on an artificial fish swarm algorithm [13]. Zuo and Liu proposed a modified gray wolf algorithm to optimize the typical square, rectangular, and circular layouts of the light source, which further reduces the fluctuations of the received optical power [14].

Although much work has been carried out on the optimization of light source layout for indoor VLC systems, there is still room for improvement. Most of the existing schemes are based on the global optimization strategy, which requires searching over a huge number of coordinates, leading to a complicated optimization procedure. Fitness functions based on the received optical power are employed in these optimization schemes, which results in a large number of computations and further reduces the optimization efficiency. Motivated by the above analysis, we propose a sub-regional optimization strategy for light source layout. The main contributions are summarized below.

1. To reduce the limitations of local optimization, we propose a sub-region partition and optimization priority selection scheme based on the principle of minimum influence among sub-regions.
2. To address the unpredictable impact of the light sources located on the latter sub-region, we propose a pre-bias compensation scheme based on equivalent mapping.
3. To reduce the computational complexity, the variance of the square value of the distance between the projection of the light source on the receiving plane and the received point is employed as the fitness function in the optimization of the single light source layout in the sub-region.

The remainder of this paper is organized as follows. The model of the indoor VLC system is described in Section 2. The design of the sub-regional light source layout optimization strategy is presented in Section 3, followed by simulation results and a comparative analysis in Section 4. Finally, the conclusions are drawn in Section 5.

2. System Model

2.1. Channel Model

In indoor VLC systems, several reflecting propagation paths, also known as non-line-of-sight (NLOS) paths, can coexist with LOS communication. These reflections can be viewed as multiple distortions. The NLOS component is the sum of the contributions from all non-LOS links that are due to one or more reflections between the wall surfaces [15]. In the following, the LOS channel, NLOS channel, and overall optical channels are discussed.

The geometry of a point-to-point VLC downlink is shown in Figure 1. According to the Lambert radiation model, the LOS channel gain between the LED and the PD can be modeled as [16]

$$h_{LOS} = \begin{cases} \frac{A(m+1)}{2\pi d^2}\cos^m(\phi)\cos(\varphi), 0 < \varphi < \Psi_c \\ 0, \qquad \varphi > \Psi_c \end{cases} \quad (1)$$

where A is the effective area of the PD, m is the order of the Lambertian emission given by $m = -\log(2)/\log(\cos(\Phi_{1/2}))$, such that $\Phi_{1/2}$ represents the semi-angle of the LED [17], d is the Euclidean distance between LED and PD, φ and ϕ are the angle of incidence and irradiance, respectively, and Ψ_c is the receiver field of view (FOV).

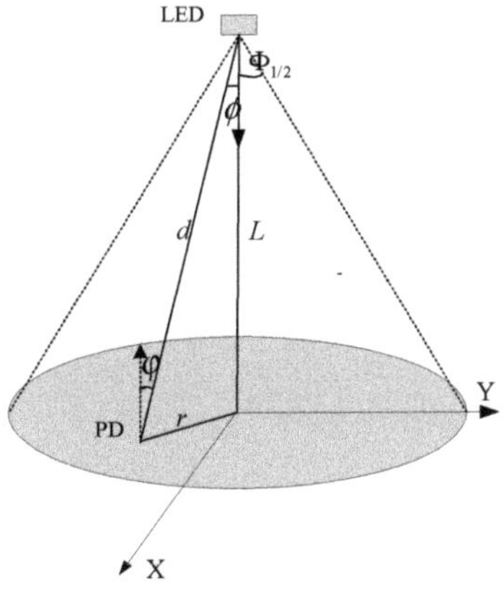

Figure 1. System model.

Assuming that the LED is placed downward and the PD is placed upward, the values of $\cos\phi$ and $\cos\varphi$ can be determined as [18]

$$\cos\phi = \cos\varphi = \frac{L}{d} = \frac{L}{\sqrt{r^2 + L^2}} \quad (2)$$

where r is the distance between the projection of the light source on the receiving plane and the PD, and L is the vertical distance between the ceiling and the receiving plane.

Substituting $C = \frac{A(m+1)L^{m+1}}{2\pi}$ and Equation (2) into Equation (1), we can obtain another expression of the LOS channel gain:

$$h = \begin{cases} \frac{C}{(r^2+L^2)^{\frac{m+3}{2}}}, & 0 < \varphi < \Psi_c \\ 0, & \varphi > \Psi_c \end{cases} \quad (3)$$

It can be seen from Equation (3) that for given m and L, h is a function of r.

The NLOS link is the sum of the contributions from all reflecting propagation links that are due to one or more reflections between the wall surfaces. A discrete reflector model was obtained by dividing the walls into small surface elements. The impulse response h_{NLOS} is approximated by the sum of the contributions from reflections up to a given order N_{ref} [15]:

$$h_{NLOS} \approx \sum_{k=1}^{N_{\text{ref}}} h_{\text{DIFF}}^{(k)}. \tag{4}$$

where $h_{\text{DIFF}}^{(k)}$ is the response after k-reflection to the LED source that is given by [15]. Studies show that the NLOS component accounts for about 7% of the total received power [11]. Moreover, there is no clear linear relationship between r and N_{ref}. Therefore, we do not consider the NLOS component in the design of the light source layout optimization strategy.

2.2. Received Optical Power Model

For a multi-source system, the optical power received by the PD is the sum of the optical power transmitted by the light sources within its FOV. To simplify the analysis, we set the transmitted optical power of each light source to 1W, and then the received optical power of the PD can be expressed as

$$P = \sum_{n=1}^{N} \frac{C}{\left(r_n^2 + L^2\right)^{\frac{m+3}{2}}} \tag{5}$$

where N is the number of light sources within the PD's FOV, and r_n is the distance between the projection of the light source n on the receiving plane and the PD.

For a single-light source system, the optical power received by the PD can be expressed as

$$P(u) = \frac{C}{\left(u + L^2\right)^{\frac{m+3}{2}}} \tag{6}$$

where $u = r^2$. We can see from Equation (6) that P is a monotonically decreasing function of u and P is completely determined by u for a given L. It can be concluded that the fluctuation of P can be approximately estimated by the fluctuation of u.

3. Design of the Sub-Regional Light Source Layout Optimization Strategy

To balance the efficiency and performance of light source layout optimization, we propose a sub-regional light source layout optimization strategy, which consists of a sub-regional partition and optimization priority selection scheme, a pre-bias compensation scheme, and a distance-based sub-regional single-source layout optimization scheme.

3.1. Sub-Region Partition and Optimization Priority Selection Scheme

In order to reduce the mutual influence between sub-regions, we propose a sub-regional partition and optimization priority selection scheme, which divides the optimization region into multiple sub-regions with different areas and optimization priorities. It is important to note that the choice of the number of sub-regions is not the focus of this paper, as this issue has been extensively studied in the relevant literature. We mainly focus on the sub-region partitioning and optimization priority selection under a given number of sub-regions. Considering that the edge sub-regions are less affected than the center sub-regions, we set larger areas and higher optimization priorities for the edge sub-regions. More specifically, the distance between the center of the sub-region and the center of the overall region is used to determine the area and the optimization order of each sub-region. The larger the distance, the smaller the area and the higher the optimization order of the sub-region. Sub-regions with the same distance have the same area but have different optimization priorities. The optimization priority of these sub-regions with the same area is designed as follows. First, anyone is selected as the reference sub-region and the highest optimization priority is set on it. Then, the other sub-regions are assigned optimization priority based on the distance between the center of those sub-regions and the center of the reference sub-region; the larger the distance, the higher the optimization priority.

Figure 2 shows two examples of the sub-region partitions for a 5 m × 5 m region. The numbers listed in Figure 2 indicate the optimization priority of the sub-regions; the smaller the number, the higher the optimization priority. Since the centers of sub-regions 1–4 are equally distant from the center of the overall region, these four sub-regions are assigned equal area and any of them can be chosen as the reference sub-region. Considering the symmetry of positions between different sub-regions, the optimization orders for the sub-regions exhibited in Figure 2a,b seem different, but they are essentially consistent. The proposed sub-region partition strategy provides a partitioning principle but does not specify the size of the region. The area of each sub-region is a flexible variable that can be set according to the size of the room. In this work, we present two examples of the sub-region partitions for a 5 m × 5 m room. In these two examples, sub-regions 1–4 are the first priority, with the same area, 1 m^2, sub-regions 5–12 are the second priority, with the same area, 1.5 m^2, and sub-regions 13–16 are the third priority, with same area, 2.25 m^2.

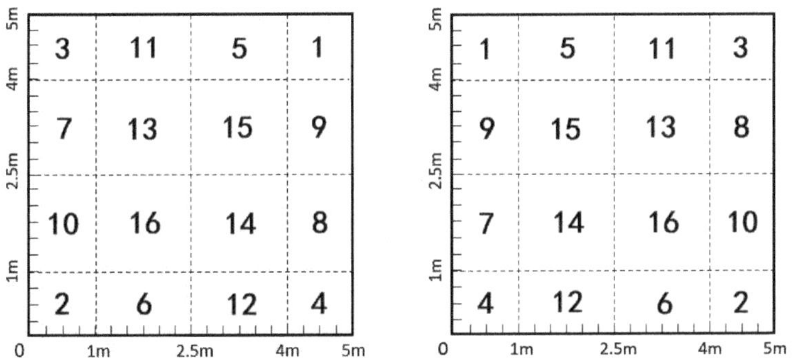

(**a**) sub-region partition (**b**) optimization priority selection

Figure 2. Examples of sub-region partition and optimization priority selection.

3.2. Pre-Bias Compensation Scheme

Since the light source layout in the sub-regions is optimized sequentially, the influence of the light source located in sub-region j on sub-region i $(i < j)$ is uncertain, as the light source layout optimization in sub-region j has not yet been implemented. In order to minimize as much as possible the optimization error due to such uncertain effects, we propose a pre-bias compensation scheme based on equivalent maps. This scheme approximately equates the effect of a light source located in the unoptimized sub-regions to that of a virtual source located in the center of the overall region, and adjusts this effect for different sub-regions with different factors. The specific implementation procedure is presented below.

When optimizing sub-region i $(i = 1, 2, \cdots, N-1)$, we assume that the other $N - i$ unoptimized sub-regions all have a virtual light source located in their centers. Since the ranges of each sub-region have been determined by the proposed sub-regional partition scheme presented in the previous subsection, the projections of their center on the receiving plane can be determined accordingly. Then the square value of the distance between the projection of these virtual light sources on the receiving plane and the center of sub-region i can be written as

$$u_{c_i} = \sum_{j=1}^{N-i} (x_{c_j} - x_{c_i})^2 + (y_{c_j} - y_{c_i})^2 \tag{7}$$

where (x_{c_j}, y_{c_j}, z) and (x_{c_i}, y_{c_i}, z) are the projections of virtual light source j and the center of sub-region i on the receiving plane, respectively.

Meanwhile, we assume that there is a virtual light source located at the center of the overall optimization region, and the square value of the distance between its projection on the receiving plane and the projection of the center of sub-region i can be written as

$$\bar{u}_{c_i} = (x_c - x_{c_i})^2 + (y_c - y_{c_i})^2 \tag{8}$$

where (x_c, y_c, z) is the projection of this virtual light source on the receiving plane.

Consequently, the pre-bias factor corresponding to the impacts of the light sources located on the other $N - i$ unoptimized sub-regions on sub-region i can be obtained as

$$Q_i = u_{c_i} / \bar{u}_{c_i} \tag{9}$$

The specific application of the pre-bias factor is described in the following subsection.

3.3. Single-Light Source Layout Optimization Scheme for Sub-Region

It can be seen from Equation (6) that the computation required by P is much greater than that of u. When the NLOS component of the channel link is considered, the computation required by P would be greatly increased. To reduce the computation complexity, we propose a single-light source layout optimization scheme, where the variance of the squared distance replaces the receiving optical power as the new fitness function. Moreover, in the optimization of the single-light source layout for the sub-region, the possible impact of the unoptimized sub-regions is simulated through a pre-bias factor.

Assuming (x_i, y_i, z) is the projection of the light source located in sub-region i on the receiving plane, then the square of the distance between this projection and the receiving point (x, y, z) can be written as

$$u_i = (x - x_i)^2 + (y - y_i)^2 \tag{10}$$

When $i = 1$, only the impacts of the latter optimized $N - i$ sub-regions need to be considered. Therefore, we revise u_i through the pre-bias factor as follows.

$$u_i = (x - x_i)^2 + (y - y_i)^2 + Q_i((x - x_c)^2 + (y - y_c)^2) \tag{11}$$

When $1 < i < N$, besides the impacts of the light sources located in the latter optimized $N - i$ sub-regions, we need to consider the impacts of the light sources located in the former $i - 1$ optimized sub-regions. Assuming that the projection of the light source located in optimized sub-region $k(k = 1, 2, \cdots, i - 1)$ on the receiving plane is (x_k, y_k, z_L), u_i can be revised as

$$u_i = (x - x_i)^2 + (y - y_i)^2 + Q_i((x - x_c)^2 + (y - y_c)^2) + \sum_{k=1}^{i-1} (x - x_k)^2 + (y - y_k)^2 \tag{12}$$

When $i = N$, only the impacts of the light sources located in the former $N - 1$ optimized sub-regions need to be considered. Therefore, u_i can be revised as

$$u_i = (x - x_i)^2 + (y - y_i)^2 + \sum_{k=1}^{N-1} (x - x_k)^2 + (y - y_k)^2 \tag{13}$$

where u_i can be seen as a random variable and its fluctuations can be described by its variance, which is calculated as

$$D(\tilde{u}_i) = \frac{\iint\limits_{C_i} (u_i - \overline{u_i})^2 ds}{S_i} \tag{14}$$

where C_i and S_i are the range and area of sub-region i, and $\bar{u}_i$ is the mean value of u_i, which can be calculated as

$$\overline{u}_i = \frac{\iint\limits_{C_i} u_i ds}{S_i} \quad (15)$$

The minimum fluctuation of P can be achieved by searching for the (x_i, y_i) that minimizes the value of $D(\tilde{u}_i)$. Thus, the optimization of the light source layout can be transformed into an optimization search problem, as presented in the following:

$$\begin{aligned} \min_{x_i, y_i} \quad & D(\tilde{u}_i) \\ s.t. \quad & (x_i, y_i) \in C_i \end{aligned} \quad (16)$$

This is a non-convex problem for which closed-form solutions cannot be obtained. However, it can be solved through optimization algorithms, such as genetic algorithms (GAs) and bat algorithms (BAs). GA is an optimization method that simulates the principles of biological inheritance and evolution in nature. By simulating the natural evolution process, it searches for the optimal solution in the global range. In the search process, there is no strict requirement on the objective function and constraints, and at the same time, it has strong parallel processing ability, global search ability, and fast convergence ability. Therefore, it has been widely used in many fields.

4. Simulation Results and Discussion

In this section, we evaluate the performance of the proposed light source layout optimization strategy (denoted as the distance-related scheme). The room size we consider is 5 m × 5 m and the total number of light sources is 16, as set in [12]. Therefore, the sub-region partition and optimization order for each sub-region are the same as shown in Figure 2. The GA is applied to solve the optimization problem presented in Equation (16), whose parameters are described in Table 1. The Python software is used to realize data processing. And the entire strategy flowchart is shown in Figure 3.

Table 1. Genetic algorithm parameters.

Population size	Maximum genetic generation	Individual length
40	50	16
Generation gap	Cross probability	Mutate probability
1	0.5	0.01

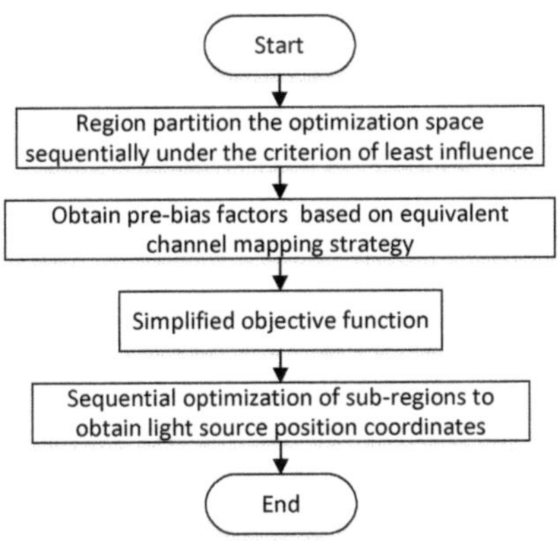

Figure 3. The flowchart of the proposed optimization strategy.

GA provides a general framework for solving a nonlinear program. It encodes the problem parameters as chromosomes, generating an initial population and then using iterative choose, cross, and mutate operations to exchange the information of the chromosomes in the population, ultimately generating a chromosome that meets the optimization objective. In genetic algorithms, chromosomes correspond to data or arrays, usually represented by a one-dimensional string structure, and each position on the string corresponds to the value of the gene.

Taking GA as an example, Figure 4 shows the process of single-light source layout optimization. It is noteworthy that the fitness function employed in the proposed distance-related optimization scheme is independent for the vertical distance between the ceiling and the receiving plane. Each step of the genetic algorithm is briefly described below:

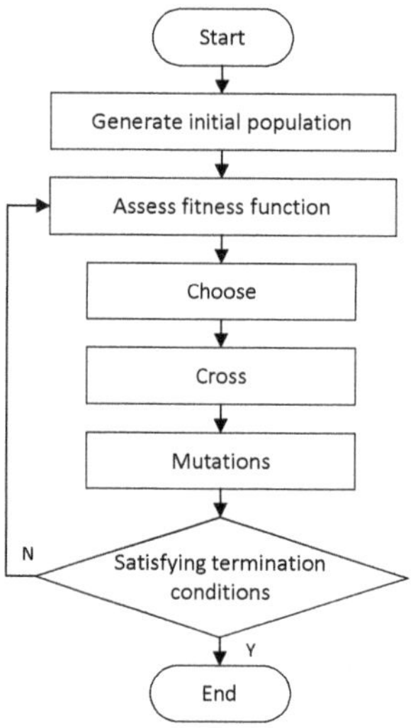

Figure 4. The process of single-light source layout optimization (GA).

- Generate initial population

  N initial coding strings are randomly generated, each coding string is called an individual, and N individuals constitute a population. The GA uses these N coding strings as the initial point to start the evolution. N is the population size.
- Assess fitness function

  The fitness indicates the superiority or inferiority of an individual or a solution. As the genetic algorithm proceeds, the quality of the solution improves, the fitness increases, and the genetic algorithm is terminated once a solution with a satisfactory fitness value is found. The design of the corresponding fitness function varies from different problems.
- Choose

  The purpose of choose is to choose the best individuals from the current population so that they have the opportunity to reproduce as parents for the next generation. Genetic algorithms embody this idea through the process of choose, which is based on the principle that a well-adapted individual has a high probability of contributing one or more offspring to the next generation. Choose reflects the Darwinian principle of survival of the fittest.
- Cross

  Cross is the most important operation in the genetic algorithm. Through the cross operation, a new generation of individuals can be obtained, and the new individuals combine the characteristics of their parents. Cross embodies the idea of information exchange. This step makes the chromosome sequence change greatly; the final chromosome mapping to the real number will also change greatly, making the population dispersed, reflecting the characteristics of the algorithm's global search.

- Mutations

  Mutation begins by randomly choosing an individual in the population, and for the chosen individual, the value of one of the strings in the coding string is randomly changed with a certain probability. This step causes the chromosome sequence to fluctuate within a small range, reflecting the local search ability. As in biology, the probability of a mutation occurring in GA is low and usually takes on small values.

4.1. *Performance Analysis*

First, we present the optimization results of the proposed optimization strategy. For comparison, the light source layout determined by the scheme proposed in [12] (denoted as immune clonal bat algorithm (ICBA)) is also presented in Figure 5. We can see that the light source layout determined by the proposed optimization strategy is different from that determined by ICBA.

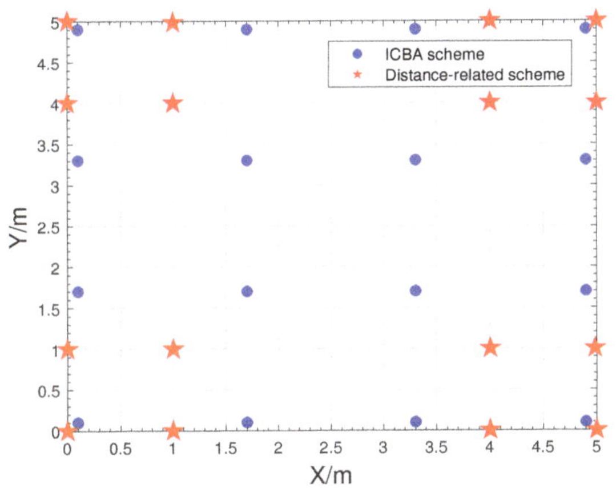

Figure 5. Light source layout of different schemes.

Next, we display the received power distribution determined by the proposed optimization strategy and the ICBA scheme in Figure 6. Of special note is that the received power consists of two component: the LOS component and NOLS component. Considering the spatial differences of rooms in different scenarios, we set the vertical distance between the ceiling and the receiving plane (denoted as L) as 2.5 m, 2.85 m, 3.15 m, and 3.5 m, respectively. For the proposed optimization strategy, the FOV is set to 90° and the semi-angle at half power of the LED is set to 60°. The parameters used for the ICBA scheme are the same as those presented in [12]. We can see that the proposed optimization strategy has a larger fluctuation in the optical power distribution fluctuation compared to the ICBA scheme when L = 2.5 m. However, the range of fluctuations decreases as the distance increases. When L = 2.85 m, the optical power distribution of the proposed optimization strategy is lower than that of the ICBA scheme. The greatest challenge for the proposed optimization strategy is the uncertain impact of the latter optimized sub-regions on the current optimized sub-region. However, this uncertain effect decreases as the vertical distance between the ceiling and the receiving plane increases. Therefore, the proposed optimization strategy performs well when $L \geq 2.85$ m. In scenarios such as airports, train stations, and shopping malls, the vertical distance between the ceiling and the receiving plane is always larger than 2.85 m. This proposed optimization strategy is recommended for these scenarios.

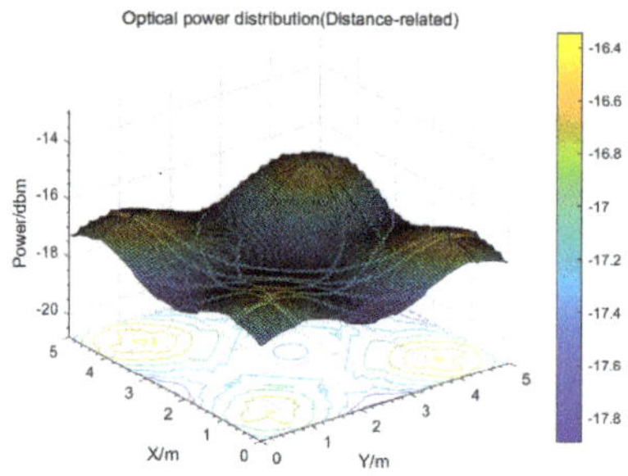

(a) The proposed scheme (*L*: 2.5 m)

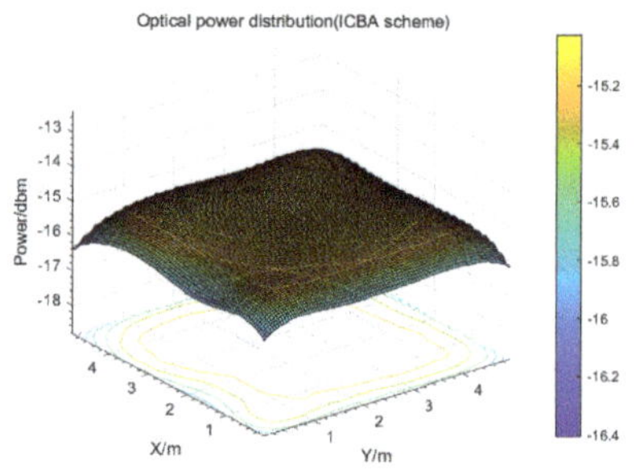

(b) ICBA scheme (*L*: 2.5 m)

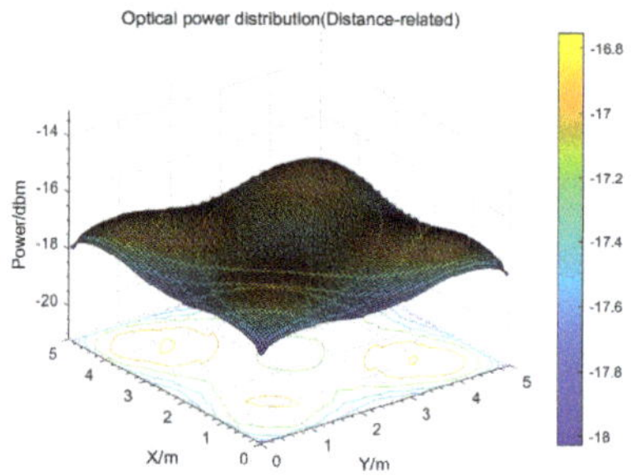

(c) The proposed scheme (*L*: 2.85 m)

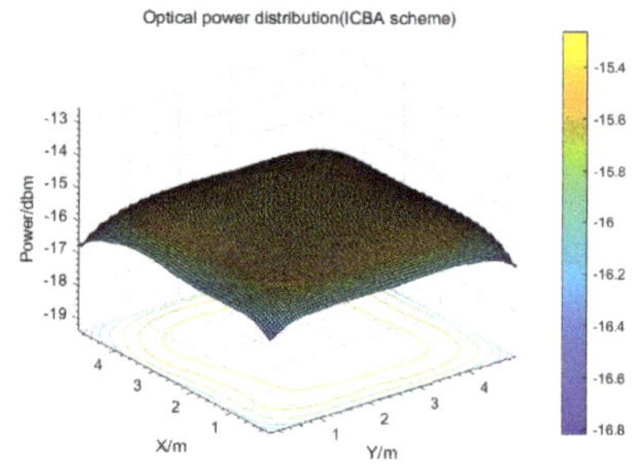

(d) ICBA scheme (*L*: 2.85 m)

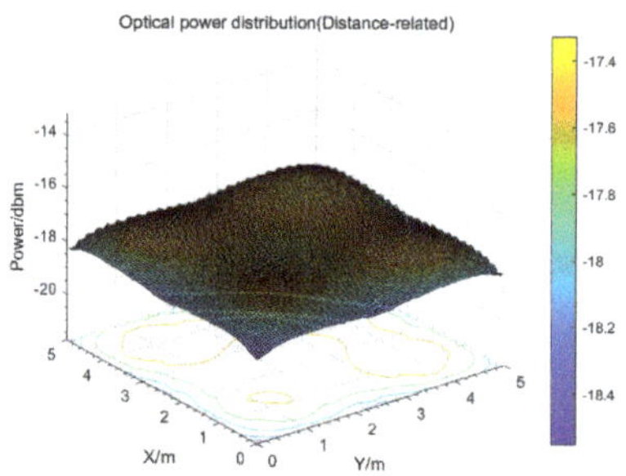

(e) The proposed scheme (*L*: 3.15 m)

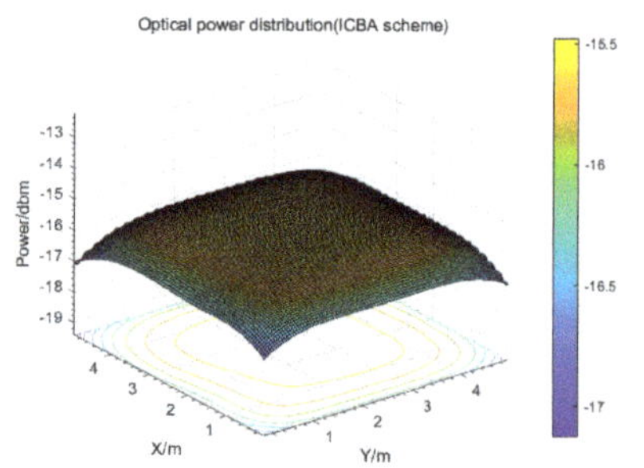

(f) ICBA scheme (*L*: 3.15 m)

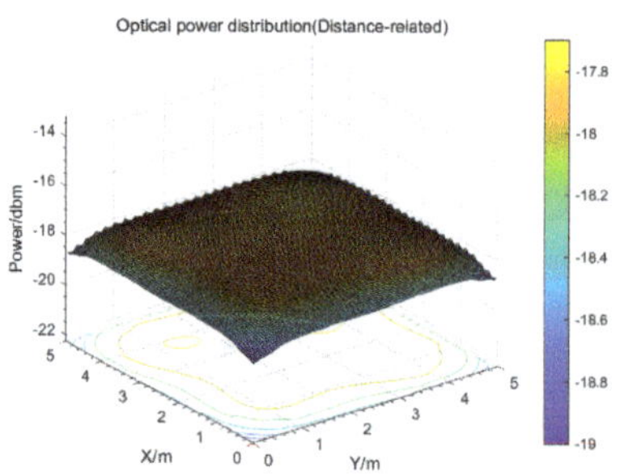

(g) The proposed scheme (*L*: 3.5 m)

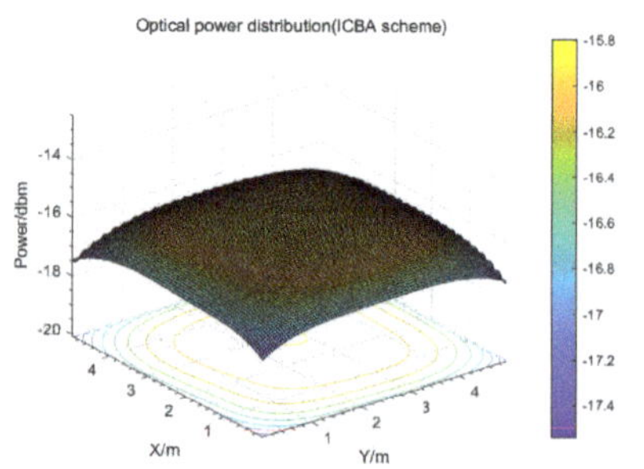

(h) ICBA scheme (*L*: 3.5 m)

Figure 6. Received optical power distribution.

In addition, Table 2 presents the power indices obtained using the two different optimization methods. It is observed that the fluctuations of the maximum and the minimum optical power value in the receiving plane at different heights in the proposed scheme are large compared to the ICBA scheme. However, the power fluctuation range in the proposed scheme becomes limited with the increase in the height, which shows that the receiving optical power distribution is improved when the vertical distance between the ceiling and the receiving plane is not less than 2.85 m, providing a better service quality for users at different positions in the given receiving plane.

Table 2. Powervalues for different schemes.

L (m)	Scheme	Maximum Power (dBm)	Minimum Power (dBm)	Power Fluctuation (dBm)	Power Variance
2.5	The Proposed scheme	−16.3448	−17.8917	1.5468	0.0888
	ICBA scheme	−15.0258	−16.4085	1.3828	0.0456
2.85	The Proposed scheme	−16.7525	−18.0286	1.2761	0.0444
	ICBA scheme	−15.2625	−16.8194	1.5569	0.0678
3.15	The Proposed scheme	−17.3268	−18.5515	1.2247	0.0318
	ICBA scheme	−15.4781	−17.1271	1.6490	0.0825
3.5	The Proposed scheme	−17.6989	−19.0047	1.3058	0.0380
	ICBA scheme	−15.7962	−17.5427	1.7466	0.1001

In addition, by traversing L in the range of [2.5 m 3.5 m] with an interval of 0.35 m, we compare the proposed optimization strategy with the ICBA scheme in terms of average performance over a 5 m × 5 m × 1 m space in Table 3. We can see that the the proposed optimization strategy is is more robust against ICBA.

Table 3. Power values over 5 m × 5 m × 1 m space with L = 2.5 m to L = 3.5 m.

Space	Scheme	Maximum Power (dBm)	Minimum Power (dBm)	Power Fluctuation (dBm)	Power Variance
5 × 5 × 1	ICBA scheme	−15.3907	−16.9744	1.5838	0.0740
	The Proposed scheme	−17.0308	−18.3691	1.3384	0.0508

To analyze the impact of sub-region partition on the proposed optimization strategy, we also evaluate the other two sub-region partitions for a 5 m × 5 m room, denoted as sub-region partition 2 and sub-region partition 3, respectively, as shown in Figure 7. In sub-region partition 2, sub-regions 1–4 are the first priority, with the same area, 1 m^2, sub-regions 5–12 are the second priority, with the same area, 1.5 m^2, and sub-regions 13–16 are the third priority, with the same area, 2.25 m^2. In sub-region partition 3, sub-regions 1–16 have the same priority, with same area, 1.5625 m^2. The power values on the receiver plane (L = 2.85 m) for different sub-region partitions are shown in Table 4, where sub-region partition 1 is the sub-region partition shown in Figure 2. We can see that sub-region partition 1 used in our work has a clear advantage over the other two.

Table 4. Power values for different sub-region partition schemes.

L (m)	Scheme	Maximum Power (dBm)	Minimum Power (dBm)	Power Fluctuation (dBm)	Power Variance
	Sub-region partition 1	−16.7525	−18.0286	1.2761	0.444
2.85	Sub-region partition 2	−17.1149	−18.9374	1.8224	0.1818
	Sub-region partition 3	−16.8796	−18.6252	1.7456	0.0644

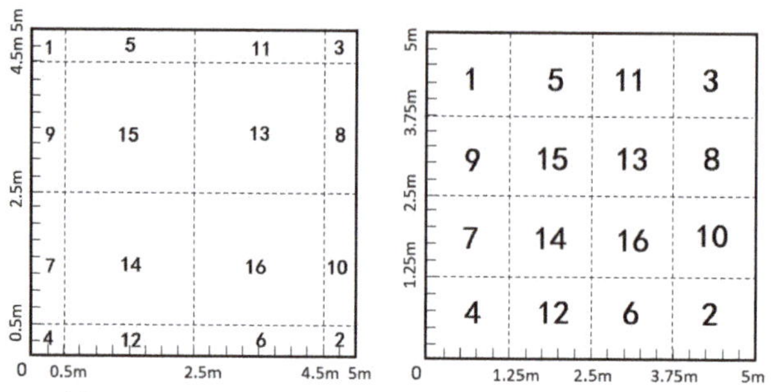

Figure 7. Two examples of sub-region partition for a 5 m × 5 m room.

4.2. Complexity Analysis

To demonstrate the low complexity and high efficiency of the proposed optimization strategy, we compare the computational complexity of the power-related fitness function and the distance-related fitness function. In the NLOS channel model, the walls in the room are divided into many small surface elements and each surface element itself acts as a radiator by reflecting the impinging light. Usually, each wall is divided into more than 50 small surfaces. The total gain for all NLOS links is the sum of the contributions from reflections up to a given order. Consequently, obtaining the NLOS component requires a significant amount of computation, which is about a hundred times more than that of u employed in the proposed optimization strategy. Therefore, the proposed optimization strategy has a simple fitness function, which reduces the optimization complexity.

In addition, we compare the convergence of different fitness functions through simulation. Figure 8 shows the iterative diagram of the distance-related fitness functions and the power-related fitness functions. It can be seen from Figure 8 that the distance-related fitness function reduces the fitness value to the target value after about 10 iterations, while the power-related fitness function reduces the fitness value to the target value after about 30 iterations. The latter is three times more than the former. In addition, we compare the computer execution time of the two fitness functions: 15.198 s for the distance-related fitness function and 49.317 s for the power-related fitness function. The former reduces the computation by about 70% compared to the latter. Moreover, the proposed optimization strategy does not require diversity in the semi-angle at half power of the LED, which helps to improve the optimization efficiency and expand the applications.

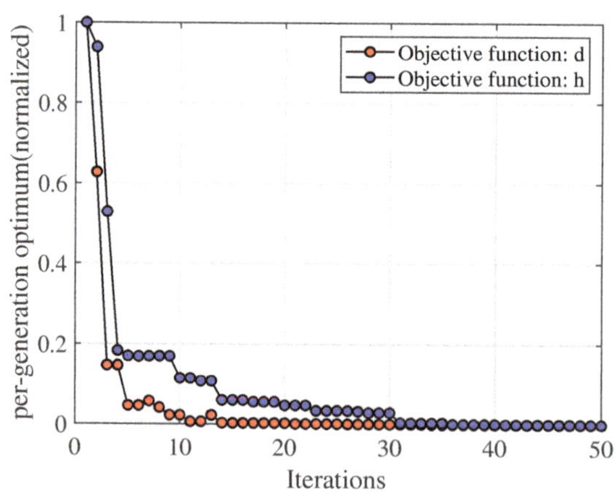

Figure 8. Convergence curves of two objective functions.

5. Conclusions

In this paper, we introduce an optimization strategy based on pre-bias compensation to improve the reliability and effectiveness of light source layout optimization for indoor VLC systems. Under this optimization strategy framework, we propose a sub-region partition scheme to minimize the mutual impacts between sub-regions as much as possible, a pre-bias compensation scheme to help offset some of the impacts of the latter optimized sub-regions on the current optimized sub-region, and a distance-based sub-regional single-light source layout optimization scheme to improve the optimization efficiency. Simulation results have shown that the proposed optimization strategy has an obvious advantage over the ICBA scheme when $L > 2.5$ m, which means that the proposed optimization strategy performs well in rooms with open spaces, such as airports, train stations, and shopping malls. Moreover, the proposed optimization strategy is low-complexity and high-efficiency, as it employees a simple distance-related fitness function and does not require diversity in the semi-angle at half power of the LED. Comparison results in terms of performance and complexity further validate the reliability and effectiveness of the proposed optimization strategy.

Author Contributions: Methodology, Q.Z.; Software, W.Z.; Data curation, J.F.; Writing—original draft, W.Z.; Writing–review & editing, Q.Z. and L.D. All authors have read and agreed to the published version of the manuscript.

Funding: This work was supported by the National Natural Science Foundation of China (62101444) and the Education Department of Shaanxi Province General Special Research Project (21JK0629).

Institutional Review Board Statement: Not applicable.

Informed Consent Statement: Not applicable.

Data Availability Statement: Data is contained within the article.

Acknowledgments: The authors would like to thank the anonymous Reviewers for their valuable comments and suggestions.

Conflicts of Interest: The authors declare no conflict of interest.

References

1. Pathak, P.H.; Feng, X.; Hu, P.; Mohapatra, P. Visible light communication, networking, and sensing: A survey, potential and challenges. *IEEE Commun. Surv. Tutor.* **2015**, *17*, 2047–2077. [CrossRef]
2. Ren, J.; Zhu, Y.; Zhang, Y.; Li, D. Optimization of multi-receiver SNRs for indoor visible light communication based on modified evolutionary algorithm. *Optik* **2021**, *228*, 166158. [CrossRef]
3. Mapunda, G.A.; Ramogomana, R.; Marata, L.; Basutli, B.; Khan, A.S.; Chuma, J.M. Indoor visible light communication: A tutorial and survey. *Wirel. Commun. Mob. Comput.* **2020**, *2020*, 8881305. [CrossRef]
4. Chvojka, P.; Zvanovec, S.; Haigh, P.A.; Ghassemlooy, Z. Channel characteristics of visible light communications within dynamic indoor environment. *J. Light. Technol.* **2015**, *33*, 1719–1725. [CrossRef]
5. Chen, C.; Zhong, W.D.; Yang, H.; Zhang, S.; Du, P. Reduction of SINR fluctuation in indoor multi-cell VLC systems using optimized angle diversity receiver. *J. Light. Technol.* **2018**, *36*, 3603–3610. [CrossRef]
6. Sun, G.; Liu, Y.; Yang, M.; Wang, A.; Liang, S.; Zhang, Y. Coverage optimization of VLC in smart homes based on improved cuckoo search algorithm. *Comput. Netw.* **2017**, *116*, 63–78. [CrossRef]
7. Liu, H.; Lin, Z.; Xu, Y.; Chen, Y.; Pu, X. Coverage uniformity with improved genetic simulated annealing algorithm for indoor Visible Light Communications. *Opt. Commun.* **2019**, *439*, 156–163. [CrossRef]
8. Komine, T.; Nakagawa, M. Fundamental analysis for visible-light communication system using LED lights. *IEEE Trans. Consum. Electron.* **2004**, *50*, 100–107. [CrossRef]
9. Wang, Z.; Yu, C.; Zhong, W.D.; Chen, J.; Chen, W. Performance of a novel LED lamp arrangement to reduce SNR fluctuation for multi-user visible light communication systems. *Opt. Express* **2012**, *20*, 4564–4573. [CrossRef] [PubMed]
10. Pei, H.; Jing, L.; Tong, Z.; Ma, M.; Cui, X. Layout and optimization of LED light source for indoor visible light communication. *Microw. Opt. Technol. Lett.* **2023**, *65*, 710–716. [CrossRef]
11. Liu, C.; Tang, Y.; Yan, W.; Bai, Y. Optimizing the Light Source Layout of the Indoor Visible Light Communication System. *IEEE Access* **2022**, *10*, 27223–27229. [CrossRef]
12. Che, H.; Wang, P.; Chi, S.; Sun, Y.; Yang, T.; Wang, Z. LED layout optimization in visible light communication system by a hybrid immune clonal bat algorithm. *Opt. Commun.* **2022**, *520*, 128532. [CrossRef]

13. Jiaan, W.; Ancheng, X.; Jintao, J.; Linyang, G. Optimization lighting layout of indoor visible light communication system based on improved artificial fish swarm algorithm. *J. Opt.* **2020**, *22*, 035701. [CrossRef]
14. Zuo, Y.; Liu, B.; Shao, K. Symmetrical indoor visible light layout optimized by a modified grey wolf algorithm. *Appl. Opt.* **2022**, *61*, 6016–6022. [CrossRef] [PubMed]
15. Lee, K.; Park, H.; Barry, J.R. Indoor Channel Characteristics for Visible Light Communications. *IEEE Commun. Lett.* **2011**, *15*, 217–219. [CrossRef]
16. Kumar, A.; Sudha, V. Optical power distribution and statistical analysis of indoor visible light communication. In Proceedings of the 2019 TEQIP III Sponsored International Conference on Microwave Integrated Circuits, Photonics and Wireless Networks (IMICPW), Tiruchirappalli, India, 22–24 May 2019; pp. 383–386.
17. Lin, S.H.; Liu, C.; Bao, X.; Wang, J.Y. Indoor visible light communications: Performance evaluation and optimization. *EURASIP J. Wirel. Commun. Netw.* **2018**, *2018*, 228. [CrossRef]
18. Liu, Y.; Lu, J.; Sun, G.; Liu, L.; Zhang, J. Optical Power Coverage Optimization for UAV-enabled Visible Light Communication. In Proceedings of the ICC 2022—IEEE International Conference on Communications, Seoul, Republic of Korea, 16–20 May 2022; pp. 1–6.

photonics MDPI

Article

Channel Characterization and Comparison in Industrial Scenario from Sub-6 GHz to Visible Light Bands for 6G

Yue Yin [1], Pan Tang [2,*], Jianhua Zhang [1], Zheng Hu [1], Tao Jiang [3], Liang Xia [3] and Guangyi Liu [3]

1 State Key Laboratory of Networking and Switching Technology, Beijing University of Posts and Telecommunications, Beijing 100876, China; yinyue18@bupt.edu.cn (Y.Y.); jhzhang@bupt.edu.cn (J.Z.); huzheng@bupt.edu.cn (Z.H.)
2 State Key Laboratory of Information Photonics and Optical Communications, Beijing University of Posts and Telecommunications, Beijing 100876, China
3 China Mobile Research Institute, Beijing 100053, China; jiangtao@chinamobile.com (T.J.); xialiang@chinamobile.com (L.X.); liuguangyi@chinamobile.com (G.L.)
* Correspondence: tangpan27@bupt.edu.cn

Abstract: The industrial scenario is indispensable for ubiquitous 6G coverage, which demands hyper-reliable and low-latency communication for full automation, control, and operation. To meet these demands, it is widely believed that it is necessary to introduce not only the conventional sub-6 GHz bands but also high-frequency technologies, such as millimeter wave (mmWave), terahertz (THz), and visible light bands. In this paper, we conduct a channel characterization and comparison in the industrial scenario from the sub-6 GHz to visible light bands. The channel characteristics, including the path loss (PL), root mean square (RMS) delay spread (DS), and angle spread (AS), were analyzed with respect to the frequency dependence and the distance dependence. On the one hand, the visible light band exhibited significant differences in channel characteristics compared to the electronic wave band. Due to the line-of-sight transmission of VLC, the visible light band had a higher path loss, and the path loss exponent reached 3.84. Due to the Lambertian radiation pattern, which has a wide range of reflection angles, the AS of the visible light band was much larger than that of the electronic wave band, which were 1.73 and 0.80 for the visible light and THz bands, respectively. On the other hand, the blockage effect of the metal instruments in the industrial scenario will greatly affect the channel characteristics. As the transceiver distance grows large, signals from both sides of the receiver will be blocked by metal instruments, resulting in a decreasing trend in the RMS DS for the electronic wave band. Moreover, the statistical characteristics of the channel properties were modeled and compared with the 3GPP TR 38.901 standard. It was found that the height of the receiver caused the difference between the proposed model and the 3GPP model and needs to be taken into account when modeling. Furthermore, we extended the 3GPP model to the THz and VLC bands and provided the statistical parameters of the channel characteristics for all frequency bands. This study can provide insights for the evaluation and standardization of multi-frequency communication technology in the industrial scenario.

Keywords: channel characteristics; comparison; industrial scenario; VLC; mmWave; THz; sub-6 GHz

1. Introduction

At the recent International Telecommunication Union Radiocommunication (ITU-R) Working Party (WP) 5D meeting, a new recommendation was approved which highlights

 https://doi.org/10.3390/photonics12030257

six usage scenarios of IMT-2030 [1]. As a typical usage case of the hyper-reliable and low-latency communication scenario, the industrial scenario has received widespread attention again. It extends the Ultra-Reliable and Low-Latency Communication (URLLC) of IMT-2020 and covers specialized use cases that are expected to have more stringent requirements for enhanced reliability, low latency, and precise positioning [2].

To satisfy the high performance metrics of IMT-2030 for the industrial scenario, communication technology continues to develop from sub-6 GHz toward high-frequency bands [3]. Firstly, sub-6 GHz bands have low propagation loss and a strong coverage capability which can provide extensive coverage for the industrial scenario, which has a larger physical size than an ordinary indoor scenario [4]. Secondly, to meet the typical business requirements of an ultra-wide bandwidth and ultra-low-latency transmission in the industrial scenario, high-frequency communication technologies such as millimeter wave (mmWave) and terahertz (THz) bands are applied [5]. Thirdly, excessive electromagnetic interference will be generated due to the operation of the abundant equipment in the industrial scenario, and the visible light band is regarded as a solution to interference problems due to the significant frequency isolation between the light and radio waves. Furthermore, an abundance of metallic scatterers in the industrial scenario produces specular reflection which results in an increasing intensity of reflection and dense multipath components (MPCs) [6]. Due to the different capabilities for the reflection, diffraction, and scattering of signals in different frequency bands, the impact of the frequency is more pronounced than in traditional indoor hotspot scenarios. Hence, it is imperative to explore the propagation characteristics for all frequency bands within an industrial scenario and elucidate the distinct channel properties across various frequency bands.

1.1. Literature Review

Looking at the existing literature, a number of works have been reported on channel characteristics and modeling at multiple bands in the industrial scenario, as shown in Table 1. In the following subsection, we will review the progress of industrial scenario channels according to frequency bands.

Table 1. Research progress in channel characteristics at multiple bands in industrial scenarios.

Frequency Band	Frequency Point	Modeling Approach	Channel Characteristics	Influence Factors	Ref.
Sub-6 GHz	2.4 GHz	Channel measurement	Path loss	Transceiver distance	[7]
	4.145 GHz	Channel measurement	Path loss, RMS DS	Environmental density	[8]
	4.1 GHz	Channel measurement	Path loss, RMS DS	Transceiver distance, LOS and NLOS cases	[9]
mmWave	28 GHz	Channel measurement	Path loss, RMS DS, AS	Antenna height	[6]
	28 GHz	Ray-tracing-based simulation	Path loss, LOS probability	Environmental density	[10]
	28 GHz	Channel measurement	Path loss	Transceiver distance, LOS and NLOS cases	[11]
	28 GHz	Channel measurement	Path loss, RMS DS, AS	LOS and NLOS cases	[12]
	105 GHz	Channel measurement	PDP, RMS DS, temporal PDP correlation coefficient	Time-varying channel	[13]
	142 GHz	Channel measurement	Path loss, RMS DS, AS	Polarization characteristics	[14]
	142 GHz	Channel measurement	Path loss, RMS DS, AS	Antenna transmission pattern, antenna height	[15]
	142 GHz	Channel measurement	A 3-D omnidirectional multipath channel model based on clusters	LOS and NLOS cases	[16]

Table 1. *Cont.*

Frequency Band	Frequency Point	Modeling Approach	Channel Characteristics	Influence Factors	Ref.
THz	220 GHz	Channel measurement	Path loss, RMS DS, AS, power delay angle profile	Environmental density	[17]
	300 GHz	Channel measurement	Path gain, path loss, RMS DS, AS	Scenario size	[18]
	304.2 GHz	Channel measurement	Path gain, PDP, RMS DS	Time-invariant and time-variant settings, antenna height	[19]
	300.7 GHz	Channel measurement	Received power, PDP	Penetration through glass, blockage by frame losses	[20]
VLC		Ray-tracing-based simulation	The channel frequency responses, channel direct current gain, RMS DS	Comparison of visible light and infrared frequency bands	[21]
		Ray-tracing-based simulation	The channel frequency responses, channel direct current gain, RMS DS	Various receiving positions	[22]
		Ray-tracing-based simulation	Path loss, RMS DS, AS	Antenna height	[23]
Multiple bands	3–4 GHz and 38–40 GHz	Channel measurement	Channel gain coefficient, RMS DS, and Ricean K factor	LOS and NLOS cases	[24]
	3.7 GHz and 28 GHz	The Volcano ray-tracing channel simulation and channel measurement	Channel gain, RMS DS, AS	LOS and NLOS cases	[25]
	6.75 GHz, 74.25 GHz, and 305.27 GHz	Channel measurement	Path loss, normalized power, DS	LOS and NLOS cases	[26]
	28 GHz, 38 GHz, 132 GHz, and 220 GHz	Channel measurement	Path loss, RMS DS, AS, K factor	Scenario size	[27]

With respect to the sub-6 GHz band, the existing channel research on the industrial scenario mainly concentrates on the path loss and root mean square (RMS) delay spread (DS), and the reference distance (CI) model is mainly used for path loss modeling. In [7], the author investigated the multi-wall–multi-floor (MWF) model of the industrial environment, and measurements were conducted at 2.4 GHz in the multistoried building. By comparing the MWF model with the path loss obtained through measurements, it was found that the MMF model can accurately characterize the path loss in industrial scenarios. In [8], the author presented channel measurements at 4.145 GHz in both a sparse industrial environment with a normal size and a dense industrial environment with an extra-large size. It was found that the line-of-sight (LOS) path loss exponents (PLEs) of the CI model were 1.24 and 1.5, and the RMS DS was 22 and 83.2 for the sparse environment and dense environment, respectively. This indicates that the path loss and RMS DS are dependent on the industrial environment. In [9], the CI path loss model and the RMS DS were investigated at 4.1 GHz in both LOS and non-line-of-sight (NLOS) scenarios. The PLEs for the LOS and NLOS cases were 1.96 and 2.72, and the majority of the RMS DSs for the LoS and NLoS cases were 172 ns and 204 ns, respectively. However, due to the lack of detailed descriptions of the measurement scenario and measurement setting in [9], it is difficult to explain why there were significant differences in the channel characteristic parameters between [8] and [9]. Moreover, research on the spatial domain is still lacking.

Regarding the mmWave band, channel research for the industrial scenario mainly focuses on 28 GHz [6,10–12] and 142 GHz [14–16], which is also known as the sub-THz frequency band. There is sufficient research on channel characteristics such as the path loss, RMS DS, and angle spread (AS), and many influencing factors, such as the antenna height, antenna polarization configurations, and time-varying effects, have been taken into account. In [6], the path loss, DS, and AS were analyzed and modeled in a factory-like scenario at 28 GHz. The path loss was modeled by both the CI model and the floating intercept (FI) model, and the FI model showed high accuracy. Moreover, the RMS DS and AS were modeled by a linear function, and the cumulative distribution function (CDF) fitting line as well as the fitting parameters were given. Furthermore, it was found that the antenna heights would have an impact on these channel characteristics. In [10], a ray-tracing-based simulation was conducted at 28 GHz in a light factory and a heavy factory. The path loss was analyzed and modeled by the CI model, and the PLEs were 1.9 and 2.1 for the light and heavy factories, respectively. In [11], the author presented an extensive measurement campaign at 28 GHz inside a factory environment. It was found that the path gain in LOS cases could be well represented by the Friis free space formula, with an RMS error of 3.6 dB. In [12], the path loss, RMS DS, and AS in the factory environment at 28 GHz were analyzed and compared to the indoor hotspot (InH) model. The author proposed a path loss model similar to the FI model and used a logarithmic function to model the RMS DS. In [13], time-varying THz channel measurements at 105 GHz were conducted in an indoor factory. The number of MPCs and the RMS DS in LOS and NLOS environments were analyzed and compared. The correlation between power delay profiles (PDPs) was described by the temporal PDP correlation coefficient. In [14–16], the author conducted channel measurements at 142 GHz and studied three different aspects of channel characteristics in the industrial scenario. Ref. [14] focused on the influence of co-polarized and cross-polarized antennas on the path loss. The omnidirectional and directional path loss with two antenna polarization configurations produced gross cross-polarization discrimination (XPD) with a mean of 27.7 dB. Ref. [15] conducted measurements with low-RX and high-RX antenna heights of 0.5 m and 1.5 m to characterize the propagation channel for close-to-floor applications. Ref. [16] proposed a sub-THz statistical channel model for the indoor factory scenario based on an extensive dataset of radio propagation characterization in factory buildings at 142 GHz. The temporal and spatial channel parameters were modeled by distribution fitting and showed satisfactory p-scores in the goodness-of-fit test. However, the channel characteristics in the industrial scenario at frequencies other than 28 GHz and 142 GHz still need further investigation.

With respect to THz bands, the research on industrial scenario channels is relatively insufficient, with the frequency points mainly concentrated at 300 GHz. In [17], the author presented channel measurement campaigns in a frequency range of 215–225 GHz in micro drilling–milling and large milling areas. The results indicated that the FI model is potentially more accurate compared to the CI model and the distribution density of scatterers in the environment has significant effects on the RMS DS and AS. In [18], channel measurements were conducted at a carrier frequency of 300 GHz with a measurement bandwidth of 2 GHz in three distinct industrial scenarios. The channel parameters of each receiver were provided, such as the path gain, path loss, RMS DS, and AS. It can be seen that there are obvious differences in the channel characteristic parameters in different scenarios. However, due to the lack of statistical analysis, the impact of scenarios on the channel parameters has not been quantified. In [19], measurements were conducted at 304.2 GHz for both time-invariant and time-variant settings. For the time-invariant setups, the evaluation results included the path gain, PDP, and RMS DS. For the time-variant setups, the evaluation results included a 3D power delay profile and the path gain versus the time for a fixed delay value. In [20], a novel ultra-wideband dual-polarized double-directional measurement was conducted at 300 GHz for the application of an access point to the inside of a machine in an industrial scenario. The results showed a sparse spatial/temporal channel with multiple paths from the different metallic objects and their influence on polarization. However, the channel characteristics in the industrial scenario at other frequencies in the THz band have not been fully investigated.

For visible light bands, in two IEEE channel model standards, manufacturing cells have been defined as typical application scenarios for visible light communication (VLC). However, the channel characteristics for visible light bands in industrial scenarios are rarely researched. In the IEEE 802.15.7 [21] released in 2018 and the IEEE 802.11bb [22], which was released in 2023, the channel frequency responses, channel direct current gain, and

RMS DS for manufacturing cells were given. However, the simulation environment was too simplified, with only two machines inside, which cannot reflect the unique channel characteristics of the industrial scenario. What is more, in [23], the simulation environment was more realistic, with abundant scatterers, and the path loss, the RMS DS, and the AS were investigated at the receiving heights of 1.5 m and 2.5 m.

Furthermore, there are works that have conducted measurements at multiple frequencies in the same industrial scenario to fairly compare the channel characteristics of different frequency bands. In [24], wideband channel characteristics, for example, the channel gain coefficient, DS, and Ricean K factor at the 3–4 GHz and 38–40 GHz bands, were analyzed and modeled. In [25], the Volcano ray-tracing channel simulation was tuned and benchmarked against field measurements at two frequencies relevant to 5G industrial networks: 3.7 GHz and 28 GHz. Large-scale parameters, such as the PDP, and the CDF of the RMS DS were explored. In [26], the author performed simultaneous multi-band ultra-wideband dual-polarized double-directional measurements at the sub-6 GHz (6.75 GHz), mmWave (74.25 GHz), and THz (305.27 GHz) bands in a small machine room. In [27], channel measurements were conducted at 28 GHz, 38 GHz, 132 GHz, and 220 GHz in four industrial scenarios. The channel characteristics were extracted and modeled, including the path loss, K factor, RMS DS, and AS. However, the existing research on the comparison of multi-band channel characteristics in industrial scenarios only considers a few electronic wave bands, and there is no fair comparison of the channel characteristics covering the sub-6 GHz, mmWave, THz, and visible light bands in the same industrial scenario. The impact of frequencies on the channel characteristics of the industrial scenario is still unknown.

To sum up, some studies have been conducted on the channel characteristics for industrial scenarios in various bands. However, the research on each band has only focused on a few specific frequency points, for example, 28 GHz and 142 GHz in the mmWave band and 300 GHz in the THz band. Moreover, there is a particular lack of research on industrial scenario channels in the visible light band. Furthermore, it can be found that some influencing factors could have a significant impact on the channel characteristics of the industrial scenario, such as the environment density and antenna height. Therefore, in order to accurately describe the impact of the frequency on the channel characteristics of industrial scenarios, it is necessary to conduct channel characteristic research covering the sub-6 GHz to visible light bands with a unified antenna height in the same industrial scenario.

1.2. Contribution of This Paper

In this paper, we conduct a comprehensive channel characterization and comparison in the industrial scenario from the sub-6 GHz to visible light bands. The main contributions are as follows:

(1) The multi-frequency channel simulations were conducted in the same industrial scenario with the same configuration for both the electronic bands and the visible light bands. The frequency range covered the sub-6 GHz, mmWave, THz (i.e., 3.5 GHz, 6 GHz, 28 GHz, 60 GHz, 220 GHz, 260 GHz, 300 GHz) and visible light bands, which could provide channel data for a comprehensive channel characterization and comparison in the industrial scenario.

(2) We made a comprehensive comparison of the channel characteristics for multiple electronic bands and visible light bands in the industrial scenario, including the path loss, shadow fading, RMS DS, and AS. The frequency dependence and the distance dependence were analyzed based on the industrial features and frequency properties.

(3) The statistical characteristics of the channel properties were modeled and compared with the 3GPP TR 38.901 standard [28]. Furthermore, we extended the 3GPP model to

the THz and visible light bands and provided the statistical parameters of the channel characteristics for all the frequency bands.

1.3. Paper Organization

The rest of this paper is organized as follows. In Section 2, the scenario and ray-tracing simulation setup are described in detail. In Section 3, the channel characteristic data processing methods for both the electronic wave band and the VLC band are given. In Section 4, the channel characteristics, including the path loss, RMS DS, and AS, are analyzed, and the statistical characteristics are investigated. Finally, conclusions are drawn in Section 5.

2. Scenario and Ray-Tracing Simulation Setup

In this section, we introduce a realistic industrial scenario and describe how we conducted a channel simulation at multiple electronic wave bands and VLC bands. Firstly, detailed information on the industrial scenarios and internal reflectors, including their position, size, and materials, is presented. Secondly, the channel modeling methodology for the electronic wave band is introduced, and the corresponding setups are provided. Thirdly, the channel modeling methodology and the corresponding setups for the VLC band are introduced.

2.1. Industrial Scenario

We chose the fiber optic cable laboratory of the Beijing University of Posts and Telecommunications to conduct the simulation in, which was a typical factory-like scenario. On the one hand, the size of the scenario was larger than the common indoor scenario, which reaches 33.3 m × 7.7 m × 3 m. On the other hand, there was a rich variety of metal reflectors in the scenario, most of which were metal instruments, and the material was defined as ferrum (Fe). There were a total of 38 metal instruments in the scenario, whose average height was 1.27 m. The layout of the scenario is shown in Figure 1. In addition, there was a small number of non-metallic reflectors which could cause the occurrence of pure Lambertian radiation for VLC, including the floor, ceiling, walls, and workspace. Among them, the walls were covered by plaster. The window was made of glass. The workspace in the bottom left corner of Figure 1 and the long office desk surrounding it were made of galvanized steel and pinewood.

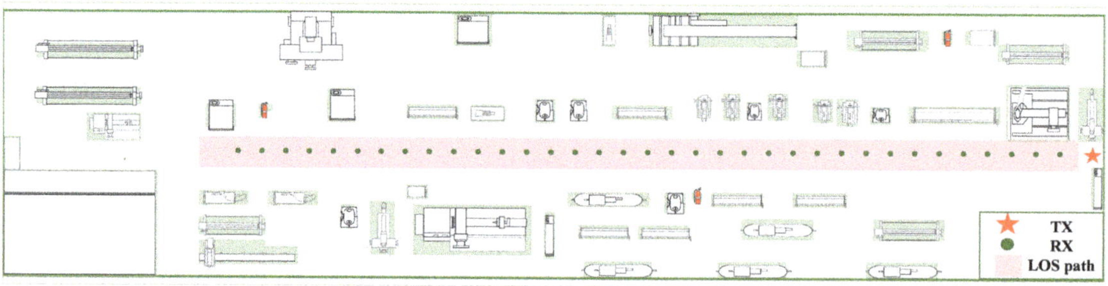

Figure 1. Layout of the industrial scenario.

2.2. Ray-Tracing Simulation Setup for Electronic Wave Bands

We used Wireless InSite® (Remcom, State College, PA, USA) software tools for channel modeling at electronic wave bands [29]. Wireless InSite® is a suite of ray-tracing models and high-fidelity electromagnetic equation solvers for the analysis of site-specific radio wave propagation. There are five main steps for channel modeling in Wireless InSite®, including features, waveforms, antennas, transmitters and receivers, and outputs. In the following subsection, we will explain the main settings for each step in the simulation study.

The first step was the feature setup, which built the three-dimensional model of the industrial environment in the software. The computer-aided design (CAD) models for both the non-metallic reflectors (floor, ceiling, and windows) and the metal instruments were imported to model the realistic industrial scenario. Then, the surface materials of each object could be selected from the database. Secondly, for the waveform setup, we used a sinusoid waveform with carrier frequencies of 3.5 GHz, 6 GHz, 28 GHz, 60 GHz, 220 GHz, 260 GHz, and 300 GHz.

For the antenna, we chose to utilize an isotropic antenna instead of an omnidirectional antenna. The reason was that an omnidirectional antenna produces an omni-pattern in the azimuthal XoY plane with a null on the Z-axis. When the transmitter and the receiver are not placed at the same vertical height, it is very likely that only a small amount of the signal can be received or even that no signal can be received. Meanwhile, an isotropic antenna provides a uniform field strength in all directions.

For the transmitter, it was placed on a passable path near the wall and is shown by the red star in Figure 1. The distance between the transmitter and the wall was 2 m, and the height of the transmitter was 1.60 m. For the receivers, these were placed on the LOS route with a height of 1.27 m, and the route is shown by the red square. The interval between two adjacent points was 0.50 m. A total of 50 points were arranged.

As outputs, Wireless Insite® could generate the received power, the path loss, and the channel impulse response (CIR) automatically by selecting those items as the intended results in the output window, which could be used to calculate the channel characteristics.

2.3. Ray-Tracing Simulation Setup for VLC Bands

We utilized the non-sequential ray-tracing software Zemax® (Zemax LLC, Kirkland, WA, USA) to obtain the VLC channel data [30]. VLC channel models developed using this non-sequential ray-tracing method were accepted as reference channel models in IEEE 802.15.13 and IEEE 802.11bb [22]. This channel modeling approach has been validated through experimental results in [31].

The first step was to construct a three-dimensional simulation environment to recreate the industrial scenario. Firstly, the floor, ceiling, and walls were constructed according to the scenario size. Then, CAD models that matched the actual length, width, and height of the metal instruments were chosen and imported into the software according to their position in the realistic industrial scenario.

Moreover, the reflection characteristics of each object were set according to those of its actual material. The reflection coefficient refers to the ratio of the intensity of the reflected light to the intensity of the incident light on the reflective material surface. It is influenced by the wavelength of the light signal. Due to the large wavelength coverage of visible light signals, the wavelength dependence of the reflection coefficients for various materials needed to be taken into consideration, which is shown in Figure 2. Commonly used materials for indoor scenarios, such as those of the ceiling, floor, and plaster wall, were obtained from [30]. The remaining materials, such as plate window glass, pinewood, galvanized steel, and Fe, were obtained from [32]. Furthermore, the radiation pattern was set based on the roughness of the object. For non-metallic reflectors, which are relatively rough, the pure Lambertian radiation model was adopted. For metal instruments, specular reflection and Lambertian radiation simultaneously occurred, with each accounting for a proportion of half of the received power.

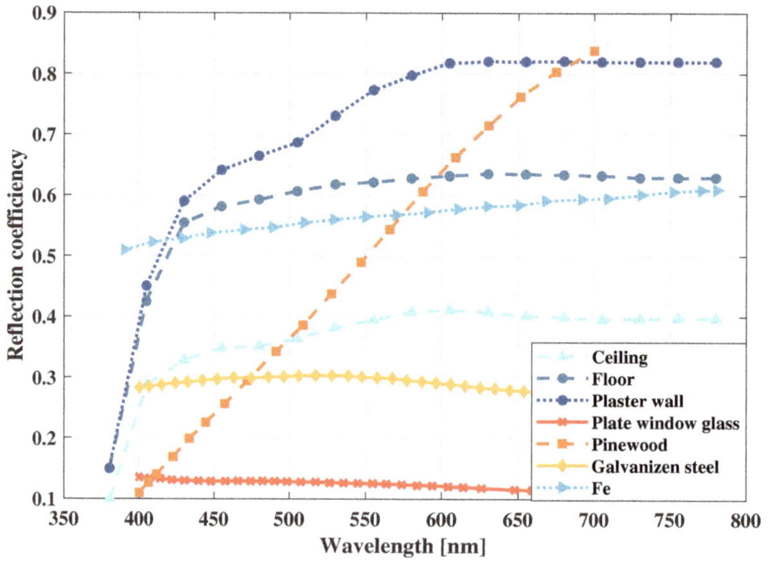

Figure 2. The reflection coefficients for various materials.

The second step was to set up the Tx and Rx in the simulation environment. We placed the Tx and Rx on the same LOS path as that in the electronic wave band simulation. On the Tx side, we utilized the method of splicing multiple LEDs to ensure that the light source and the isotropic antenna in the electronic wave band had the same transmission pattern. Nine commercially available light sources (PAR20) that had a viewing angle of 40° were used to achieve 360° coverage, with different lights pointing from 0° to 360° (interval of 40°) at TX. Furthermore, the relative spectral power distribution of the light source followed that of typical white light distribution, as shown in Figure 3. On the Rx side, we used rectangular detectors that were facing in the direction of the Tx. All the key parameters can be found in Table 2.

Table 2. Parameters of simulation setup.

	Electronic Wave Band	Visible Light Band
Scenario size	33.3 m × 7.7 m × 3 m	
Average height of metal instruments	1.27 m	
Central frequency	3.5/6/28/60/220/260/300 GHz	White light (380–780 THz)
TX height	1.60 m	
TX pattern	Isotropic	Isotropic-like
Analysis rays	10^7	
Minimum relative ray intensity	10^{-3}	
Transmitter power	1 W	
Radiation pattern	Specular reflection	Specular reflection 50% Lambert radiation 50%
RX height	1.27 m	
RX pattern	Isotropic	Rectangular (10 cm)

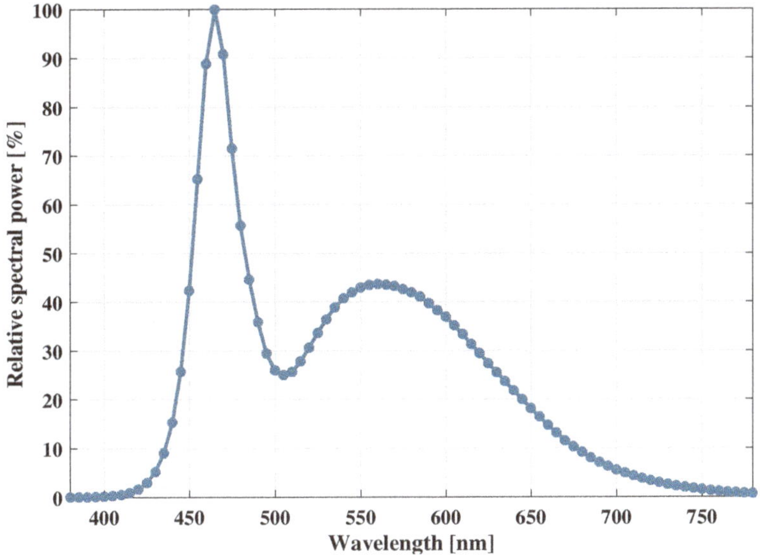

Figure 3. The relative spectral power distribution of the light source.

3. Data Processing

In this section, we introduce the channel characteristic parameter extraction method based on the electronic wave band channel data from Wireless InSite® and the VLC channel data from Zemax®. Three types of channel characteristics will be presented, including the path loss, RMS DS, and AS.

For the electronic wave band, the CIR can be given by

$$h(\tau) = a_1\delta(\tau - \tau_1) + \sum_{n=2}^{N} a_n e^{j\phi_n}\delta(\tau - \tau_n), \tag{1}$$

where a_1 presents the deterministic (typically LOS) component with the excess delay τ_1. The parameters a_n, τ_n, and ϕ_n are the amplitude, excess delay, and initial phase of the nth stochastic (typically NLOS) component, respectively. N denotes the number of multiple paths.

For the visible light band, the CIR can be given by

$$h(\tau) = \sum_{n=1}^{N} a_n\delta(\tau - \tau_n). \tag{2}$$

In the following equation, the CIR is utilized to obtain the channel characteristics. The path loss is the reduction in the power density of a radio wave as it propagates, which is of high importance for coverage prediction and interference analysis. According to the definition of the path loss, it can be defined by [33]

$$PL[dB] = -10\log_{10}(\sum_{n=1}^{N} |h(\tau_n)|^2), \tag{3}$$

The RMS DS describes the time dispersion of the MPCs, which determines the length of the guard interval in the OFDM system. It can be expressed by

$$\tau_{RMS} = \sqrt{\frac{\sum_{n=1}^{N}(\tau_n - \tau_{mean})^2|h(\tau_n)|^2}{\sum_{n=1}^{N}|h(\tau_n)|^2}}, \tag{4}$$

in which τ_{mean} is the average excess delay, which can be derived by

$$\tau_{mean} = \frac{\sum_{n=1}^{N}\tau_n|h(\tau_n)|^2}{\sum_{n=1}^{N}|h(\tau_n)|^2}. \tag{5}$$

The AS can characterize the dispersion degree of MPCs in the angle dimension, which can be used to represent the gain brought by MIMO technology [34]. In this paper, we analyze the angle spread of the arrival for the azimuth dimension, which can be expressed by

$$\sigma_{AS} = \sqrt{-2\ln\left(\frac{\sum_{n=1}^{N} e^{j\theta_n}|h(\tau_n)|^2}{\sum_{n=1}^{N}|h(\tau_n)|^2}\right)}, \tag{6}$$

where θ_n is the angle of the MPCs in radians.

4. Results and Analysis

In this section, we analyze the channel characteristics, present the channel characteristic models, and compare the channel characteristic model parameters between the sub-6 GHz, mmWave, THz, and visible light bands. Firstly, the path loss is analyzed and the distance dependency of the path loss is characterized by the FI model. Secondly, the RMS delay spread is analyzed with regard to the distance dependency, and the CDF of the RMS delay spread is modeled. Finally, the azimuth AS is analyzed, and the CDF model of the AS is presented.

4.1. Path Loss

Figure 4 presents the path loss results for all the frequency bands. It was observed that along the LOS path, the path loss for all the frequency bands showed an obvious distance dependence, and the path loss increased with an increasing distance from the transceiver with a non-linear relationship. Moreover, it could be observed that the path loss of the visible light band exhibited more significant variation with the transceiver distance compared with the electronic wave bands, and the path loss variation for various electronic wave bands was similar. For example, when the transceiver distance increased from 3 m to 9 m, the variation in the path loss for the visible light band was 15.5 dB, while the path loss variation for the other electronic wave bands was 6.43, 5.92, 6.49, 6.98, 5.41, 5.12, and 4.55 dB, respectively. This can be attributed to the LOS transmission of VLC. That is, in the received signal, LOS components account for a majority of the received power, while NLOS components only account for a small portion. As the LOS component mainly varies depending on the transceiver distance, the path loss for the visible light channel varies significantly with the transceiver distance.

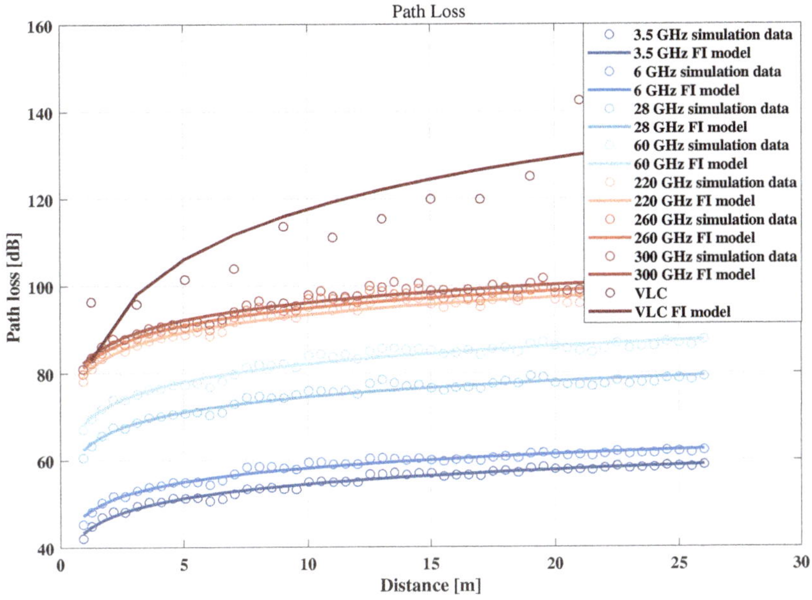

Figure 4. Path loss vs. transceiver distance.

In addition, at the same transceiver distance, the path loss increased with an increase in the frequency. Specifically, the path loss of the visible light channel was much higher than that of the electronic wave channel in general. At a small transceiver distance, the difference in the path loss between the electronic wave and visible light wave was relatively small. As the transceiver distance increased, the difference in the path loss between the visible light and electronic wave bands grew larger. For example, at a transceiver distance of 2.8 m, the path loss difference between the visible light signal and the 300 GHz signal, which was in the THz band, was only 8.37 dB. When the transceiver distance was 23 m, the path loss difference reached 27.10 dB. This shows that in short-range communication, there is not much difference in performance between THz communication and VLC. For long-range communication, electronic wave communication possesses great advantages over VLC.

In the THz bands, a large frequency change can only bring about small changes in the path loss, and in the mmWave and sub-6 GHz bands, a smaller frequency change will lead to a large path loss difference. For instance, at a transceiver distance of 15 m, the path loss difference between 300 GHz and 220 GHz was only 3.13 dB, while the path loss difference between 60 GHz and 39 GHz was 8.58 dB, and the path loss difference between 6 GHz and 3.5 GHz was 3.07 dB. Therefore, there will be significant differences in the coverage capability when applying different frequency bands, and a strict frequency selection must be made based on the propagation characteristics during deployment.

To accurately characterize the path loss, the FI model was utilized to fit the path loss for all the frequency bands, which can be expressed as

$$PL^{FI}(d) = \alpha + 10\beta\log_{10}d + X_{\sigma}, \tag{7}$$

where α is the floating intercept, β is the slope of the model, which is the PLE, and X_{σ} is a zero-mean Gaussian variable with a standard derivation, σ, representing the shadow effect. The fitting parameters for all the frequency bands are shown in Table 3.

Table 3. FI model fitting parameters.

	Frequency/GHz	α	β	σ
Sub-6 GHz	3.5	43.73	1.078	0.603237
	6	47.63	1.054	0.696649
mmWave	28	62.82	1.174	0.980498
	60	68.66	1.343	1.04594
THz	220	80.13	1.321	1.11161
	260	81.79	1.304	1.22165
	300	82.83	1.33	1.19503
	VLC	79.25	3.84	10.1258
3GPP	0.5–100 GHz	-	2.15	4.00

In 3GPP TR 38.901, the channel parameters of the factory scenario are presented, which can support the frequency bands for 0.5–100 GHz. The path loss model parameters in the factory scenario are distinguished based on the density of scatterers and the height of the transmitter. In our research, the simulation scenario had dense metal instruments and high transmitters, which corresponds to the DH scenario with dense clutter and high base stations in 3GPP TR 38.901, and the PLE in 3GPP TR 38.901 is 2.19, which is much higher than that of the electronic wave bands obtained in our research.

This can be attributed to the inconsistent height of the receiver. In 3GPP TR 38.901, the height of the receiver is described as embedded, which represents that the receiver height is the same as the average height of the reflector in the scenario, while in this research, the receiver was located lower than the average height of the metal instruments, which meant that there were a large number of metal reflectors distributed around the receiver. The abundant metal reflectors could lead to various NLOS signals. Therefore, in the received signal, the ratio of the LOS signal was relatively small, resulting in a smaller PLE. This can also be verified by the K factor. In 3GPP TR 38.901, the average K factor for the factory scenario is 7 dB, while in this research, the average K factor for the 3.5 GHz channel was −7.74 dB. The smaller value of the K factor represents a small proportion of LOS power in this research. Therefore, the height of the receiver will also greatly affect the parameters of the path loss model in industrial scenarios, which must be taken into account during path loss modeling.

Furthermore, due to the abundant NLOS signals, the PLE for the industrial scenario is lower than that for the common indoor hotspot scenario for all the electronic wave bands. For instance, for 0.5–100 GHz bands, the PLE for the indoor office scenario is 1.73 in 3GPP TR 38.901. Moreover, for the VLC band, the PLE for the industrial scenario is much higher than that for the common indoor hotspot scenario. For example, in [35], the PLE is 1.70 for the indoor conference scenario, while the PLE for the industrial scenario was 3.84 in this research, which is a similar value to that in [23]. This indicates that the light signals will suffer a faster decay in the industrial scenario. This may be due to the combined effect of the large size of the industrial scenario and the special specular reflection on metal surfaces, which would not happen in an indoor office. On the one hand, the large size makes the reflected light rays take longer paths to the RX. On the other hand, specular reflection reduces the angle of the reflected signal compared to pure Lambertian radiation. Therefore, the difficulty for the reflected signals in reaching the receivers is greatly increased. As the transceiver distance increases, the reflected signals that can reach the receiver decrease significantly, resulting in a significant attenuation of the received power with respect to the transceiver distance.

4.2. RMS Delay Spread

Figure 5 illustrates the RMS DS results in the industrial scenario. Firstly, as the transceiver distance increased, the RMS DS for the electronic wave bands showed an overall trend of first increasing and then decreasing, with the boundary point at a transceiver distance of 13 m. At the receiver closest to the transmitter, only a few NLOS components could be received, and the LOS component power accounted for a large proportion of the received power; thus, the RMS DS was at its minimum. As the transceiver distance increased, the number of NLOS components received from different directions increased, resulting in an increased RMS DS. When the transceiver distance continued to increase and exceeded 13 m, the RMD DS did not increase continuously as in most cases but instead decreased with an increasing distance. This can be attributed to the height of the transceivers, which was lower than the height of most metal instruments on both sides. The high-order reflection MPCs from both sides were obstructed by the metal instruments and could not reach the receivers, and only some low-order reflection rays from the transmitter direction could be received, leading to a decrease in the RMS DS. Therefore, in industrial scenarios, the height of the transmitter and receiver relative to that of the metal equipment can have a great impact on the RMS DS for electronic wave channels. Especially when the scenario size is large, a lower height of the transceiver will lead to a decreasing trend in the RMS DS with an increasing transceiver distance.

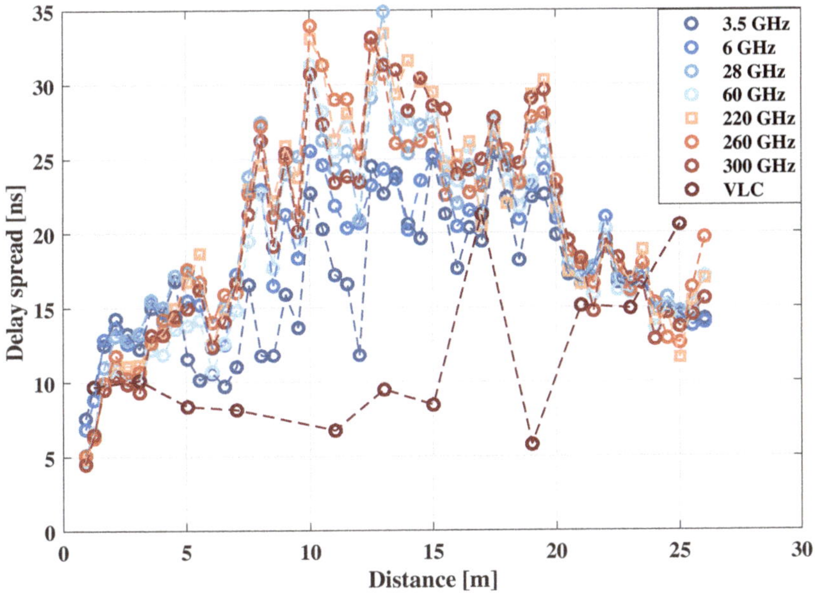

Figure 5. RMS DS vs. transceiver distance.

For the visible light band, the RMS DS increased with an increase in the transceiver distance. Even when the distance was greater than 13 m, there was no decreasing trend as in the electronic wave bands. This is because visible light signals exhibit Lambertian radiation on the surface of metal objects, and the reflected optical signals will have a wide range of angles. Although the metal instruments on both sides of the receivers may block some of the optical signals from propagating into the surrounding space, multiple reflections will still occur between the two rows of metal instruments. As the distance increases, the received NLOS path increases, resulting in an increase in the RMS DS.

Thirdly, from a frequency perspective, the RMS DS of the visible light band was significantly lower than that of the electronic wave band, while in the electronic wave band, the impact of the frequency on the RMS DS was not significant. On the one hand, the low RMS DS of the visible light band is attributed to the LOS transmission of VLC. The LOS components that have a low time delay account for a large proportion of the received signal power; thus, the RMS DS is relatively small. On the other hand, for the electronic wave band, the distribution of the MPCs that reaches the receiver is similar, which leads to a similar RMS DS. In specular reflection on the surface of metal instruments, the attenuation of signals is relatively small in various frequency bands. There is a low probability that the MPCs experience excessive power attenuation after multiple reflections due to a frequency increase which prevents it from reaching the receivers. Therefore, for various electronic wave bands, there is an obvious frequency dependence of the RMS DS.

Finally, it was found that the RMS DS of both the electronic wave and visible light bands could be fitted well by a lognormal distribution, as shown in Figure 6. The fitting parameters are given in Table 4. For the mean of the RMS DS fitting model, that of the visible light channel was obviously lower than that of the electronic wave channel, which was $-7.93\ \log_{10}(\mathrm{s})$. In the electronic wave bands, the value of μ was not significantly changed with the frequency, ranging from -7.72 to $-7.79\ \log_{10}(\mathrm{s})$, with an average value of $-7.74\ \log_{10}(\mathrm{s})$. In 3GPP TR 38.901, the mean of the RMS DS is also a fixed value, which is expressed as

$$\mu_{DS}^{3GPP} = \log_{10}(26 \times \frac{V}{S} + 14) - 9.53, \tag{8}$$

where V is the hall volume in m^3 and S is the total surface area of the hall in m^2 (walls + floor + ceiling). By substituting the volume and area values of the researched scenario into Equation (8), the mean was obtained as $-7.74\ \log_{10}(\mathrm{s})$, which is the same as the average value of μ in our research. This indicates that the 3GPP model of the RMS DS for the factory scenario can be extended to apply to THz bands. For the standard deviation, the value for the visible light channel was much larger than that of the electronic wave channel, which was 0.23. For electronic wave channels, the standard deviation increased with the frequency, which can be modeled as

$$\sigma_{DS} = 0.034 \times \log_{10}(1 + f_c) + 0.10, \tag{9}$$

where f_c is the frequency. The root mean square error (RMSE) was 0.009, indicating that the proposed model is in good agreement with the simulated data.

Comparing the RMS DS of the industrial scenario with that of indoor hotspot scenarios, it was found that there was only a small difference in the electronic wave band. For example, at 28 GHz, the mean values of the RMS DS for the industrial scenario and the indoor office scenarios were $-7.72\ \log_{10}(\mathrm{s})$ and $-7.71\ \log_{10}(\mathrm{s})$, respectively. Furthermore, for the visible light bands, the RMS DS in the industrial scenario was greater than that in the indoor conference scenario, which was $-7.93\ \log_{10}(\mathrm{s})$ and $-8.13\ \log_{10}(\mathrm{s})$, respectively. This is because the obstruction of the metal instruments limited the propagation environment of most electronic wave signals to the space between the metal instruments on both sides, and the impact of the scenario on the signals was not significant. In VLC, the optical signal had a larger angle range due to Lambertian radiation, and the effect of obstruction on the signal was relatively small. The influence of the scenario size on the channel characteristics became increasingly prominent. Thus, the industrial scenario, which had a much larger size than that of the indoor conference scenario, had a greater RMS DS.

Table 4. Fitting parameters of lognormal distribution for RMS DS ($\log_{10}$(s)).

Frequency/GHz		μ	σ
Sub-6 GHz	3.5	−7.79	0.12
	6	−7.75	0.12
mmWave	28	−7.72	0.16
	60	−7.75	0.18
THz	220	−7.72	0.18
	260	−7.72	0.17
	300	−7.73	0.19
VLC		−7.93	0.23

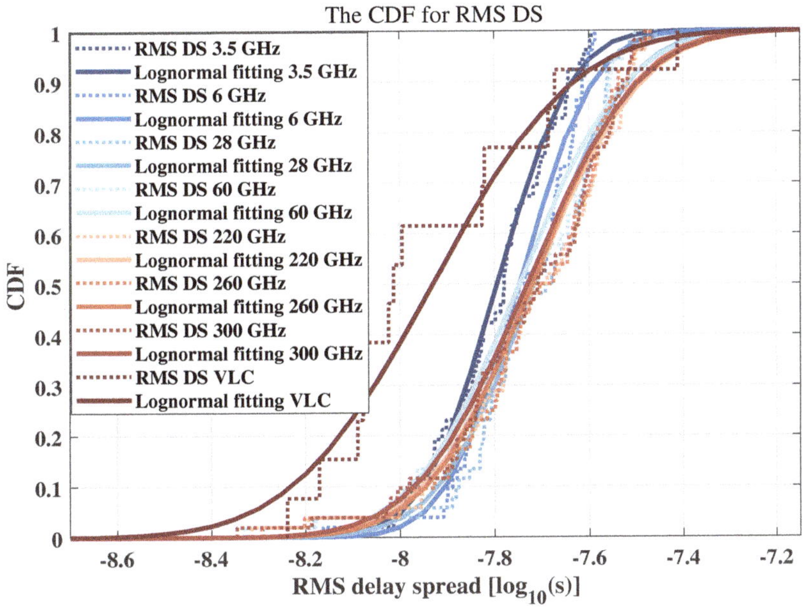

Figure 6. CDF of the RMS DS.

4.3. Angle Spread

Figure 7 illustrates the AS for all the frequency bands in the industrial scenario. Firstly, the AS of the visible light band was much larger than that of the electronic wave bands. This was due to the different reflection patterns on the surface of metal reflectors for the visible light and electronic wave bands. There are two kinds of reflection patterns for visible light signals on non-smooth metal surfaces: specular reflection and Lambertian radiation. Meanwhile, electronic wave signals mainly undergo specular reflection. In specular reflection, the direction of the reflected signal is singular, while in Lambertian radiation, the direction of the received signal exists within the range of 0–360°. Therefore, the AS for the visible light channel always maintains a high value.

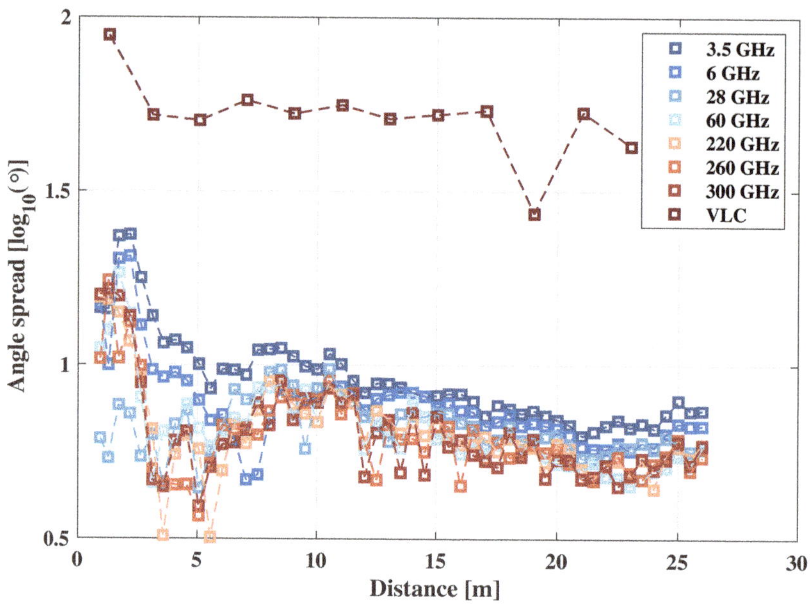

Figure 7. AS v.s. transceiver distance.

Moreover, in the electronic wave band, the AS slightly decreased with an increase in the frequency bands, and within each frequency band, the difference in the AS was not significant. For example, at a distance of 21.50 m, for the sub-6 GHz bands, the AS for 3.5 GHz and 6 GHz was 0.81 and 0.76 $\log_{10}(°)$, respectively. For the mmWave bands, the AS for 28 GHz and 60 GHz was 0.73 and 0.72 $\log_{10}(°)$, respectively. For the THz bands, the AS for 220, 260, and 300 GHz was 0.68, 0.68, and 0.69 $\log_{10}(°)$, respectively. This is because as the frequency increases, the reflection ability of the signal decreases, and the power of the NLOS signals decreases, resulting in a decrease in the AS.

Secondly, it can be observed that the AS for the electronic wave band channels experienced four stages as the transceiver distance increased. Firstly, at the receiver closest to the transmitter, the power of the LOS path accounted for a large proportion, and the AS was small, as shown in Figure 8a. As the transceiver distance increased to 2 m, the AS increased significantly with an increase in the transceiver distance. On the one hand, the number of NLOS components increased, which led to an increase in the AS. On the other hand, there was a metal instrument with a large size located at the position with a transceiver distance of 2 m, resulting in a series of strong NLOS signals. The strong NLOS signals caused by this large metal instrument can be clearly seen from the power angular spectrum (PAS), as shown in Figure 8b, which caused an increase in the AS by a large margin. Within the transceiver distance range of 2–5 m, the strong NLOS signals caused by this large metal instrument disappeared, and the AS suddenly decreased. Thirdly, as the transceiver distance increased from 5 m to 10 m, the power of the LOS component decreased greatly, and the NLOS components from the metal instruments on both sides gradually intensified, as shown in Figure 8d, resulting in a slow growth in the AS. Finally, when the transceiver distance was greater than 10 m, the AS decreased as the transceiver distance increased. It can be clearly observed from Figure 8e that the angle range of the received signal gradually converged towards the direction of the transmitter. Due to the lower height of the receivers, the blockage effect of the metal instruments on both sides caused the angle of the received signals to continue to be concentrated toward the transmitter, resulting in a continuous decrease in the AS. Therefore, in the industrial scenario, both ultra-large metal equipment and the height of the receivers can have a significant impact on the AS.

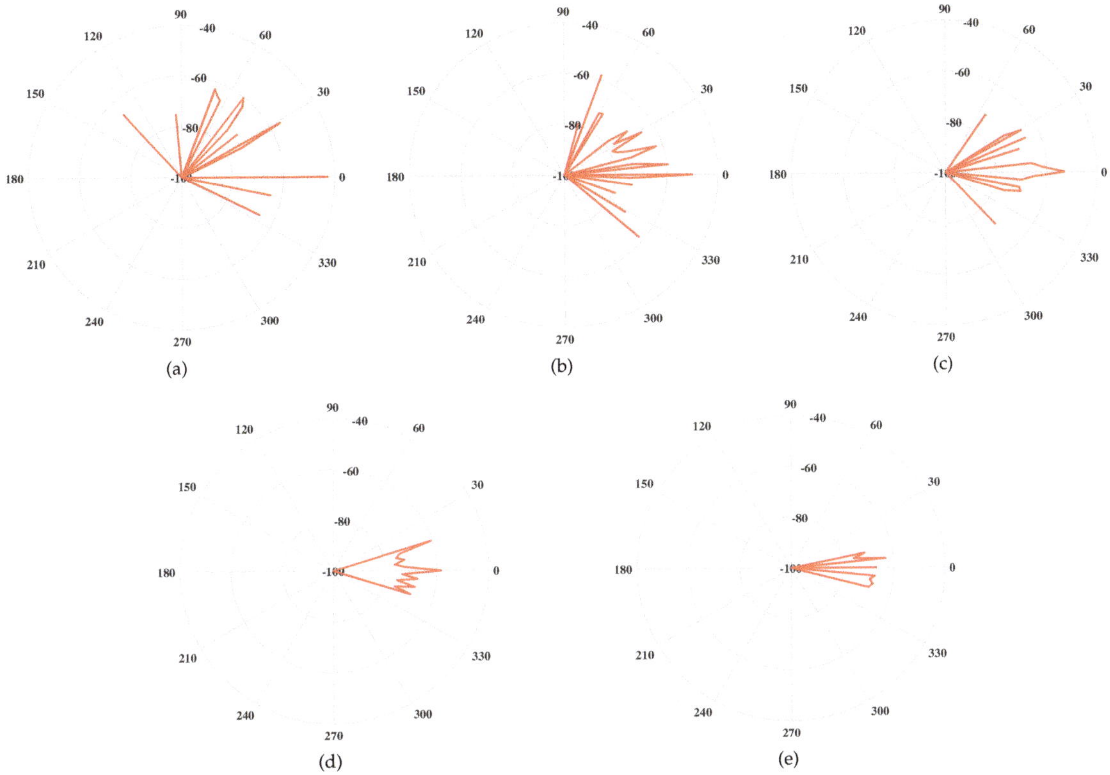

Figure 8. PAS of the electronic wave channel: (**a**) d = 1 m, (**b**) d = 2 m, (**c**) d = 5 m, (**d**) d = 10 m, (**e**) d = 27 m.

For the visible light channel, the AS almost remained constant with an increasing transceiver distance and was mainly distributed between 1.6 and 1.8 $\log_{10}(^{\circ})$. It can be seen from Figure 9 that there were MPCs with high power distributed randomly in the range of 0–360° at each receiver for the visible light band, which kept the AS in a stable state.

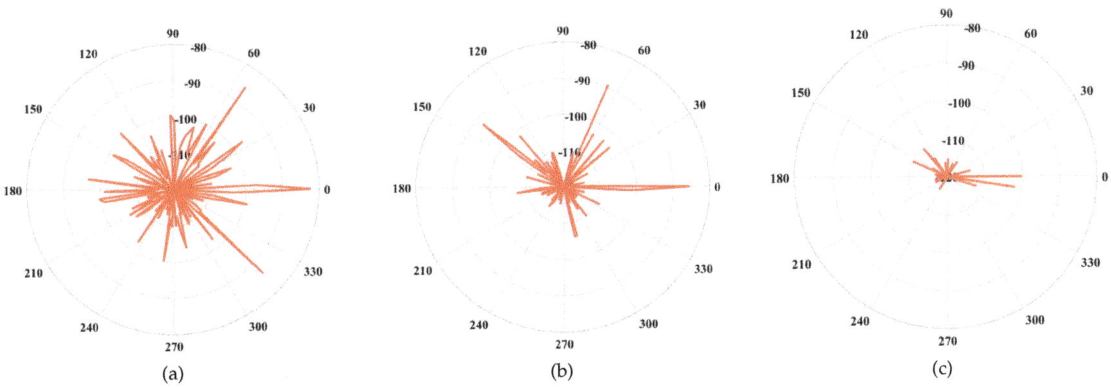

Figure 9. PAS of the VLC channel: (**a**) d = 5 m, (**b**) d = 10 m, (**c**) d = 21 m.

Furthermore, we used curve-fitting tools to fit the AS data and found that the AS in both the visible light and electronic wave channels followed a lognormal distribution, as shown in Figure 10. The fitting parameters are shown in Table 5.

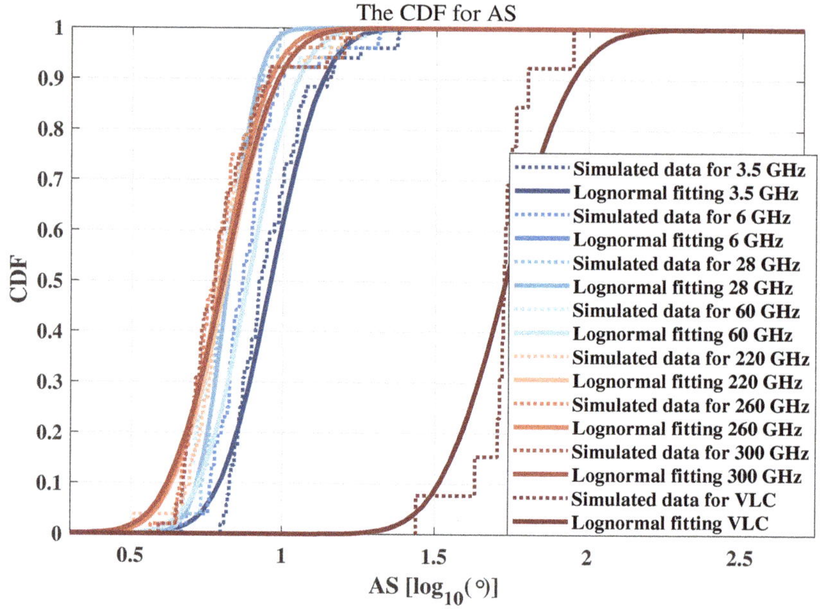

Figure 10. CDF of the AS.

Table 5. Fitting parameters of the lognormal distribution for the AS.

	Frequency/GHz	μ	σ
Sub-6 GHz	3.5	0.96	0.13
	6	0.89	0.12
mmWave	28	0.82	0.08
	60	0.82	0.12
THz	220	0.81	0.14
	260	0.79	0.13
	300	0.80	0.14
VLC		1.73	0.17

For the mean of the AS fitting model, the value for the visible light band was much higher than that for the electronic wave band, which was 1.73 ($\log_{10}(°)$). In the electronic wave band, the mean had a decreasing trend with an increasing frequency, as shown in Figure 11, which can be modeled as

$$\mu_{AS}^{ele} = -0.07 \times \log_{10}(1 + f_c) + 0.96. \quad (10)$$

The RMSE was 0.031, indicating that the proposed model has a goodness of fit to the simulated data. For the standard deviation, the value of the visible light band was higher than that of the electronic wave band, which was 0.17. In the electronic wave band, the minimum value of the standard deviation was 0.07 at 28 GHz. At other electronic wave frequencies, the standard deviation values varied between 0.12 and 0.14, with little difference and an average of 0.1234.

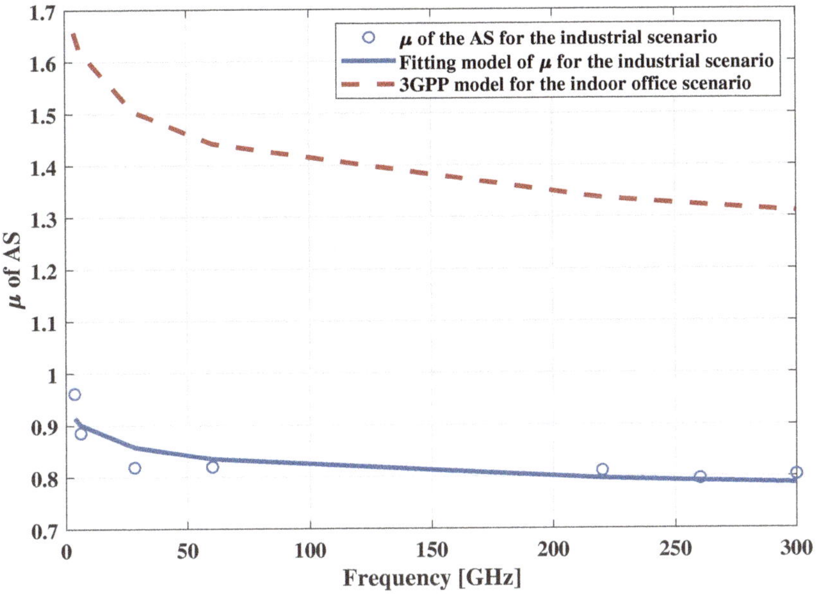

Figure 11. The mean value of the lognormal fitting for the AS.

Comparing the AS model of the industrial scenario with that of indoor hotspot scenarios in 3GPP TR 38.901, it was found that for the electronic wave band, the μ of the industrial scenario was much lower than that of the indoor office scenario. This was due to the blockage effect of the metal instruments caused by the lower height of the receivers. For the visible light band, there is no proposed model of the AS in the existing study. In [36], the median of the measured AS in the office was 1.40 ($\log_{10}(°)$), which is lower than the μ in the industrial scenario. This is due to the fact that the abundant metal instruments generated rich reflection signals which improved the AS.

5. Conclusions

In this paper, we conducted a comprehensive comparison of the channel characteristics for all the frequency bands, including sub-6 GHz, mmWave, THz, and VLC, in the industrial scenario. Firstly, the channel characteristics, including the path loss, RMS DS, and AS, were analyzed with respect to the frequency dependence and the distance dependence. On the one hand, the visible light band exhibited significant differences in channel characteristics compared to the electronic wave band. Due to the LOS transmission of VLC, the visible light band had a higher path loss, and the path loss exponent reached 3.84. Due to the Lambertian radiation pattern, which has a wide range of reflection angles, the AS of the visible light band was much larger than that of the electronic wave band, which was 1.73 ($\log_{10}(°)$) and 0.80 ($\log_{10}(°)$) for the visible light and THz bands, respectively. On the other hand, the blockage effect of the metal instruments in the industrial scenario will greatly affect the channel characteristics. As the transceiver distance grows large, signals from both sides of the receiver will be blocked by metal instruments, resulting in a decreasing trend in the RMS DS for the electronic wave band.

Secondly, the statistical characteristics of the channel properties were analyzed and modeled. It was found that both the RMS DS and the AS could be well fitted by the lognormal function for all the frequency bands. For the mean of the RMS DS fitting model, the value for the visible light band was -7.93 ($\log_{10}(s)$), which was obviously lower than that of the electronic wave bands. In the electronic wave bands, the value of the mean was not significantly changed with the frequency and could be modeled as a constant with the value of -7.74 ($\log_{10}(s)$). For the mean of the AS fitting model, the value for the visible light band was 1.73 ($\log_{10}(°)$), which was much higher than that for the electronic wave band. In the electronic wave band, the mean of the AS had a decreasing trend with an increasing frequency, which could be modeled as a logarithmic function.

Finally, the proposed statistical parameters of the channel characteristics were compared with the 3GPP TR 38.901 standard. It was found that the height of the receiver caused a larger PLE and a lower AS in the electronic wave bands in our research compared with those in 3GPP TR 38.901. For the RMS DS, the 3GPP model parameter for the factory scenario can be extended to apply to THz bands. What is more, the channel characteristics of the industrial scenario were compared with those of the indoor hotspot scenario. For the electronic wave bands, the channel characteristics of the industrial scenario exhibited a lower PLE, similar RMS DS, and lower AS. For the visible light bands, the channel characteristics of the industrial scenario exhibited a greater PLE, greater RMS DS, and larger AS. This study can provide insights for the evaluation and standardization of multi-frequency communication technology in industrial scenarios.

Author Contributions: P.T., J.Z., and Z.H. are my instructors and gave me full guidance and English correction during the writing process. Then, Y.Y. carried out the related work, including the construction of the simulation environment, the analysis of the simulation data, and the writing of the original draft. T.J., L.X., and G.L., as experts in VLC channel modeling in the company, also provided a lot of guidance on modeling during the data analysis. All authors have read and agreed to the published version of the manuscript.

Funding: This research was supported in part by the National Key Research and Development Program of China (2023YFB2904805), National Natural Science Foundation of China (62201086), Beijing Municipal Natural Fund (L243002), National Natural Science Foundation of China (62341128), Beijing University of Posts and Telecommunications China Mobile Research Institute Joint Innovation Center (CMYJY-202300305), and State Key Laboratory of Information Photonics and Optical Communications (Beijing University of Posts and Telecommunications) (No. IPOC2023ZT02), China.

Institutional Review Board Statement: Not applicable.

Informed Consent Statement: Not applicable.

Data Availability Statement: The data are contained within the article.

Acknowledgments: The authors would like to thank Baobao Liu, Yu Tong, and Shuo Liu for their help in preparing and writing this paper.

Conflicts of Interest: The authors Tao Jiang, Liang Xia, and Guangyi Liu were employed by the China Mobile Research Institute. The remaining authors declare that the research was conducted in the absence of any commercial or financial relationships that could be construed as a potential conflict of interest.

References

1. ITU-R. Framework and Overall Objectives of the Future Development of IMT for 2030 and Beyond. 2023. Available online: https://www.itu.int/dms_pub/itu-d/oth/07/31/D07310000090015PDFE.pdf (accessed on 20 January 2025).
2. Tang, P.; Yin, Y.; Tong, Y.; Liu, S.; Li, L.; Jiang, T.; Wang, Q.; Chen, M. Channel Characterization and Modeling for VLC-IoE Applications in 6G: A Survey. *IEEE Internet Things J.* **2024**, *11*, 34872–34895. [CrossRef]

3. Tang, P.; Zhang, J.; Xu, H.; Miao, H.; Liu, X. Preliminary perspectives on 3GPP standardization of the propagation channel model for FR3 bands for NR. *Sci. China Inf. Sci.* **2025**, *68*, 137301. [CrossRef]
4. Zhang, J.; Zhang, Y.; Yu, Y.; Xu, R.; Zheng, Q.; Zhang, P. 3-D MIMO: How Much Does It Meet Our Expectations Observed from Channel Measurements? *IEEE J. Sel. Areas Commun.* **2017**, *35*, 1887–1903. [CrossRef]
5. Miao, H.; Zhang, J.; Tang, P.; Tian, L.; Zhao, X.; Guo, B.; Liu, G. Sub-6 GHz to mmWave for 5G-Advanced and Beyond: Channel Measurements, Characteristics and Impact on System Performance. *IEEE J. Sel. Areas Commun.* **2023**, *41*, 1945–1960. [CrossRef]
6. Wang, Y.; Jiang, T.; Tang, P.; Song, Q.; Zhao, X.; Tian, L.; Zhang, J.; Liu, B. Measurement-based analysis and modeling of channel characteristics in an industrial scenario at 28 GHz. In Proceedings of the 2021 IEEE 94th Vehicular Technology Conference (VTC2021-Fall), Virtual Conference, 27 September–28 October 2021; pp. 1–5.
7. Aung, S.M.Y.; Pattanaik, K.K. Path loss measurement for wireless communication in industrial environments. In Proceedings of the 2020 International Conference on Computer Science, Engineering and Applications (ICCSEA), Gunupur, India, 13–14 March 2020; pp. 1–5.
8. Cebecioglu, B.; Mo, Y.K.; Dinh-Van, S.; Fowler, D.S.; Evans, A.; Sivanathan, A.; Kampert, E.; Ahmad, B.; Higgins, M.D. Sub-6 GHz channel modeling and evaluation in indoor industrial environments. *IEEE Access* **2022**, *10*, 127742–127753. [CrossRef]
9. Kim, J.; Kim, C.-S.; Hong, J.-Y.; Lim, J.-S.; Chong, Y.-J. Propagation characteristics of an industrial environment channel at 4.1 GHz. In Proceedings of the 2021 International Conference on Information and Communication Technology Convergence (ICTC), Jeju Island, Republic of Korea, 20–22 October 2021; pp. 1530–1532.
10. Solomitckii, D.; Orsino, A.; Andreev, S.; Koucheryavy, Y.; Valkama, M. Characterization of mmWave channel properties at 28 and 60 GHz in factory automation deployments. In Proceedings of the 2018 IEEE Wireless Communications and Networking Conference (WCNC), Barcelona, Spain, 15–18 April 2018; pp. 1–6.
11. Chizhik, D.; Du, J.; Valenzuela, R.A.; Otterbach, J.; Fuchs, R.; Koppenborg, J. Path loss and directional gain measurements at 28 GHz for factory automation. In Proceedings of the 2019 IEEE International Symposium on Antennas and Propagation and USNC-URSI Radio Science Meeting, Atlanta, GA, USA, 7–12 July 2019; pp. 2143–2144.
12. Ito, S.; Hayashi, T.; Yamazaki, K.; Nakao, M.; Sakai, S.; Kurosawa, Y.; Matsunaga, A.; Yoshida, K. Measurement and Modeling of Propagation Characteristics for an Indoor Environment in the 28 GHz-band. In Proceedings of the 2019 IEEE 30th Annual International Symposium on Personal, Indoor and Mobile Radio Communications (PIMRC), Istanbul, Turkey, 8–11 September 2019; pp. 1–6.
13. Qin, Y.; Tang, P.; Tian, L.; Lin, J.; Chang, Z.; Liu, P.; Zhang, J.; Jiang, T. Time-Varying Channel Measurement and Analysis at 105 GHz in an Indoor Factory. In Proceedings of the 2024 18th European Conference on Antennas and Propagation (EuCAP), Glasgow, UK, 17–22 March 2024; pp. 1–5.
14. Ju, S.; Xing, Y.; Kanhere, O.; Rappaport, T.S. Sub-terahertz channel measurements and characterization in a factory building. In Proceedings of the ICC 2022-IEEE International Conference on Communications, Seoul, Republic of Korea, 16–20 May 2022; pp. 2882–2887.
15. Ju, S.; Shakya, D.; Poddar, H.; Xing, Y.; Kanhere, O.; Rappaport, T.S. 142 GHz sub-terahertz radio propagation measurements and channel characterization in factory buildings. *IEEE Trans. Wirel. Commun.* **2023**, *23*, 7127–7143. [CrossRef]
16. Ju, S.; Rappaport, T.S. Statistical channel model of Wideband sub-THz radio propagation in indoor factories at 142 GHz: Towards 6G industrial wireless networks. *IEEE Trans. Wirel. Commun.* **2024**, *23*, 16316–16331. [CrossRef]
17. Liao, X.; Fan, L.; Wang, Y.; Yu, Z.; Wang, G.; Chen, Y.; Zhang, J. Measurement-Based Channel Characterization in Indoor IIoT Scenarios at 220 GHz. In Proceedings of the 2024 IEEE Wireless Communications and Networking Conference (WCNC), Dubai, United Arab Emirates, 21–24 April 2024; pp. 1–6.
18. Schultze, A.; Schmieder, M.; Wittig, S.; Klessig, H.; Peter, M.; Keusgen, W. Angle-resolved THz channel measurements at 300 GHz in an industrial environment. In Proceedings of the 2022 IEEE 95th Vehicular Technology Conference: VTC2022-Spring, Helsinki, Finland, 19–22 June 2022; pp. 1–7.
19. Elesina, V.V.; Reinhardt, C.E.; Kürner, T. Channel Measurements in Workspace with Robotic Manipulators at 300 GHz and Recent Results. In Proceedings of the 2024 18th European Conference on Antennas and Propagation (EuCAP), Glasgow, UK, 17–22 March 2024; pp. 1–5.
20. Dupleich, D.; Ebert, A.; Völker-Schöneberg, Y.; Löser, L.; Boban, M.; Thomä, R. Spatial/temporal characterization of propagation and blockage from measurements at sub-THz in industrial machines. In Proceedings of the 2023 17th European Conference on Antennas and Propagation (EuCAP), Florence, Italy, 26–31 March 2023; pp. 1–5.
21. Uysal, M.; Miramirkhani, F.; Narmanlioglu, O.; Baykas, T.; Panayirci, E. IEEE 802.15.7r1 Reference Channel Models for Visible Light Communications. *IEEE Commun. Mag.* **2017**, *55*, 212–217. [CrossRef]
22. Miramirkhani, F.; Baykas, T.; Elamassie, M.; Uysal, M. IEEE 802.11BB Reference Channel Models For Light Communications. *IEEE Commun. Stand. Mag.* **2023**, *7*, 84–89. [CrossRef]
23. Tong, Y.; Tang, P.; Zhang, J.; Liu, S.; Yin, Y.; Liu, B.; Xia, L. Channel Characteristics and Link Adaption for Visible Light Communication in an Industrial Scenario. *Sensors* **2023**, *23*, 3442. [CrossRef] [PubMed]

24. Wang, Y.; Lv, Y.; Yin, X.; Duan, J. Measurement-based experimental statistical modeling of propagation channel in industrial IoT scenario. *Radio Sci.* **2020**, *55*, 1–14. [CrossRef]
25. Bhatia, G.; Corre, Y.; Di Renzo, M. Tuning of ray-based channel model for 5G indoor industrial scenarios. In Proceedings of the 2023 IEEE International Mediterranean Conference on Communications and Networking (MeditCom), Dubrovnik, Croatia, 4–7 September 2023; pp. 311–316.
26. Dupleich, D.; Ebert, A.; Völker-Schöneberg, Y.; Sitdikov, D.; Boban, M.; Samara, L.; Del Galdo, G.; Thomä, R. Characterization of Propagation in an Industrial Scenario from Sub-6 GHz to 300 GHz. In Proceedings of the 2023 IEEE Globecom Workshops (GC Wkshps), Kuala Lumpur, Malaysia, 4–8 December 2023; pp. 1475–1480.
27. Wang, Y.; Wang, C.; Zheng, X.; Hao, X.; Liao, X. Multi-Frequency Wireless Channel Measurements and Characterization in Indoor Industrial Scenario. *IEEE Internet Things J.* **2024**, *11*, 38455–38468. [CrossRef]
28. 3rd Generation Partnership Project (3GPP). Technical Specification Group Radio Access Network; Study on Channel Model for Frequencies from 0.5 to 100 GHz. 3GPP TR 38.901. 2019. Available online: https://www.3gpp.org (accessed on 20 March 2024).
29. Yuan, Z.; Lyu, Y.; Zhang, J.; Tang, P.; Liu, X.; Lin, J.; Petersen, S.S.; Fan, W. Sub-THz Ray Tracing Simulation and Experimental Validation for Indoor Scenarios. In Proceedings of the 2023 IEEE International Mediterranean Conference on Communications and Networking (MeditCom), Dubrovnik, Croatia, 4–7 September 2023; pp. 7–11.
30. Miramirkhani, F.; Uysal, M. Channel Modeling and Characterization for Visible Light Communications. *IEEE Photonics J.* **2015**, *7*, 1–16. [CrossRef]
31. Eldeeb, H.B.; Uysal, M.; Mana, S.M.; Hellwig, P.; Hilt, J.; Jungnickel, V. Channel modelling for light communications: Validation of ray tracing by measurements. In Proceedings of the 2020 12th International Symposium on Communication Systems, Networks and Digital Signal Processing (CSNDSP), Porto, Portugal, 20–22 July 2020; pp. 1–6.
32. ASTER Spectral Library—Version 2.0. Available online: http://speclib.jpl.nasa.gov (accessed on 7 September 2023).
33. Yin, Y.; Tang, P.; Zhang, J.; Hu, Z.; Xia, L.; Liu, G. Multi-Wavelength Path Loss Model for Indoor VLC with Mobile Human Blockage. *Electronics* **2023**, *12*, 5036. [CrossRef]
34. Tang, P.; Zhang, J.; Miao, H.; Wei, Q.; Zuo, W.; Tian, L.; Jiang, T.; Liu, G. XL-MIMO channel measurement, characterization, and modeling for 6G: A survey. *Front. Inf. Technol. Electron. Eng.* **2024**, *25*, 1627–1650. [CrossRef]
35. Liu, B.; Tang, P.; Zhang, J.; Yin, Y.; Liu, G.; Xia, L. Propagation Characteristics Comparisons between mmWave and Visible Light Bands in the Conference Scenario. *Photonics* **2024**, *9*, 228. [CrossRef]
36. Liu, S.; Tang, P.; Zhang, J.; Yin, Y.; Tong, Y.; Liu, B.; Liu, G.; Xia, L. Statistical channel modeling for indoor VLC communications based on channel measurements. *China Commun.* **2024**, *21*, 131–147. [CrossRef]

photonics MDPI

Article

Integrated Mobile Visible Light Communication and Positioning Systems Based on Decision Feedback Channel Estimation

Ruoxuan Wang [1], Yuzhe Sun [1], Zhongxu Liu [2], Mingyi Gao [1] and Xiaodi You [1,*]

[1] School of Electronic and Information Engineering, Soochow University, Suzhou 215000, China; 2128410028@stu.suda.edu.cn (R.W.); 20215228065@stu.suda.edu.cn (Y.S.); mygao@suda.edu.cn (M.G.)
[2] Department of Electrical and Electronic Engineering, The Hong Kong Polytechnic University, Hong Kong; zhongxu.liu@connect.polyu.hk
* Correspondence: xdyou@suda.edu.cn

Abstract: Visible light communication (VLC) and visible light positioning (VLP) systems are usually designed separately to prevent mutual interference, while terminal mobility often introduces challenges that can degrade their performance. In this paper, we propose an integrated visible light communication and positioning (VLCP) scheme designed for mobile scenarios, encompassing both multiple-input–single-output (MISO) and multiple-input–multiple-output (MIMO) configurations. This scheme integrates both functionalities into a unified system. Utilizing decision feedback channel estimation (DFCE), we effectively estimate the dynamic channel state information (CSI) for both VLC and VLP, thereby potentially enhancing both communication and positioning performance. Simulation results across various routes verify the effectiveness of the proposed scheme. It is observed that when the terminal moves at 1 m/s, the VLCP scheme with DFCE can maintain reliable transmission quality, ensuring bit error rates (BERs) consistently below 1.3×10^{-2}. Additionally, the mean positioning errors remain within the centimeter range in different routes, not exceeding 4.3 cm and 15.5 cm in the MISO and MIMO scenarios, respectively.

Keywords: visible light communication and positioning (VLCP); decision feedback channel estimation (DFCE); multiple-input–single-output (MISO); multiple-input–multiple-output (MIMO)

1. Introduction

With the development of mobile Internet and industrial Internet of Things, there has been a growing demand for high-efficiency indoor data access and high-accuracy positioning. Indoor visible light communication (VLC) based on light-emitting diodes (LEDs) has emerged as a promising solution, leveraging the advantages of visible light for lighting, communication, and positioning purposes [1–3], providing an additional communication channel in indoor environments where radio frequency (RF) communication may face challenges like data insecurity, interference, and limited spectrum availability. Specifically, VLC can be applied to indoor positioning, often referred to as visible light positioning (VLP) [4], offering benefits such as high accuracy, low cost, and immunity to electromagnetic interference [5,6].

However, conventional VLC and VLP systems are typically designed and deployed independently. When deployed within the same indoor environment to provide both communication and positioning services simultaneously and share similar bandwidth resources, these systems may struggle to deliver high-quality performance in their respective functions due to frequency interference. Thus, integrated visible light communication and positioning (VLCP) systems using multiplexing techniques like time division multiplexing (TDM), frequency division multiplexing (FDM), and code division multiplexing (CDM) can address this issue. TDM, the simplest method, segregates VLC and VLP signals into different time slots [7,8]. With FDM, VLC and VLP signals occupy distinct frequency subcarriers for simultaneous transmission and positioning [9–13]. CDM assigns unique spreading codes to VLC and VLP signals, which are then differentiated through despreading codes [8,14,15]. However, commercial LEDs typically offer a narrow bandwidth, often just a few megahertz. Consequently, in scenarios where both communication and positioning are equally vital, effectively managing limited bandwidth resources to ensure reliable communication and accurate positioning poses a significant challenge for integrated VLCP design. The mentioned schemes may fail to allocate sufficient communication resources for both VLC and VLP simultaneously. To optimize the utilization of constrained bandwidth resources, researchers have proposed realizing VLP based on the channel state information (CSI) during VLC [16–18]. In [16], the estimated frequency domain transfer function (FDTF) of VLC channels was utilized for ranging, enabling VLP. In [17], the distances between LEDs and the PD were estimated through the received signal strength (RSS) of VLC channels, followed by trilateration to achieve VLP. However, the above studies focus on the scenario where the terminal was stationary and do not consider the case of terminal movement.

From the user's perspective, terminal mobility poses a significant and challenging issue in mobile VLCP systems. On one hand, in mobile VLC, terminal mobility causes fluctuations in the optical channel, impacting the VLC performance [19]. Various strategies have been proposed to enhance the link performance. In [20], a decision feedback channel estimation (DFCE) scheme was proposed to estimate the time-varying CSI in mobile VLC systems. In [21,22], space–time block coding (STBC) was employed to enhance the robustness of mobile VLC against time-varying CSI. However, these studies mainly focus on single-input–single-output (SISO) or multiple-input–single-output (MISO) systems, overlooking the potential of multiple-input–multiple-output (MIMO) systems, which have been demonstrated to boost the capacity and reliability of VLC systems [23,24]. On the other hand, in mobile VLP, the optical channel fluctuation also affects the positioning performance. While studies have targeted mobile VLP, they often involve fixing terminals at discrete locations to receive VLP signals [25,26]. Few studies have considered the continuous movement of terminals. For a mobile receiver, when conducting VLP discretely (e.g., using the RSS scheme), the positioning accuracy along the moving trajectory will inevitably deteriorate due to channel variations. To enhance accuracy, reducing the time interval of positioning is necessary. However, this entails transmitting more VLP signals, leading to reduced spectrum efficiency. Some studies address this by utilizing CSI as a feature fingerprint and analyzing it through machine learning [27–29]. Nevertheless, this approach demands significant data processing and computational power from mobile terminals. It also requires extensive data collection for model training, and any environmental changes necessitate repeating the training steps, which hampers system portability. Hence, new approaches are needed to enhance mobile VLP. In summary, current research on mobile VLC and VLP has limitations, with studies on integrated mobile VLCP still being relatively scarce.

In this paper, we propose an integrated mobile VLCP scheme designed for both MISO and MIMO scenarios, leveraging DFCE to efficiently estimate dynamic CSI for both VLC and VLP. The scheme begins with using training symbols to acquire initial CSI at discrete locations. Subsequently, it transitions to a decision feedback mode to continuously track and estimate subsequent CSI as the receiver moves. For the MISO scenario, specific groups of LEDs are selected from multiple LEDs to transmit identical VLC signals simultaneously, ensuring sufficient signal-to-noise ratios (SNRs). These groups of LEDs vary in each time slot within a cycle, allowing for the extraction of CSI related to VLC from multipath channels. This information is then used, via DFCE, to determine the range between each LED and the PD for VLP purposes. For the MIMO scenario, we utilize STBC to enhance the bit error rate (BER) performance. The improved transmission quality contributes to more accurate VLC-based CSI via DFCE, thereby further enhancing positioning accuracy. Our extensive simulations evaluate both communication and positioning performance as the receiver moves through different routes within a 5 m $\times$ 5 m $\times$ 3 m indoor space.

We analyze the effects of LED transmit power and receiver velocities on communication performance. For positioning accuracy, we investigate the impacts of observation block length, receiver velocities, and other relevant factors. The results indicate that in both MISO and MIMO scenarios, implementing DFCE maintains the BERs between approximately 10^{-4} and 10^{-2} for mobile VLC, while successfully achieving centimeter-level mobile VLP. Note that part of this work was previously presented in our short conference paper [30], which only focused on the MIMO scenario. However, in this expanded study, we conduct a comprehensive investigation covering both MISO and MIMO scenarios, as well as addressing additional issues.

The rest of the paper is organized as follows: Section 2 introduces the principle of the proposed integrated mobile VLCP scheme, considering both MISO and MIMO scenarios. Simulation results and discussions are presented in Section 3. Finally, we conclude the paper in Section 4.

2. Proposed Integrated Mobile VLCP Scheme

Figure 1 illustrates the schematic diagram of an integrated mobile VLCP system model, considering both MISO and MIMO configurations. The system comprises N_T LEDs serving as transmitting antennas installed in the ceiling to transmit VLC signals. The coordinate of the ith LED ($i = 1, 2, \ldots, N_T$) is known and denoted as (x_i, y_i, z_i). In this work, we focus on the scenario where $N_T = 4$. The mobile receiver is equipped with N_R PDs serving as receiving antennas, positioned in the receiving plane. For the MISO scenario, $N_R = 1$, whereas for MIMO, $N_R = 2$ is assumed. Each PD is upward-oriented to receive VLC signals. The three-dimensional (3D) coordinate of the jth PD ($j = 1, 2, \ldots, N_R$) is (x_{rj}, y_{rj}, z_{rj}), with its two-dimensional (2D) horizontal coordinate (x_{rj}, y_{rj}) being unknown, and its vertical coordinate (z_{rj}) known. At the receiver, channel estimation is conducted to recover the VLC signal. The acquired CSI is then utilized for receiver localization, enabling simultaneous VLC and VLP within the same system.

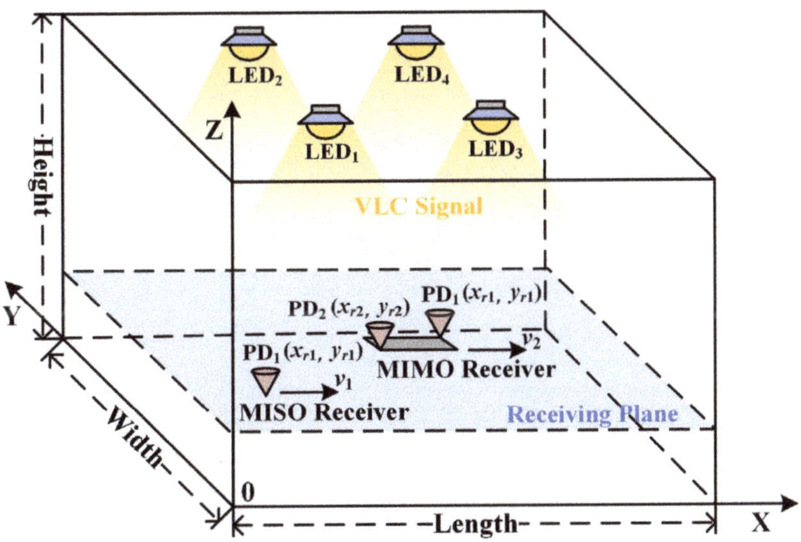

Figure 1. Schematic diagram of an integrated mobile VLCP system model.

2.1. Proposed Scheme for MISO Configuration

We first consider the MISO scenario. Figure 2 depicts the block diagram of the proposed integrated mobile VLCP scheme for MISO. At the transmitter, serial source bits are converted into quadrature amplitude modulation (QAM) symbols $X(k)$, where k

represents the subcarrier channel index (k = 1, 2, 3, ...). These QAM symbols are then modulated into the DC-biased optical OFDM (DCO-OFDM) signal $x(t)$. Next, $x(t)$ enters the LED selection stage, where the selector determines the groups of LEDs transmitting the identical $x(t)$ simultaneously. In our VLCP system, the total number of LEDs is four, which is commonly used to provide adequate and uniform light distribution in the receiving plane. From a communication perspective, adopting more LEDs can enhance SNRs for better link performance. Based on this, a simple way is to adopt all LEDs to send the same signal. However, from a positioning perspective, to enable trilateration, at least three LEDs are required, and the distances between each LED and the PD need to be estimated separately. Using a single-LED setup is inadequate to support trilateration as it does not provide enough distance measurements. For our four-LED setup, using all four LEDs to transmit signals simultaneously does not allow for the establishment of sufficient equations to calculate the distances between LEDs and the PD. To optimize both communication and positioning, different groups of LEDs can be used to transmit signals and calculate the CSI from multipath channels for VLP. For each group, a subset of N_{used} LEDs out of four LEDs is selected to transmit the same signal. In the case where N_{used} = 1, each of the four LEDs transmits the signal in one slot, but this configuration may not optimize the SNR, leading to reduced DFCE accuracy. The same issue arises for N_{used} = 2. Since involving more LEDs can enhance the SNR, we focus on N_{used} = 3, dividing the four LEDs into four groups: Group 1 (LED1, LED2, LED3), Group 2 (LED1, LED2, LED4), Group 3 (LED1, LED3, LED4), and Group 4 (LED2, LED3, LED4). This division is the optimized solution that ensures both signal recovery and optimized link performance at the receiver. The number of groups equals the total slots per cycle. After each cycle, the CSI can be estimated for VLCP and fed back for the next cycle. The four slots form a VLCP data block, and for the nth VLCP data block, the LED selection is determined by the LED selection matrix:

$$\mathbf{X}^n(t) = \begin{bmatrix} x_1^{n,1}(t) & x_2^{n,1}(t) & x_3^{n,1}(t) & 0 \\ x_1^{n,2}(t) & x_2^{n,2}(t) & 0 & x_4^{n,2}(t) \\ x_1^{n,3}(t) & 0 & x_3^{n,3}(t) & x_4^{n,3}(t) \\ 0 & x_2^{n,4}(t) & x_3^{n,4}(t) & x_4^{n,4}(t) \end{bmatrix}. \quad (1)$$

Here, $x_i^{n,w}(t)$ represents the wth OFDM symbol emitted by the ith LED in the nth VLCP data block, where i = 1, 2, 3, 4 and w = 1, 2, 3, 4. With the same w, $x_i^{n,w}(t)$ in each row of the matrix represents the same signals transmitted during the same slot, and different i values represent different expressions of the same signals emitted by different LEDs. After passing through multipath channels, the VLC signal arrives at the mobile receiver equipped with a single PD, denoted as PD1. Following light-to-current inversion, DC filtering, and time-to-frequency transformation, the electrical signal of the nth VLCP data block received by PD1 in the frequency domain can be expressed as follows:

$$\mathbf{Y}^n(k) = \mathbf{X}^n(k)\mathbf{H}^n(k) + \mathbf{N}^n(k). \quad (2)$$

Here, $\mathbf{X}^n(k)$ is the frequency representation of $\mathbf{x}^n(t)$, implying that each element $X_i^{n,l}(k)$ (l = 1, 2, 3, 4) in $\mathbf{X}^n(k)$ is the frequency representation of $x_i^{n,w}(t)$ in $\mathbf{x}^n(t)$ after Fourier transform. Thus, for the nth VLCP data block, $X_i^{n,l}(k)$ is the kth transmitted subcarrier symbol of the lth OFDM symbol at the ith LED, $\mathbf{Y}^n(k) = [Y^{n,1}(k), Y^{n,2}(k), Y^{n,3}(k), Y^{n,4}(k)]^{\mathrm{T}}$, where $Y^{n,l}(k)$ is the kth received subcarrier symbol of the lth OFDM symbol at PD1, $\mathbf{H}^n(k) = [H_1^n(k), H_2^n(k), H_3^n(k), H_4^n(k)]^{\mathrm{T}}$, where $H_i^n(k)$ is the FDTF of the channel between the ith LED and PD1, and $\mathbf{N}^n(k)$ is the noise matrix. Next, to recover the transmitted signal, we equalize the multipath channel. After zero-forcing equalization, the kth OFDM subcarrier symbol in the nth VLCP data block can be expressed as follows:

$$\hat{\mathbf{S}}^n(k) = \left[\frac{Y^{n,1}(k)}{\hat{H}^n_{T1\_total}(k)}, \frac{Y^{n,2}(k)}{\hat{H}^n_{T2\_total}(k)}, \frac{Y^{n,3}(k)}{\hat{H}^n_{T3\_total}(k)}, \frac{Y^{n,4}(k)}{\hat{H}^n_{T4\_total}(k)}\right]^T, \quad (3)$$

where $\hat{H}^n_{Tw\_total}(k)$ (w = 1, 2, 3, 4) is the estimated FDTF of the channel between all LEDs activated in the wth time slot and PD1.

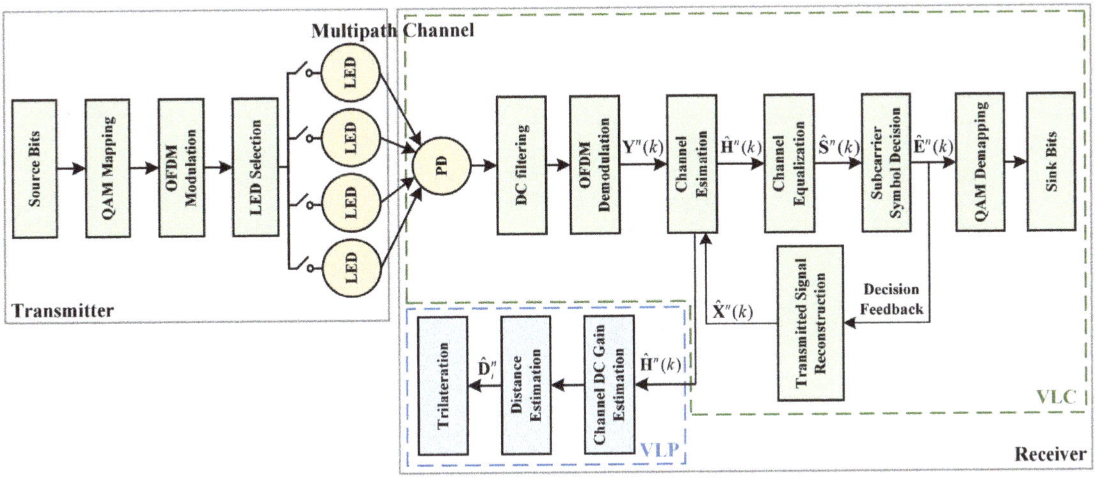

Figure 2. Block diagram of the integrated mobile VLCP scheme for MISO configuration.

According to Equation (3), before channel estimation, the receiver needs to have a priori knowledge of the CSI. Conventionally, training symbols at the beginning of each OFDM packet are employed for CSI estimation, assuming that the channel remains constant throughout the entire packet. However, in a mobile VLCP terminal, the channel characteristics vary over time. Consequently, the CSI estimated from training symbols may differ from that experienced by subsequent data symbols, leading to errors in CSI estimation and potentially affecting system performance. To effectively track the dynamic channel and achieve more accurate CSI, we implement the DFCE scheme in our system [20]. This scheme initially employs training symbols at the beginning of each OFDM packet to estimate the initial CSI. Subsequently, the estimated CSI is utilized to calculate $\hat{\mathbf{S}}^n(k)$ using Equation (3). Based on $\hat{\mathbf{S}}^n(k)$, the receiver can derive the decision value $\hat{\mathbf{E}}^n(k)$ to recover the original transmitted signal. The feedback of $\hat{\mathbf{E}}^n(k)$ is utilized to reconstruct the estimated transmitted signal matrix $\hat{\mathbf{X}}^n(k)$. Finally, $\hat{\mathbf{X}}^n(k)$ and $\mathbf{Y}^n(k)$ are used to update the CSI for subsequent channel estimation in the next VLCP data block, given by the following:

$$\hat{\mathbf{H}}^{n+1}(k) = \frac{1}{L}\sum_{p=n-L+1}^{n} \mathbf{Y}^p(k)\left[\hat{\mathbf{X}}^p(k)\right]^{-1}, \tag{4}$$

where L is the observation block length for VLCP data blocks. Equation (4) averages out noise by incorporating multiple blocks. This process is repeated to recover subsequent VLCP data blocks until the end of each OFDM packet. By doing so, we effectively track and mitigate channel variations between neighboring VLCP data blocks, thereby enhancing the channel estimation accuracy. It is worth mentioning that, based on the LED selection matrix in Equation (1), $\hat{\mathbf{X}}^n(k)$ is always invertible, ensuring the implementation of Equation (4).

Meanwhile, benefiting from the enhanced channel estimation accuracy, the CSI, specifically $\hat{H}^n_{Tw\_total}(k)$, can further be utilized to estimate the location of the mobile receiver. Based on Equation (3) and the LED selection in Equation (1), the specific expansion of the FDTFs for the multipath channels can be derived as follows:

$$\begin{cases} \hat{H}^n_{T1\_total}(k) = \hat{H}^n_{DC,1}(0)e^{-j2\pi k\Delta f t_1} + \hat{H}^n_{DC,2}(0)e^{-j2\pi k\Delta f t_2} + \hat{H}^n_{DC,3}(0)e^{-j2\pi k\Delta f t_3} \\ \hat{H}^n_{T2\_total}(k) = \hat{H}^n_{DC,1}(0)e^{-j2\pi k\Delta f t_1} + \hat{H}^n_{DC,2}(0)e^{-j2\pi k\Delta f t_2} + \hat{H}^n_{DC,4}(0)e^{-j2\pi k\Delta f t_4} \\ \hat{H}^n_{T3\_total}(k) = \hat{H}^n_{DC,1}(0)e^{-j2\pi k\Delta f t_1} + \hat{H}^n_{DC,3}(0)e^{-j2\pi k\Delta f t_3} + \hat{H}^n_{DC,4}(0)e^{-j2\pi k\Delta f t_4} \\ \hat{H}^n_{T4\_total}(k) = \hat{H}^n_{DC,2}(0)e^{-j2\pi k\Delta f t_2} + \hat{H}^n_{DC,3}(0)e^{-j2\pi k\Delta f t_3} + \hat{H}^n_{DC,4}(0)e^{-j2\pi k\Delta f t_4} \end{cases}. \tag{5}$$

Here, $\hat{H}^n_{DC,i}(0)$ is the estimated channel DC gain between the ith LED and PD1, Δf is the subcarrier frequency interval, and t_i is the propagation time of the signal from the ith LED to PD1. By squaring both sides simultaneously and setting $k = 1$, we can obtain the estimated channel DC gain from each LED to PD1:

$$\begin{cases} \hat{H}^n_{DC,1}(0) \approx \frac{1}{3}\left[\left|\hat{H}^n_{T1\_total}(1)\right| + \left|\hat{H}^n_{T2\_total}(1)\right| + \left|\hat{H}^n_{T3\_total}(1)\right| - 2\left|\hat{H}^n_{T4\_total}(1)\right|\right] \\ \hat{H}^n_{DC,2}(0) \approx \frac{1}{3}\left[\left|\hat{H}^n_{T1\_total}(1)\right| + \left|\hat{H}^n_{T2\_total}(1)\right| + \left|\hat{H}^n_{T4\_total}(1)\right| - 2\left|\hat{H}^n_{T3\_total}(1)\right|\right] \\ \hat{H}^n_{DC,3}(0) \approx \frac{1}{3}\left[\left|\hat{H}^n_{T1\_total}(1)\right| + \left|\hat{H}^n_{T3\_total}(1)\right| + \left|\hat{H}^n_{T4\_total}(1)\right| - 2\left|\hat{H}^n_{T2\_total}(1)\right|\right] \\ \hat{H}^n_{DC,4}(0) \approx \frac{1}{3}\left[\left|\hat{H}^n_{T2\_total}(1)\right| + \left|\hat{H}^n_{T3\_total}(1)\right| + \left|\hat{H}^n_{T4\_total}(1)\right| - 2\left|\hat{H}^n_{T1\_total}(1)\right|\right] \end{cases}. \tag{6}$$

In our VLCP scheme, without loss of generality, we consider a VLC channel model where LEDs are assumed to follow a Lambertian radiation pattern [31]. Whether phosphor LEDs or RGB LEDs, as long as they follow this radiation pattern, they can be used in our system. Furthermore, for other types of radiation patterns, as long as the radiation pattern is known and the distances between LEDs and PDs can be determined from the estimated channel DC gains, these types of LEDs can also be used. Considering the Lambertian radiation model, the distance between the ith LED and PD1 can be estimated using the RSS scheme:

$$\hat{D}^n_i = \sqrt[m+3]{\frac{(m+1)AT_s(\varphi)g(\varphi)h^{m+1}}{2\pi\hat{H}^n_{DC,i}(0)}}, \tag{7}$$

where m is the Lambertian emission order, A is the effective area of the PD, $T_s(\varphi)$ is the optical filter gain, $g(\varphi)$ is the concentrator gain, and h is the vertical distance between the LEDs and the receiving plane.

Based on Equation (7), we can select three from the four estimated distances to estimate the location of the receiver. For example, using $\hat{D}^n_1$, $\hat{D}^n_2$, and $\hat{D}^n_3$, the horizontal coordinate of PD1 $(\hat{x}_{r1}, \hat{y}_{r1})$ can be estimated with trilateration by solving the following:

$$\begin{cases} \left(\hat{D}^n_1\right)^2 = (\hat{x}_{r1} - x_1)^2 + (\hat{y}_{r1} - y_1)^2 + h^2 \\ \left(\hat{D}^n_2\right)^2 = (\hat{x}_{r1} - x_2)^2 + (\hat{y}_{r1} - y_2)^2 + h^2 \\ \left(\hat{D}^n_3\right)^2 = (\hat{x}_{r1} - x_3)^2 + (\hat{y}_{r1} - y_3)^2 + h^2 \end{cases}. \tag{8}$$

In the MISO scenario, $(\hat{x}_{r1}, \hat{y}_{r1})$ can simply be assumed as the receiver's horizontal coordinate $(\hat{x}_R, \hat{y}_R)$. To enhance positioning accuracy, we can select different groups of $\hat{D}^n_i$ to obtain multiple solutions, i.e., $(\hat{x}^q_{r1}, \hat{y}^q_{r1})$ (q = 1, 2, 3, 4), and then take their average value $((\hat{x}^1_{r1} + \hat{x}^2_{r1} + \hat{x}^3_{r1} + \hat{x}^4_{r1})/4, (\hat{y}^1_{r1} + \hat{y}^2_{r1} + \hat{y}^3_{r1} + \hat{y}^4_{r1})/4)$ as the final estimated horizontal coordinate of the receiver.

It is worth noting that Equation (4) assumes equal weight coefficients for the VLCP data blocks within the observation block. When the receiver is static, maintaining the same weight coefficient does not affect CSI accuracy. However, for a mobile receiver, where the channel varies over time, the estimated channel from different VLCP data blocks varies as well. Specifically, the closer the VLCP data blocks, the greater their impact on the channel estimation of the current block. Therefore, optimizing the weight coefficients can lead to better performance. Generally, closer blocks are assigned higher weights, with the neighboring block potentially having the highest weight. By considering the unequal weight coefficients, the CSI used for subsequent channel estimation in the next VLCP data block can be optimized as follows:

$$\hat{\mathbf{H}}^{n+1}(k) = \sum\nolimits_{p=n-L+1}^{n} \frac{\alpha^{n-p}(1-\alpha)}{1-\alpha^L} \mathbf{Y}^p(k)\left[\hat{\mathbf{X}}^p(k)\right]^{-1}, \tag{9}$$

where α represents a forgotten factor ($0 < \alpha < 1$).

2.2. Proposed Scheme for MIMO Configuration

Next, we consider the MIMO scenario. Figure 3 illustrates the block diagram of the proposed indoor integrated mobile VLCP scheme for MIMO. At the transmitter, serial source bits are converted into multiple parallel data streams, mapped into QAM symbols, and encoded by STBC. Here, STBC is introduced to ensure the feasibility of DFCE in the MIMO scenario, with the encoding matrix depending on the number of transmitting antennas N_T. In this work, we focus on a 4×2 MIMO system, where the receiver has two PDs (PD1 and PD2). Thus, similar to the MISO case, in this MIMO configuration, there are also four slots in a VLCP data block. For the nth VLCP data block, the STBC encoding matrix is expressed as [32]:

$$\mathbf{X}^n(k) = \begin{bmatrix} X_1^{n,1}(k) & X_2^{n,2}(k) & X_3^{n,3}(k) & 0 \\ -X_1^{n,2*}(k) & X_2^{n,1*}(k) & 0 & X_4^{n,3}(k) \\ X_1^{n,3*}(k) & 0 & -X_3^{n,1*}(k) & X_4^{n,2}(k) \\ 0 & X_2^{n,3*}(k) & -X_3^{n,2*}(k) & -X_4^{n,1}(k) \end{bmatrix}, \tag{10}$$

where * is the conjugate operation and $X_i^{n,l}(k)$ ($i = 1, 2, 3, 4$ and $l = 1, 2, 3, 4$) is the kth transmitted subcarrier symbol of the lth OFDM symbol at the ith LED. Each row of the matrix corresponds to signals transmitted during the same slot. These STBC symbols are modulated into DCO-OFDM symbols and transmitted by each LED. Upon traversing the MIMO channel, the VLC signal arrives at the mobile receiver. Following light-to-current inversion, DC filtering, and OFDM demodulation, the signal received at the jth ($j = 1, 2$) PD in the nth STBC block is represented in the frequency domain as follows:

$$\mathbf{Y}_j^n(k) = \mathbf{X}^n(k)\mathbf{H}_j^n(k) + \mathbf{N}_j^n(k). \tag{11}$$

Here, $\mathbf{Y}_j^n(t) = [Y_j^{n,1}(k), Y_j^{n,2}(k), Y_j^{n,3}(k), Y_j^{n,4}(k)]^{\mathrm{T}}$, where $Y_j^{n,l}(k)$ is the kth received subcarrier symbol of the lth OFDM symbol at the jth PD, $\mathbf{H}_j^n(k) = [H_{j1}^n(k), H_{j2}^n(k), H_{j3}^n(k), H_{j4}^n(k)]^{\mathrm{T}}$, where $H_{ji}^n(k)$ is the FDTF of the channel between the ith LED and the jth PD, and $\mathbf{N}_j^n(k)$ is the noise matrix. The parallel OFDM data are then decoded by STBC, demapped, and finally converted into serial sink bits. Following STBC decoding, the kth OFDM subcarrier symbol in the nth STBC matrix is decoded by [32]:

$$\hat{\mathbf{S}}^n(k) = \frac{\begin{bmatrix} \hat{H}_{j1}^{n*}(k)Y_j^{n,1}(k) + \hat{H}_{j2}^{n}(k)Y_j^{n,2*}(k) - \hat{H}_{j3}^{n*}(k)Y_j^{n,3*}(k) - \hat{H}_{j4}^{n*}(k)Y_j^{n,4}(k) \\ \hat{H}_{j2}^{n*}(k)Y_j^{n,1}(k) - \hat{H}_{j1}^{n}(k)Y_j^{n,2*}(k) + \hat{H}_{j4}^{n*}(k)Y_j^{n,3}(k) - \hat{H}_{j3}^{n}(k)Y_j^{n,4*}(k) \\ \hat{H}_{j3}^{n*}(k)Y_j^{n,1}(k) + \hat{H}_{j4}^{n}(k)Y_j^{n,2}(k) + \hat{H}_{j1}^{n}(k)Y_j^{n,3*}(k) + \hat{H}_{j2}^{n}(k)Y_j^{n,4*}(k) \end{bmatrix}}{\sum_{i=1}^{4}\left|\hat{H}_{ji}^{n}(k)\right|^2}. \tag{12}$$

Based on Equation (12), prior to STBC decoding, the receiver requires a priori knowledge of CSI. To track time-varying channels, we employ DFCE to obtain the CSI, similar to the MISO scenario. The training sequence is used to acquire the initial CSI, resulting in $\hat{\mathbf{S}}^n(k)$. Then, based on $\hat{\mathbf{S}}^n(k)$, the receiver derives the decision value $\hat{\mathbf{E}}^n(k)$, which is encoded by STBC to reconstruct the estimated transmitted signal matrix $\hat{\mathbf{X}}^n(k)$. Finally, $\hat{\mathbf{H}}_j^{n+1}(k)$ is updated for subsequent calculations of $\hat{\mathbf{S}}^{n+1}(k)$. With equal weight coefficients, $\hat{\mathbf{H}}_j^{n+1}(k)$ can be expressed as follows:

$$\hat{\mathbf{H}}_j^{n+1}(k) = \frac{1}{L}\sum\nolimits_{p=n-L+1}^{n} \mathbf{Y}_j^p(k)[\hat{\mathbf{X}}^p(k)]^{-1}. \tag{13}$$

With unequal weight coefficients, $\hat{\mathbf{H}}_j^{n+1}(k)$ can be expressed as follows:

$$\hat{\mathbf{H}}_j^{n+1}(k) = \sum\nolimits_{p=n-L+1}^{n} \frac{\alpha^{n-p}(1-\alpha)}{1-\alpha^L}\mathbf{Y}_j^p(k)[\hat{\mathbf{X}}^p(k)]^{-1}. \tag{14}$$

Based on the encoding matrix of STBC, $\hat{\mathbf{X}}^n(k)$ is always invertible, ensuring the implementation of Equations (13) and (14).

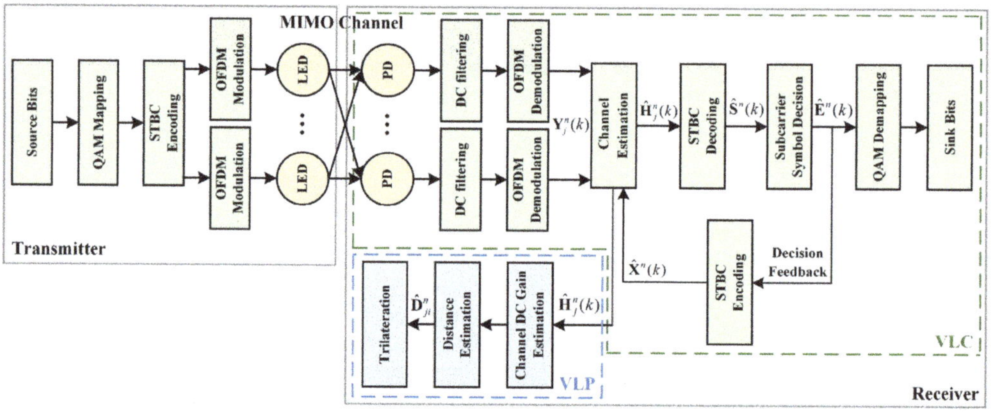

Figure 3. Block diagram of the integrated mobile VLCP scheme for MIMO configuration.

The acquired CSI can be further employed to determine the location of the mobile receiver. By setting $k = 1$ in the estimated FDTF $\hat{H}_{ji}^n(k)$ and referring to Equation (7), the distance between the ith LED and the jth PD can be estimated as follows:

$$\hat{D}_{ji}^n = \sqrt[m+3]{\frac{(m+1)AT_s(\varphi)g(\varphi)h^{m+1}}{2\pi\left|\hat{H}_{ji}^n(1)\right|}}. \tag{15}$$

With trilateration, we select different groups of the estimated distances to obtain multiple solutions for PD1, denoted as $(\hat{x}_{r1}^q, \hat{y}_{r1}^q)$ $(q = 1, 2, 3, 4)$, and for PD2, denoted as $(\hat{x}_{r2}^q, \hat{y}_{r2}^q)$, similar to the MISO system. Then, the multiple solutions of the horizontal coordinate of the receiver are calculated as the midpoint between PD1 and PD2, i.e., $((\hat{x}_{r1}^q + \hat{x}_{r2}^q)/2$, $(\hat{y}_{r1}^q + \hat{y}_{r2}^q)/2)$. Taking the average among them can finally lead to the final positioning result.

3. Results and Discussion

In this section, we consider a 5 m × 5 m × 3 m indoor integrated mobile VLCP system model, depicted in Figure 1. The transmitter consists of four LEDs, while the mobile receiver moves at a constant velocity. As illustrated in Figure 4, we investigate several test routes for the mobile receiver: Route 1 from (1, 2.5, 0.85) to (4, 2.5, 0.85), Route 2 from (1, 3.5, 0.85) to (4, 3.5, 0.85), Route 3 from (1, 2.5, 0.85) to (2, 2.5, 0.85), and Route 4 from (1, 3.5, 0.85) to (2, 3.5, 0.85). Throughout these routes, we calculate the average BERs and positioning errors of the receiver. We explore both MISO and MIMO scenarios to analyze the system performance.

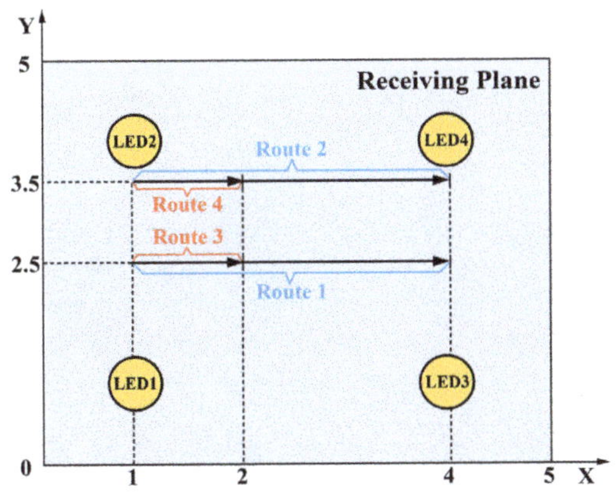

Figure 4. Bird's eye view of different test routes in the receiving plane.

In the MISO scenario, 4-QAM symbols are modulated into DCO-OFDM signals, which are then loaded onto four LEDs in the form of the LED selection matrix. All active LEDs transmit identical DCO-OFDM signals simultaneously, propagating through multipath channels. The mobile receiver, equipped with a single PD, utilizes DFCE to detect and recover the VLC signal. The CSI derived from the multipath channels is then employed for VLP. In the MIMO scenario, we consider a 4×2 MIMO system model, with a mobile receiver equipped with two PDs positioned 30 cm apart, maintaining a fixed pointing direction. 4-QAM symbols are generated, encoded with STBC, and then modulated into DCO-OFDM signals. These signals are transmitted through the mobile MIMO channel. At the receiver, the VLC signal is detected and recovered using DFCE with the dual-PD setup. Simultaneously, the CSI obtained from the VLC signal is utilized for VLP.

The OFDM symbol duration is standardized to 10 μs, with 512 subcarrier channels and a 20% cyclic prefix. For the MISO scenario, we assume that each OFDM packet consists of 2500 VLCP data blocks, with 35 blocks serving as training data. For the MIMO scenario, each OFDM packet also consists of 2500 VLCP data blocks, but with 20 blocks serving as training data. Consequently, the total duration of an OFDM packet is 0.1 s, and each OFDM packet conveys 10,000 OFDM symbols for MISO and 7500 OFDM symbols for MIMO. We then compare the communication and positioning performance of two schemes. The first scheme utilizes only OFDM training symbols at the start of each packet to estimate the CSI during terminal movement, without employing DFCE. In contrast, the second scheme integrates DFCE along with training symbols for enhanced CSI estimation. Except for Figure 12, equal weight coefficients are considered for DFCE. Additionally, the time interval for mobile VLP is set to 10 ms. For proof-of-concept, we only focus on line-of-sight (LOS) channels. When non-LOS channel conditions occur, the positioning accuracy can potentially be improved with machine learning [33], but this is beyond the scope of this paper. Other key parameters are provided in Table 1. The distribution of LEDs can affect the performance of the VLCP system. For the MISO scenario, if LEDs are positioned closely together and centrally in the ceiling, the multipath effect of the VLC channel becomes less obvious, making it difficult for the receiver to distinguish multipath components and accurately estimate the CSI for VLCP. For the MIMO scenario, closely placed LEDs in the center of the ceiling lead to a high correlation between channels, which can negatively affect the demodulation performance for VLCP. To ensure both effective communication and accurate positioning, while also maintaining uniform light distribution, we keep a certain distance between LEDs. Thus, we distribute the LEDs at the corners of the ceiling, with their coordinates listed in Table 1.

Table 1. Configuration parameters of mobile VLCP system.

Parameters	Values
Room size (length × width × height)	5 m × 5 m × 3 m
Transmit power per lamp (except for Figure 5)	25 Watt
Locations of LED1, LED2, LED3, and LED4	(1, 1, 3), (1, 4, 3), (4, 1, 3), (4, 4, 3)
Initial locations of PD1 and PD2 (for Routes 1 and 3)	(1.11, 2.61, 0.85), (0.89, 2.39, 0.85)
Initial locations of PD1 and PD2 (for Routes 2 and 4)	(1.11, 3.61, 0.85), (0.89, 3.39, 0.85)
Modulation index	0.25
Height of receiving plane	0.85 m
Physical area of PDs	10^{-4} m^2
Refractive index	1.5
PDs' field of view	85 deg.
Responsivity of PDs	0.35 A/W
Receiver velocity (except for Figures 6, 11, and 12)	1 m/s
Effective data rate for the MISO/MIMO scenario	51.2 Mbps/38.4 Mbps
Observation block length (except for Figure 10)	35 or 20

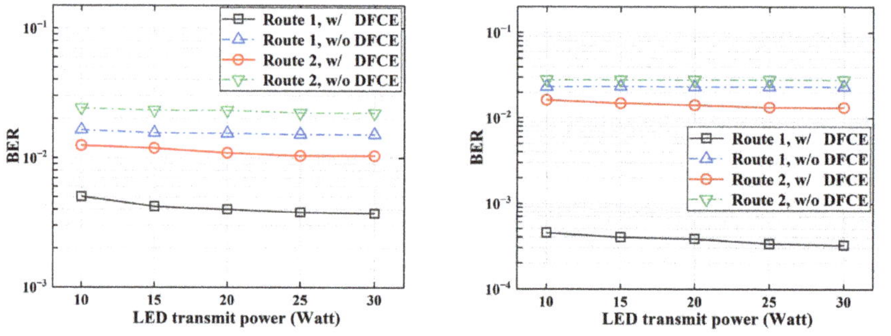

(a) (b)

Figure 5. Comparison of system BER performance with and without DFCE under varying LED transmit power levels: (**a**) MISO scenario; (**b**) MIMO scenario.

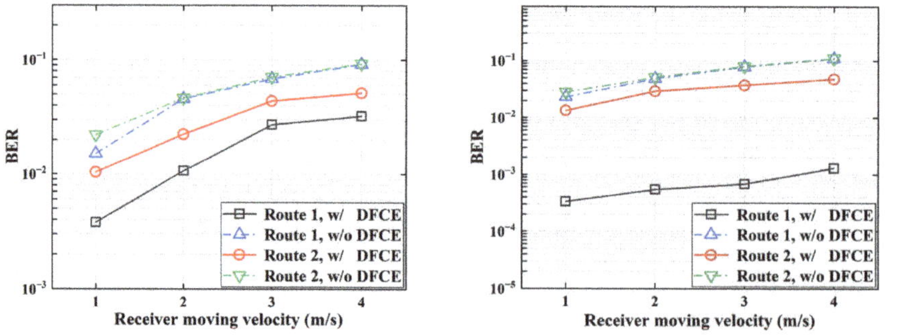

(a) (b)

Figure 6. Comparison of system BER performance with and without DFCE at different receiver moving velocities: (**a**) MISO scenario; (**b**) MIMO scenario.

We first evaluate the communication performance of the mobile VLCP systems. Figure 5 compares the BERs for the MISO and MIMO systems under different LED transmit powers. In Figure 5a, we observe a significant improvement in BER with the implementation of DFCE in the mobile MISO system. For instance, at a transmit power of 25 W, the BERs without DFCE are 1.5×10^{-2} and 2.2×10^{-2} for receiver movement along Route 1 and Route 2, respectively. After using DFCE, the BERs improve to 3.8×10^{-3} and 1.0×10^{-2} for Route 1 and Route 2, respectively. This enhancement is attributed to the increased channel estimation accuracy of DFCE, enabling effective tracking of the time-varying channel and improved signal recovery. Additionally, we observe a slight reduction in BER as the transmit power increases, although the effect is not significant. However, the influence of different routes on BER is more evident. Route 1 exhibits notably better performance than Route 2, primarily due to its closer proximity to the room center, resulting in higher SNRs. In Figure 5b, the MIMO system exhibits similar trends to the MISO system in Figure 5a. For example, in Route 1 at a transmit power of 25 W, the BER improves from 2.3×10^{-2} to 3.3×10^{-4} with the adoption of DFCE. Thus, DFCE demonstrates its capability to enhance the communication quality of the mobile MIMO system as well.

Figure 6 compares the BER performance of the MISO and MIMO systems at varying receiver moving velocities. As illustrated in Figure 6a, increased velocity leads to BER degradation across all schemes for MISO. For instance, with the velocity increasing from 1 m/s to 4 m/s, the BER without DFCE rises from 1.5×10^{-2} to 9.2×10^{-2} and from 2.2×10^{-2} to 9.3×10^{-2} for receiver movement along Route 1 and Route 2, respectively. This deterioration is a result of higher velocities inducing quicker channel variations, which negatively impact channel estimation accuracy, particularly in scenarios without feedback. However, after using DFCE, the BER of the MISO system is improved. For example, at velocities of 1 m/s and 4 m/s in Route 1, the BER improves to 3.8×10^{-3} and 3.2×10^{-2}, respectively. In Figure 6b, similar trends are observed for the MIMO scenario. For instance, in Route 1, the BER with DFCE improves from 4.9×10^{-2} to 5.4×10^{-4} and from 1.1×10^{-1} to 1.3×10^{-3} at velocities of 2 m/s and 4 m/s, respectively. In Route 2, at 1 m/s, the BER with DFCE improves from 2.8×10^{-2} to 1.3×10^{-2}. These results highlight the effectiveness of DFCE in accurately estimating the time-varying channel and recovering the transmitted signal, especially at higher velocities.

Next, we assess the positioning performance of the mobile VLCP systems. Figure 7 displays the positioning results for the MISO and MIMO systems when the receiver moves along different routes. In both scenarios and routes, when DFCE is not utilized, the positioning results are only obtained at discrete locations near the receiver's moving trajectories. This limitation arises from solely using training symbols for channel estimation, resulting in a constant CSI estimation throughout the entire OFDM packet. Consequently, the positioning results in each packet overlap at the same locations. After employing DFCE, continuous positioning results become apparent, being in much closer alignment with the actual locations. This improvement is attributed to the ability of DFCE to dynamically estimate the channel, thereby yielding more accurate CSI. This accurate CSI, in turn, contributes to enhanced positioning accuracy within the VLCP system. Furthermore, in Figure 7b, due to the receiver's higher SNRs in Route 1, the positioning accuracy appears to be superior compared to Route 2 for the MIMO system. Therefore, the VLCP scheme with DFCE can leverage the advantages of higher SNRs to enhance positioning accuracy.

In Figure 8, we compare the cumulative distribution functions (CDFs) of positioning errors calculated from the results in Figure 7, for both the MISO and MIMO systems. For the MISO system in Figure 8a, before using DFCE, 90% of positioning errors are less than 9.5 cm and 9.4 cm when the receiver moves along Route 1 and Route 2, respectively. After using DFCE, these errors decrease to 5.8 cm and 7.5 cm, respectively. Similar trends are observed in the MIMO system in Figure 8b. Along Route 1, before employing DFCE, 90% of positioning errors are less than 8.2 cm. However, this error reduces to 4.0 cm with the adoption of DFCE. The advantage of DFCE becomes more obvious along Route 2. Without DFCE, 90% of positioning errors are less than 22.4 cm. However, this error significantly reduces to 16.5 cm when DFCE is employed. These findings validate the effectiveness of the feedback mechanism in enhancing mobile positioning accuracy.

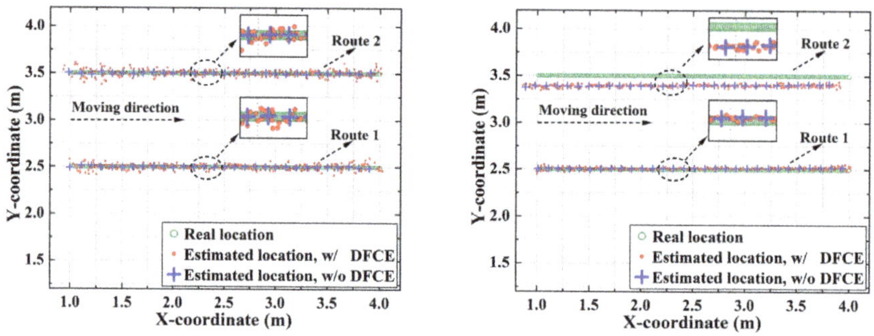

(**a**) (**b**)

Figure 7. Results for mobile VLP with and without DFCE as the receiver moves along different test routes: (**a**) MISO scenario; (**b**) MIMO scenario.

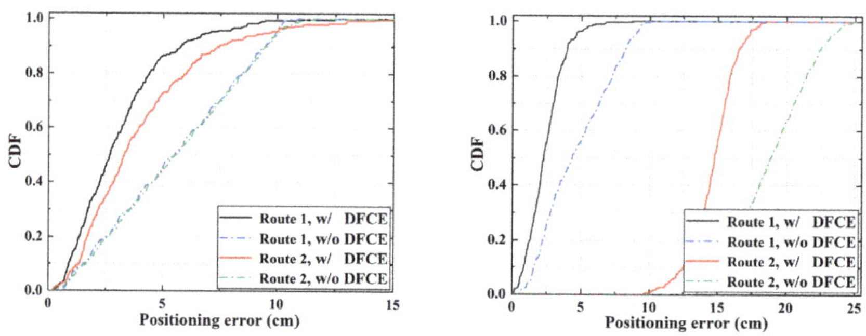

(**a**) (**b**)

Figure 8. Comparison of CDFs for positioning errors with and without DFCE: (**a**) MISO scenario; (**b**) MIMO scenario.

It is worth mentioning that the BERs and positioning errors mentioned above are average values observed while the receiver moves along different routes. For VLC, the BERs are linked to the performance of all subcarrier channels. When channel characteristics degrade, error propagation due to decision feedback, which is a weakness of DFCE, impacts overall BER performance. This issue is particularly pronounced when light intensity is low. However, the CSI used for positioning remains accurate as it only depends on the estimated FDTF of the first subcarrier channel, as shown in Equations (6) and (15). At $k = 1$, subcarrier symbols experience a higher channel response and greater noise robustness compared to others. Therefore, even at overall BER levels higher than 10^{-3} as in Figures 5 and 6, positioning errors are still maintained at the centimeter level. To further explain this, Figure 9 compares the eye diagrams of the real part for the received 4-QAM symbols in different subcarrier channels equalized with DFCE in Route 2. In both MISO and MIMO scenarios, we can clearly see that for $k = 1$, due to higher noise robustness, the eye diagram is clear and open. Conversely, for $k = 100$, due to lower noise robustness and the high possibility of error propagation from decision feedback, the eye diagram is closed, indicating poor performance. To address this issue, a possible solution is to use machine learning to mitigate the error propagation of DFCE and improve the channel equalization performance [34].

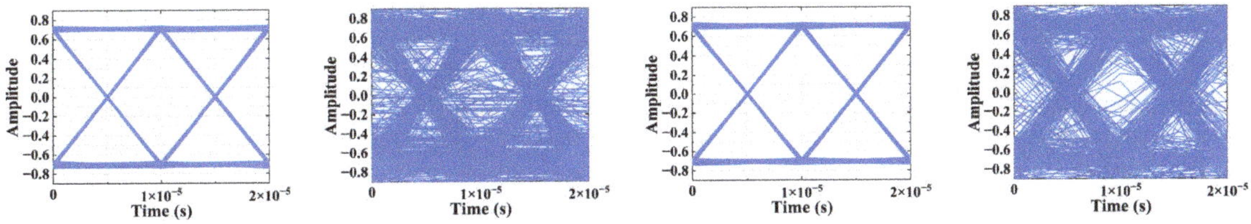

(a) (b) (c) (d)

Figure 9. Eye diagrams of the real part for the received 4-QAM symbols equalized with DFCE in Route 2: (**a**) MISO scenario, $k = 1$; (**b**) MISO scenario, $k = 100$; (**c**) MIMO scenario, $k = 1$; (**d**) MIMO scenario, $k = 100$.

Next, we examine the impact of observation block length L on positioning performance. Figure 10 compares the mean positioning errors obtained in the VLCP system using different L values for the MISO and MIMO systems. In both scenarios, we generally observe improved positioning accuracy with increasing L. For instance, for the MISO system in Figure 10a, along Route 1, the mean positioning error with DFCE decreases from 3.4 cm to 3.1 cm when L is increased from 25 to 35. This improvement is attributed to the longer observation block length effectively equalizing the channel and averaging out noise, resulting in more accurate CSI for mobile VLP. However, further increasing L does not lead to significant enhancement in positioning accuracy, particularly when L exceeds 40 for the MISO system. Similarly, for the MIMO system in Figure 10b, there is no significant enhancement when L exceeds 20. Because longer observation block lengths result in higher computing and storage requirements at the receiver, to balance system complexity and positioning accuracy, an optimal L value could be selected around 35 and 20 for MISO and MIMO systems, respectively, when DFCE is employed. This selection yields mean positioning errors of 3.1 cm for the MISO system and 2.4 cm for the MIMO system in Route 1.

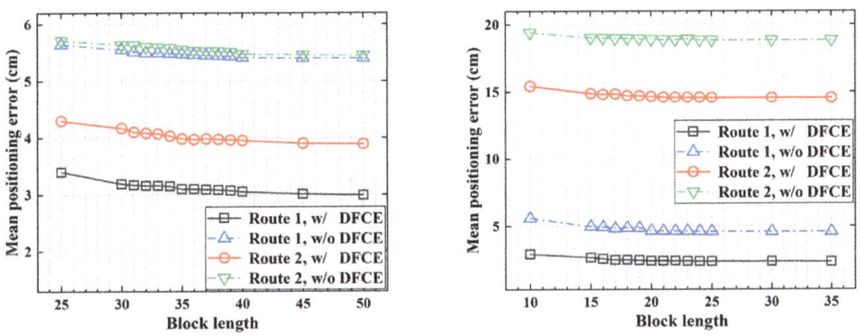

(a) (b)

Figure 10. Comparison of mean positioning errors with and without DFCE for different observation block lengths: (**a**) MISO scenario; (**b**) MIMO scenario.

Figure 11 compares the positioning performance for the MISO and MIMO scenarios at different receiver moving velocities. For the MISO system in Figure 11a, we observe a significant increase in mean positioning errors with increasing velocity when DFCE is not employed. For instance, in Route 1, the mean positioning error increases from 5.5 cm to 21.3 cm when the velocity increases from 1 m/s to 4 m/s. Similarly, in Route 2, the mean positioning error increases from 5.6 cm to 21.5 cm with the same change in velocity. This is due to the faster channel variations caused by higher velocities, which result in reduced channel estimation accuracy. However, when DFCE is employed to track and estimate the time-varying channel, the positioning errors remain relatively stable, ranging from 3.1 cm to 3.6 cm in Route 1 and from 4.0 cm to 4.3 cm in Route 2. In the MIMO system, the results exhibit a similar trend to the MISO system, with errors ranging from 2.4 cm to 2.5 cm in Route 1 and from 14.6 cm to 14.9 cm in Route 2. This indicates the effectiveness of DFCE in maintaining stable positioning performance regardless of the terminal moving speed.

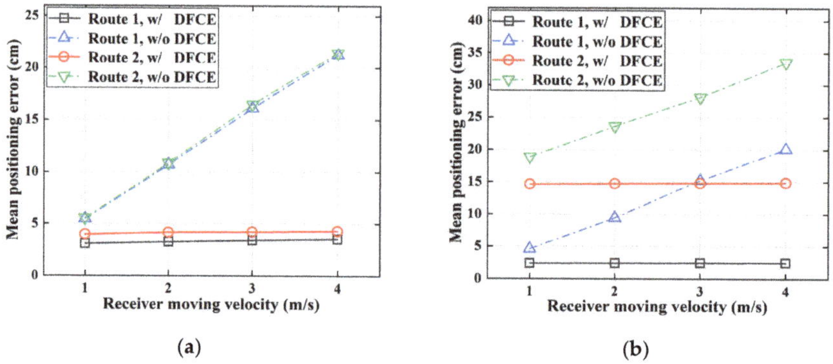

Figure 11. Comparison of mean positioning errors with and without DFCE at different receiver velocities: (**a**) MISO scenario; (**b**) MIMO scenario.

Next, we assess the impact of different weight coefficients on the positioning performance with DFCE. Considering various values of the forgotten factor α in Equations (9) and (14), Figure 12 compares the mean positioning errors under different weight coefficients for the MISO and MIMO systems. Here, the terminal velocity is set at 4 m/s. For the MISO system in Figure 12a, with equal weight coefficients, an increase in the number of subcarriers leads to higher mean positioning errors. For example, as the number of subcarriers increases from 128 to 1024, the mean positioning error increases by 1.2 cm and 1.8 cm when the receiver moves along Route 3 and Route 4, respectively. This occurs because, under a constant data rate, more subcarriers result in longer OFDM symbol durations, extending the decision feedback duration and causing more significant channel variations. However, more appropriate weight coefficients can improve positioning performance. For instance, in Route 4, when the number of subcarriers is 1024, the mean positioning error is optimized with α = 0.7, resulting in 4.8 cm, compared to 6.0 cm with equal weight coefficients. Increasing the weight for estimated CSI in VLCP data blocks adjacent to the current channel can enhance the accuracy of channel estimation due to their high correlation. However, higher weights also amplify the impact of noise, underscoring the importance of selecting appropriate weights. Similarly, in the MIMO system, optimal weight coefficients occur when α = 0.7 in both Route 3 and Route 4.

Finally, considering the variation in terminal size in our MIMO-based VLCP system, we assess the positioning performance by adjusting the distance between the two PDs. Figure 13 compares the mean positioning errors with and without DFCE at different PD distances. Along Route 1, where the SNR is sufficient, increasing the distance between PDs does not notably enhance the positioning performance. However, in Route 2, without DFCE, the mean positioning error can be reduced from 20.0 cm to 17.5 cm as the distance between PDs increases from 20 cm to 50 cm. After employing DFCE, the mean positioning error can be further reduced from 15.5 cm to 12.5 cm. This improvement is attributed to the decreased correlation between channels as the distance between PDs increases, resulting in reduced interference between signals and less additional noise being introduced.

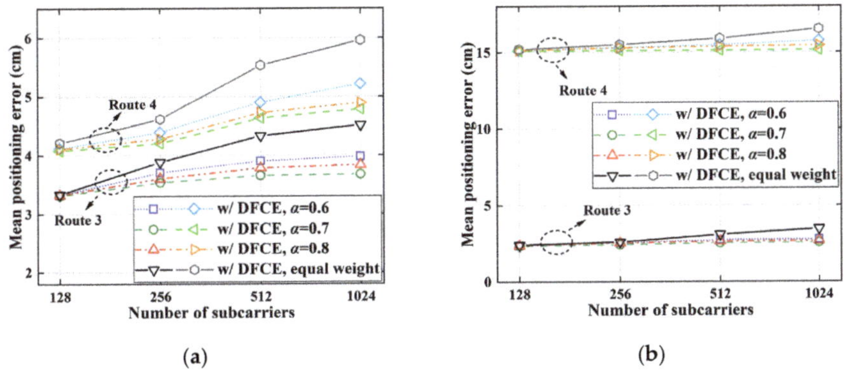

Figure 12. Comparison of mean positioning errors with DFCE when using different weight coefficients: (**a**) MISO scenario; (**b**) MIMO scenario.

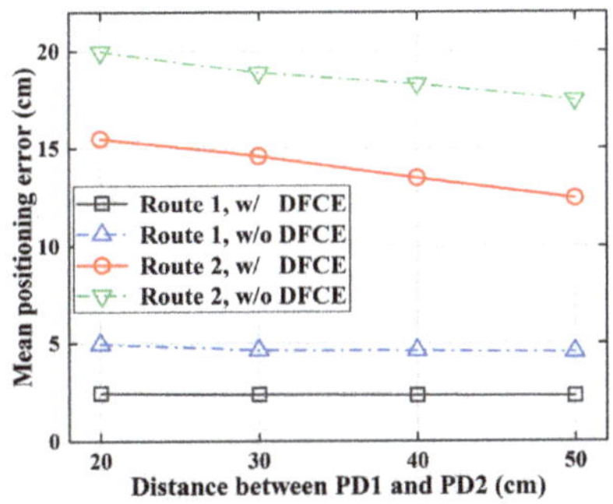

Figure 13. Comparison of mean positioning errors in MIMO with different PD distances.

4. Conclusions

We proposed an integrated mobile VLCP scheme capable of simultaneously integrating VLC and VLP into a single system. By employing DFCE for accurate estimation of time-varying CSI, we enhanced the performance of both communication and positioning under mobile receiver scenarios. Our approach considers VLCP for both MISO and MIMO scenarios, utilizing multipath channels for CSI extraction in MISO and employing STBC for CSI estimation in MIMO. Comprehensive simulations were conducted to evaluate performance. The results showed that in the MISO system, when utilizing DFCE, the BER is 3.8×10^{-3} and 1.0×10^{-2}, with 90% of positioning errors less than 5.8 cm and 7.5 cm when the receiver moves along Route 1 and Route 2, respectively. In the MIMO system, with DFCE employed, the BER is 3.3×10^{-4} and 1.3×10^{-2}, with 90% of positioning errors less than 4.0 cm and 16.5 cm in Route 1 and Route 2, respectively. Furthermore, optimizing weight coefficients and increasing PD distance for MIMO receivers can effectively enhance positioning accuracy. In future work, we will consider 3D positioning and receiver tilt.

Author Contributions: Conceptualization, X.Y.; methodology, R.W., Y.S. and X.Y.; software, R.W., Y.S. and Z.L.; validation, R.W. and Y.S.; formal analysis, R.W. and Y.S.; investigation, R.W. and Z.L.; resources, M.G. and X.Y.; data curation, R.W.; writing—original draft preparation, R.W. and X.Y.; writing—review and editing, M.G. and X.Y.; visualization, R.W. and Z.L.; supervision, X.Y.; project administration, X.Y.; funding acquisition, X.Y. All authors have read and agreed to the published version of the manuscript.

Funding: This research was funded by the National Natural Science Foundation of China (62001319), Suzhou Science and Technology Bureau-Technical Innovation Project in Key Industrial (SYG202112), and Open Fund of IPOC (BUPT) (IPOC2020A009).

Institutional Review Board Statement: Not applicable.

Informed Consent Statement: Not applicable.

Data Availability Statement: The raw data supporting the conclusions of this article will be made available by the authors on request.

Conflicts of Interest: The authors declare no conflicts of interest.

References

1. Liu, X.; Guo, L.; Wei, X. Indoor visible light applications for communication, positioning, and security. *Wirel. Commun. Mob. Com.* **2021**, *2021*, 1730655. [CrossRef]
2. Karunatilaka, D.; Zafar, F.; Kalavally, V.; Parthiban, R. LED based indoor visible light communications: State of the art. *IEEE Commun. Surv. Tutor.* **2015**, *17*, 1649–1678. [CrossRef]
3. Zvanovec, S.; Chvojka, P.; Haigh, P.A.; Ghassemlooy, Z. Visible light communications towards 5G. *Radioengineering* **2015**, *24*, 1–9. [CrossRef]
4. Zhuang, Y.; Hua, L.; Qi, L.; Yang, J.; Cao, P.; Cao, Y.; Wu, Y.; Thompson, J.; Haas, H. A survey of positioning systems using visible LED lights. *IEEE Commun. Surv. Tutor.* **2018**, *20*, 1963–1988. [CrossRef]
5. Pang, M.; Shen, G.; Yang, X. Achieving reliable underground positioning with visible light. *IEEE Trans. Instru. Meas.* **2022**, *71*, 1–15. [CrossRef]
6. Maheepala, M.; Kouzani, A.Z.; Joordens, M.A. Light-based indoor positioning systems: A review. *IEEE Sens. J.* **2020**, *20*, 3971–3995. [CrossRef]
7. Lin, X.; Zhang, L. Intelligent and practical deep learning aided positioning design for visible light communication receivers. *IEEE Commun. Lett.* **2020**, *24*, 577–580. [CrossRef]
8. Chen, J.; You, X. Visible light positioning and communication cooperative systems. In Proceedings of the 2017 16th International Conference on Optical Communications and Networks (ICOCN), Wuzhen, China, 7–10 August 2017; pp. 1–3. [CrossRef]
9. Yang, H.; Chen, C.; Zhong, W.; Alphones, A.; Zhang, S.; Du, P. Demonstration of a quasi-gapless integrated visible light communication and positioning system. *IEEE Photon. Technol. Lett.* **2018**, *30*, 2001–2004. [CrossRef]
10. Yang, H.; Zhong, W.; Chen, C.; Alphones, A.; Du, P. QoS-driven optimized design-based integrated visible light communication and positioning for indoor IoT networks. *IEEE Internet Things J.* **2020**, *7*, 269–283. [CrossRef]
11. Xu, Y.; Wang, Z.; Liu, P.; Chen, J.; Han, S.; Yu, C.; Yu, J. Accuracy analysis and improvement of visible light positioning based on VLC system using orthogonal frequency division multiple access. *Opt. Express.* **2017**, *25*, 32618–32630. [CrossRef]
12. Yang, H.; Zhong, W.; Chen, C.; Alphones, A.; Du, P.; Zhang, S.; Xie, X. Coordinated resource allocation-based integrated visible light communication and positioning systems for indoor IoT. *IEEE Trans. Wirel. Commun.* **2020**, *19*, 4671–4684. [CrossRef]
13. Ma, S.; Yang, R.; Du, C.; Li, H.; Wu, Y.; Al-Dhahir, N.; Li, S. Robust power allocation for integrated visible light positioning and communication networks. *IEEE Trans. Commun.* **2023**, *71*, 4764–4777. [CrossRef]
14. Chen, D.; Fan, K.; Wang, J.; Lu, H.; Jin, J.; Feng, L.; Chen, H.; Xue, Z.; Wang, Y. Integrated visible light communication and positioning CDMA system employing modified ZCZ and Walsh code. *Opt. Express* **2022**, *30*, 40455–40469. [CrossRef] [PubMed]
15. Liu, Z.; Yu, C. Multi-user visible light communication and positioning system based on dual-domain multiplexing scheme. *Photonics* **2023**, *10*, 306. [CrossRef]
16. Liu, Z.; You, X.; Chen, J.; Yu, C. Two-LED indoor visible light positioning method based on channel estimation with a mirror. In Proceedings of the 2019 Asia Communications and Photonics Conference (ACP), Chengdu, China, 2–5 November 2019; pp. 1–2. Available online: https://ieeexplore.ieee.org/document/8989700 (accessed on 10 February 2020).
17. Wang, K.; Liu, Y.; Hong, Z. RSS-based visible light positioning based on channel state information. *Opt. Express* **2022**, *30*, 5683–5699. [CrossRef]
18. Deng, L.; Fan, Y.; Zhao, Q.; Wu, P. Indoor visible light positioning based on channel estimation and cramér–rao low bound analysis with random receiving orientation of user equipment. *Photonics* **2023**, *10*, 812. [CrossRef]
19. Hong, Y.; Chen, L.K. Toward user mobility for OFDM-based visible light communications. *Opt. Lett.* **2016**, *41*, 3763–3766. [CrossRef] [PubMed]
20. You, X.; Zhong, Y.; Chen, J.; Yu, C. Mobile channel estimation based on decision feedback in vehicle-to-infrastructure visible light communication systems. *Opt. Commun.* **2020**, *462*, 125261. [CrossRef]
21. Shao, Y.; Hong, Y.; Deng, R.; Chen, L.K. Performance enhancement by spatial diversity for robust VLC systems with fast-moving terminals. In Proceedings of the 2019 Optoelectronics and Communications Conference (OECC) and 2019 International Conference on Photonics in Switching and Computing (PSC), Fukuoka, Japan, 7–11 July 2019; pp. 1–3. [CrossRef]
22. Li, C.; Yi, Y.; Lee, K.; Lee, K. Performance analysis of visible light communication using the STBC-OFDM technique for intelligent transportation systems. *Int. J. Electron.* **2013**, *101*, 1117–1133. [CrossRef]

23. Souto, V.D.P.; Dester, P.S.; Facina, M.S.P.; Silva, D.G.; de Figueiredo, F.A.P.; Tejerina, G.R.D.; Santos, J.C.S.; Ferreira, J.S.; Mendes, L.L.; Souza, R.D.; et al. Emerging MIMO technologies for 6G networks. *Sensors* **2023**, *23*, 1921. [CrossRef]
24. Khwandah, S.A.; Cosmas, J.P.; Lazaridis, P.I.; Zaharis, Z.D.; Chochliouros, I.P. Massive MIMO systems for 5G communications. *Wirel. Pers. Commun.* **2021**, *120*, 2101–2115. [CrossRef]
25. Zheng, H.; Xu, Z.; Yu, C.; Gurusamy, M. A 3-D high accuracy positioning system based on visible light communication with novel positioning algorithm. *Opt. Commun.* **2017**, *396*, 160–168. [CrossRef]
26. Huang, N.; Gong, C.; Luo, J.; Xu, Z. Design and demonstration of robust visible light positioning based on received signal strength. *J. Light. Technol.* **2020**, *38*, 5695–5707. [CrossRef]
27. Tran, H.Q.; Ha, C. Improved visible light-based indoor positioning system using machine learning classification and regression. *Appl. Sci.* **2019**, *9*, 1048. [CrossRef]
28. Wei, T.; Liu, S.; Du, X. Visible light integrated positioning and communication: A multi-task federated learning framework. *IEEE Trans. Mobile Comput.* **2023**, *22*, 7086–7103. [CrossRef]
29. Wei, Y.; Liu, X.; Wang, Y.; Ma, S.; Zhou, F.; Dobre, O.A. Channel and location estimation enabled a novel BDCNP network for massive MIMO VLCP systems. *IEEE Wirel. Commun. Lett.* **2024**, *13*, 218–222. [CrossRef]
30. Sun, Y.; You, X.; Chen, J.; Yu, M.; Gao, M.; Shen, G. Decision feedback channel estimation for integrated mobile VLCP based on STBC-MIMO. In Proceedings of the 2023 Asia Communications and Photonics Conference/2023 International Photonics and Optoelectronics Meetings (ACP/POEM), Wuhan, China, 4–7 November 2023; pp. 1–4. [CrossRef]
31. Gfeller, F.R.; Bapst, U. Wireless in-house data communication via diffuse infrared radiation. *Proc. IEEE* **1979**, *67*, 1474–1486. [CrossRef]
32. Damen, M.O.; Gamal, H.E.; Beaulieu, N.C. Linear threaded algebraic space-time constellations. *IEEE Trans. Inf. Theory* **2003**, *49*, 2372–2388. [CrossRef]
33. Abou-Shehada, I.M.; AlMuallim, A.F.; AlFaqeh, A.K.; Muqaibel, A.H.; Park, K.H.; Alouini, M.S. Accurate indoor visible light positioning using a modified pathloss model with sparse fingerprints. *J. Light. Technol.* **2021**, *39*, 6487–6497. [CrossRef]
34. Xu, J.; Li, L.; Zheng, L.; Liu, L. Detect to learn: Structure learning with attention and decision feedback for MIMO-OFDM receive processing. *IEEE Trans. Commun.* **2024**, *72*, 146–161. [CrossRef]

photonics

MDPI

Article

Single-Source VLCP System Based on Solar Cell Array Receiver and Right-Angled Tetrahedron Trilateration VLP (RATT-VLP) Algorithm

Dawei Xie [1], Zhongxu Liu [1,*] and Changyuan Yu [1,2]

1 Department of Electrical and Electronic Engineering, The Hong Kong Polytechnic University, Kowloon, Hong Kong 999077, China; dawei.xie@connect.polyu.hk (D.X.); changyuan.yu@polyu.edu.hk (C.Y.)
2 Shenzhen Research Institute, The Hong Kong Polytechnic University, Shenzhen 518000, China
* Correspondence: zhongxu.liu@connect.polyu.hk

Abstract: A significant deployment limitation for visible light communication and positioning (VLCP) systems in energy- and light-source-restricted scenarios is the reliance of photodetectors (PDs) on external power supplies, compromising sustainability and complicating receiver charging. Solar cells (SCs), capable of harvesting and converting environmental light into electrical energy, offer a promising alternative. Consequently, we first propose an indoor VLCP system that utilizes an SC array as the receiver, alongside a right-angled tetrahedron trilateration visible light positioning (RATT-VLP) algorithm based on a single light source and multiple receivers. The proposed system uses an SC array in place of PDs, utilizing binary phase shift keying (BPSK) signals for simultaneous communication and positioning. In experiments, we verified the system's error-free communication rate of 1.21 kbps and average positioning error of 3.40 cm in a 30 cm × 30 cm area, indicating that the system can simultaneously satisfy low-speed communication and accurate positioning applications. This provides a viable foundation for further research on SC-based VLCP systems, facilitating potential applications in environments like underwater wireless communication, positioning, and storage tank inspection.

Keywords: visible light positioning; visible light communication; visible light communication and positioning; solar cell

1. Introduction

As the scarcity of spectrum resources in traditional radio frequency (RF) communication methods approaches critical levels [1], researchers and engineers are increasingly turning their attention to optical wireless communication (OWC) technologies. Among the various categories of OWC, visible light communication (VLC) stands out due to its advantages, such as no spectrum license being required, minimal interference, high security, low system complexity, low cost, high data transmission rates, and a wide range of potential applications [2]. Additionally, VLC can simultaneously perform illumination and data transmission, making it a promising complement to traditional wireless local area network WLAN technology [3]. Within the various applications derived from VLC, visible light positioning (VLP) technology shows considerable potential for development. Compared with traditional RF-based positioning technologies, VLP provides several advantages, including higher spatial resolution due to the excellent directionality of visible light, which allows for centimeter-level accuracy in confined spaces [4]. Additionally, VLP's compatibility with existing LED lighting infrastructure and power line communication (PLC) technologies simplifies its deployment [5]. As Internet of Things (IoT) technologies and the sixth generation of wireless communication networks (6G) continue to evolve, standalone VLC and VLP systems struggle to meet the fundamental requirements for the simultaneous communication and positioning functions of emerging IoT devices such as robotic vacuum cleaners, smart streetlights, wireless underwater exploration equipment, etc. Therefore, developing systems that integrate both VLC and VLP capabilities, i.e., visible light communication and positioning (VLCP), is crucial for advancing the development of 6G IoT.

Previous studies have proposed various design strategies for VLCP systems tailored to different application scenarios. For example, in terms of channel reuse, Yu's group proposed a multi-user simultaneous indoor VLCP system scheme and solutions based on channel optimization and frame structure design [6,7]; in terms of underwater applications, Fu's group proposed a full-duplex VLCP solution for underwater wireless sensor networks [8]; Shao et al. relied on backscatter communication in the RF field to propose a retroreflective-VLCP scheme in the optical domain [9]; and Yang et al. proposed a VLCP system design for modulation technology optimization based on filter bank multicarrier [10]. In all of these VLCP systems, photodiodes (PDs) serve as receivers to capture the light signals and convert them into a photocurrent proportional to the received light power. Nevertheless, a major drawback of using PDs as receivers is their need for external power to support the conversion from photocurrent to voltage and to maintain the PDs at their ideal operating point, which limits the independent operation ability and the miniaturization of the receivers. This limitation is disadvantageous for the deployment of devices in various IoT applications, such as smart homes and smart factories, and restricts their use in energy-constrained environments like underwater and storage tanks.

To address the above issue, research into integrating photovoltaic (PV) technology into indoor VLCP could be highly beneficial. Several studies have verified the feasibility of using solar cells (SCs) to replace photodiodes (PDs) in VLC systems for simultaneous data and energy reception [11–13], introducing the concept of simultaneous lightwave information and power transfer (SLIPT) [14]. A vital advantage of this concept is that solar cells, as passive devices, do not require external power to achieve photovoltaic conversion, thereby enabling self-powered operation [15]. Additionally, compared with PDs, which typically have a few square millimeters of effective receiving area, solar cells generally feature a much larger effective receiving area, ranging from several tens to hundreds of square centimeters. This larger area not only provides a broader receiving angle but also helps mitigate issues like partial fading. However, most research into SLIPT in the VLC domain primarily focuses on enhancing transmission rates [16,17], often using laser light sources [18,19]. As for the field of indoor VLP, studies integrating it with PV technology remain scarce. To date, only Hsu et al. have proposed a positioning system using PV as the receiver, which accomplishes positioning relative to three LED sources through identity positioning and radio frequency carrier allocation, without discussing the system's communication capabilities [20]. This does not address the challenges faced by VLCP system receivers in environments with limited light sources and energy supplies, such as underwater, unmanned storage, and unmanned factories, where synchronous communication and positioning are required. In this context, studies of indoor VLCP systems and algorithms based on PV receivers represent a promising and largely unexplored research direction within the VLCP field.

In this paper, we propose an indoor VLCP system utilizing one LED transmitter and an array of multiple SC receivers, alongside a right-angled tetrahedron trilateration VLP (RATT-VLP) algorithm. The system achieves VLC through the modulation and demodulation of BPSK signals. Concurrently, using the received signal's strength (RSS) extracted from the demodulation process as an indicator, the signal transmission distances between the LED and the receivers can be determined, given the known coordinates of the LED and the relative positions of the SC receivers. Then, the position of the devices is ascertained through the RATT-VLP algorithm based on the LED's coordinate, the configuration of the receivers, and signal transmission distances, thereby facilitating VLP. By only employing one transmitter, our system offers deployment flexibility under limited light source conditions. Our experimental results demonstrate that this system can simultaneously perform signal transmission and receiver positioning, showcasing the feasibility of applying SCs to indoor VLCP systems and the potential for implementing SLIPT in such systems.

2. System Mechanism and Setup

2.1. VLCP System Mechanism

The indoor VLCP system based on one LED and an array of multiple SC receivers proposed in this paper is depicted in Figure 1a and structurally divided into three sections: the transmitter, the free-space VLC channel, and the receiver. For the user data to be transmitted, the maximum length sequence (MLS or m-sequence) is initially attached to its head and tail for signal synchronization and extraction. Then, the combined data, modulated by binary phase shift keying (BPSK), are converted into an alternating current (AC) analog signal by a digital-to-analog converter (DAC). Subsequently, an optical driver supplies both direct current (DC) and AC signals to the LED, achieving the required drive voltage to convert electrical energy into light. The optical signal carrying the VLCP information exits the transmitter and propagates through the free space VLC channel, ultimately arriving at the effective detection surface of the SC array, where it undergoes photovoltaic conversion. The received electrical signal is then reconverted into a digital signal by an analog-to-digital converter (ADC). After utilizing the MLS for signal alignment and extraction, on the one hand, the data are demodulated through threshold judgment for VLC; on the other hand, the RSS is calculated using the Fourier transform method, thereby calculating the signal transmission distances to determine the receiver's location via the RATT-VLP algorithm for VLP. Thus, the system is capable of simultaneous signal transmission and receiver-positioning functions. We will elaborate on the principles of implementing VLCP in the following sections.

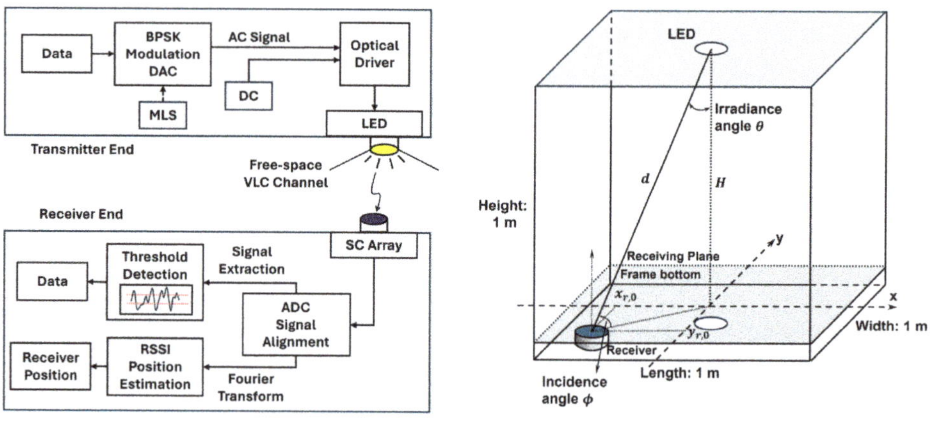

Figure 1. (**a**) Schematic diagram of indoor SC-array-based VLCP system and (**b**) diagram of the indoor VLCP system model.

Firstly, we consider the user data $b[n] \in \{-1, 1\}$; thus, the n-th symbol is expressed as

$$s(t) = b[n], \qquad nT_s \le t < (n+1)T_s, \tag{1}$$

where T_s is the symbol period. The BPSK-modulated signal can be expressed as

$$S(t) = s(t) \cdot \cos(2\pi f_c t), \tag{2}$$

where f_c denotes the carrier frequency. Now, the transmitted VLCP signal can be written as

$$T(t) = P_t(\alpha S(t) + 1), \tag{3}$$

where P_t is the DC transmit power of LED, and α is defined as the ratio of the maximum current variation of the modulated signal to the LED bias current [21,22].

Then, the VLCP signal propagates through the VLC channel, as depicted in Figure 1b. The intensity of the system's LED transmitter is assumed to follow the Lambertian emission model, expressed as $I(\theta) = I(0)\cos^m(\theta)$, where θ is the angle of emission. Given that existing research indicates that the Non-Line-of-Sight (NLoS) component of VLC channels is significantly smaller than the Line-of-Sight (LoS) component [23] and considering the integrative effect, which mitigates multipath interference [5], this paper disregards the channel's NLoS components. It is assumed that the LED is parallel to the receiving plane, and its vertical distance from the receiving plane is H. Under these assumptions, the DC gain of the channel in this VLCP model can be determined by the following equation:

$$G(0) = \begin{cases} \frac{(m+1)AH^{m+1}}{2\pi d^{m+3}} G_F(\phi) g_c(\phi) & 0 \le \phi \le \Phi_c \\ 0 & \phi > \Phi_c \end{cases}, \tag{4}$$

where m is the Lambertian emission order related to the semi-angle at half-power (or the viewing half angle) of the LED. ϕ is the angle of incidence; A is the effective detection area of the detector; d is the distance between the LED and the detector; $G_F(\phi)$ is the gain of the optical filter; and $g_c(\phi)$ is the gain of the optical concentrator, which is related to the refractive index of the detector , n, and the width of the field of view, Φ_c, at the receiver, expressed as

$$g_c(\phi) = \begin{cases} \frac{n^2}{\sin^2 \Phi_c} & 0 \le \phi \le \Phi_c \\ 0 & 0 \ge \Phi_c \end{cases}. \tag{5}$$

Therefore, the received VLCP signal with the DC component removed can be expressed as

$$R(t) = G(0) \cdot P_t \alpha S(t) = \frac{(m+1)AH^{m+1}}{2\pi d^{m+3}} G_F(\phi) g_c(\phi) \cdot P_t \alpha S(t) + N. \tag{6}$$

where N denotes the noise during the signal transmission process.

For VLC, through the method of threshold judgment accomplished by comparing the received signal value with the threshold, ξ, the received signal can be demodulated to obtain the user data. The recovered n-th symbol, r_n, of user data can be described as

$$r(t) = \begin{cases} 1, & R(t) \ge \xi \\ -1, & R(t) < \xi \end{cases}. \tag{7}$$

For VLP, we define the maximum power spectral density (PSD) of a received signal as the RSS, which is calculated by performing a Fourier transform on the signal and then taking the square of the magnitude of the resulting frequency components. Given that the incident angle, ϕ, is less than or equal to the field of view, Φ_c, the RSS at the i-th SC receiver, $P_{r,i}$, can be represented as follows:

$$P_{r,i} = \max|\mathcal{F}\{R_i(t)\}|^2 = \frac{(m+1)A_i H^{m+1}}{2\pi d_i^{m+3}} G_{F,i}(\phi) g_{c,i}(\phi) \cdot P_t + N, \tag{8}$$

With the fixed parameters associated with specific SC receivers and structures defined as $\chi_i = (m+1)A_i G_{F,i}(\phi) g_{c,i}(\phi) H^{m+1} P_t / 2\pi$, the signal transmission distance between the i-th SC receiver and the LED can be derived from the following equation:

$$d_i = (\chi)^{[\frac{1}{m+3}]} \cdot (P_{ri} - N)^{-[\frac{1}{m+3}]}, \tag{9}$$

After obtaining the signal transmission distances between the LED and all SC receivers, we use the RATT-VLP algorithm to locate the receiver. Figure 2 depicts the spatial positioning and distribution of the array of SC receivers in the proposed VLCP system. The array consists of three solar cells, SC1, SC2, and SC3, arranged in an L-shaped configuration with SC1 at the vertex and separated from SC2 and SC3 by distance, d_p. With SC1's coordinates at (x_0, y_0), SC2 is located at $(x_0 + d_p, y_0)$, and SC3 at $(x_0, y_0 + d_p)$. The distances from the LED to SC1, SC2, and SC3 can be calculated as follows:

$$\begin{cases} d_1^2 = x_0^2 + y_0^2 + H^2 \\ d_2^2 = (x_0 + d_p)^2 + y_0^2 + H^2 = x_0^2 + y_0^2 + H^2 + d_p^2 + 2x_0 d_p, \\ d_3^2 = x_0^2 + (y_0 + d_p)^2 + H^2 = x_0^2 + y_0^2 + H^2 + d_p^2 + 2y_0 d_p \end{cases} \tag{10}$$

The coordinate of SC1 can be derived as

$$\begin{cases} x_0 = \frac{d_2^2 - d_1^2 - d_p^2}{2d_p} \\ y_0 = \frac{d_3^2 - d_1^2 - d_p^2}{2d_p}, \end{cases} \tag{11}$$

Thus, the location of the receiver that we define as the coordinate of SC1 can be determined.

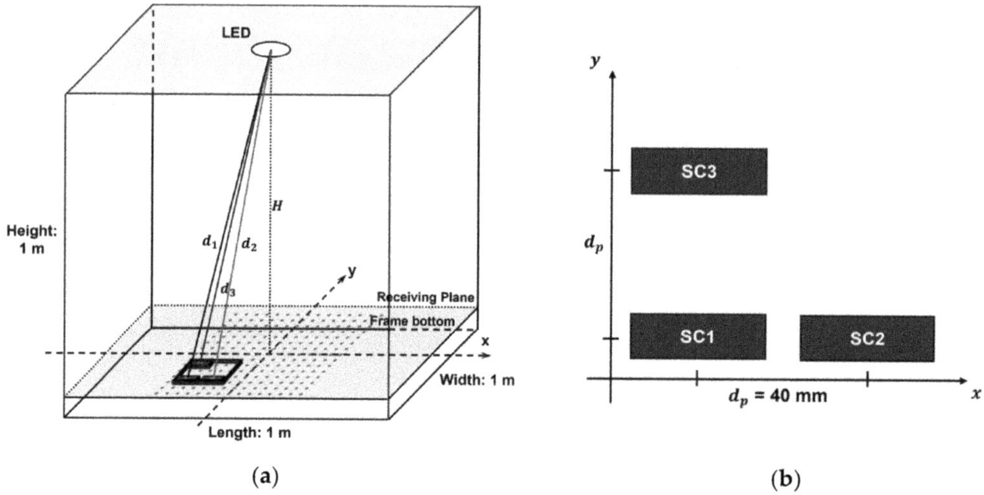

Figure 2. (**a**) Spatial positioning of the SC array receiver within the SC-array-based VLCP system; (**b**) configuration of the SC array.

2.2. Experimental Setup

The experimental setup is shown in Figure 3a, featuring a frame measuring 1 m × 1 m × 1 m. Centrally mounted on the top of the frame is a commercial LED with an output power of 3W, while a homemade SC receiver is positioned on the bottom receiving plane. The vertical separation between the LED and the receiving plane is 92 cm. The L-shaped SC array is composed of three commercially available ANYSOLAR KXOB25-14X1F rectangular monocrystalline silicon solar cells, each boasting a typical power conversion efficiency (PCE) of 25%. These solar cells each have an effective reception area of 21.5 mm × 6.59 mm and are spaced 40 mm apart from each other.

Figure 3b is the schematic block diagram of the process for transmitting and receiving signals. Initially, a random BPSK data sequence is generated, and m-sequences are appended to its head and tail to facilitate data alignment at the receiver. The m-sequence exhibits excellent autocorrelation properties, which aids in extracting the data sequences through autocorrelation calculations in signal processing. Subsequently, the transmission signal is modulated using a RIGOL DG2102 arbitrary waveform generator (AWG), and a DC power supply is provided to the LED by setting a bias voltage. The peak-to-peak voltage (V_{pp}) of the signal is set at 2 V, the bias voltage (V_{dc}) is 6 V, and the sampling rate is 50 kSa/s. Next, the SC array receiver is employed to capture the transmitted signal using a RIGOL MSO5354 oscilloscope (OSC) with a sampling rate of 5 MSa/s. A photograph of the homemade SC array receiver is also presented in the lower right corner of Figure 3a. To enhance the electrical signals output by the three SCs, a low-noise JFET quad operational amplifier, STMicroelectronics TL074CDT, is employed within the SC array receiver. The amplified signals are subsequently transmitted to an oscilloscope through a 50-ohm RF adapter for signal acquisition. The captured signals are then transferred to a computer for further processing.

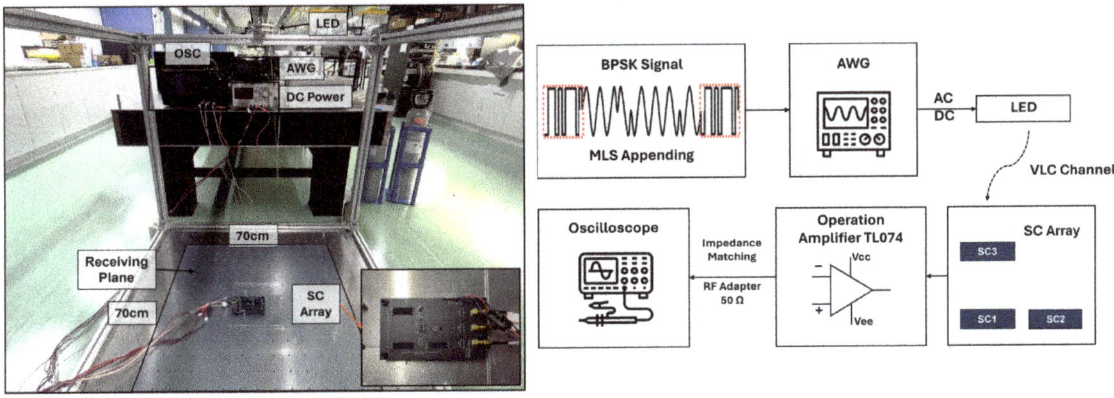

(a) (b)

Figure 3. (**a**) Diagram of the experimental platform for the SC-array-based VLCP system and (**b**) schematic block diagram of the SC array receiver.

To evaluate the system's communication performance, the bias voltage of the LED is adjusted, and the sampling rate of the AWG is tuned under stable conditions to achieve the highest possible rate. At the receiving end, the received signals are autocorrelated with the original m-sequence according to its autocorrelation property. Signal alignment is then performed based on the position of the peak in the autocorrelation coefficient, followed by the demodulation of the signals using the threshold judgment method.

To assess the positioning ability of the system, the SC array receiver moves parallel across the receiving plane, forming a uniformly distributed grid of 15 × 15 test points centered around the origin of the receiving surface. Each adjacent test point is spaced 5 cm apart, yielding a total of 225 positioning data points. Finally, the received signals are processed using a computer to extract the corresponding signal segments. By utilizing the autocorrelation property of the m-sequence, the initially attached m-sequence is auto-correlated with the received signal sequence. The peak value of the computed autocorrelation coefficient indicates the starting point of the m-sequence within the received signal sequence, allowing for synchronization and data extraction. Then, the PSDs are calculated using Fourier transform methods, as with the RSS. Here, due to the complexity of the experimental environment, the experimental results may deviate from theoretical calculations. To minimize experimental errors, we used an exponential nonlinear curve-fitting method to model the relationship between RSS and signal transmission distance, as shown in Equation (9). The results of these experiments will be discussed in the following section.

3. Results and Discussion

Figure 4 shows the downsampled sinusoidal waveforms received by SC1, SC2, and SC3 at three selected points on the diagonal of the receiving plane: point A (0 cm, 0 cm), point B (15 cm, 15 cm), and point C (25 cm, 25 cm). The received sine signals are downsampled according to the ratio of the sampling rates of the oscilloscope (OSC) and the arbitrary waveform generator (AWG), and their AC components are extracted. The waveforms at each point, as illustrated in the figure, contain noise and exhibit amplitude variations across different points. These will be processed later to ascertain specific signal strengths. Furthermore, Figure 5 showcases the waveforms of the BPSK signals received at point A under varying LED DC drive voltages, alongside the results of their demodulation and a comparison with the original data. All waveforms were successfully demodulated without errors, achieving a maximum data rate of 1.21 kbps. This data rate demonstrates that this SC-based VLCP system possesses adequate capabilities for supporting common low-speed VLC schemes, which may include applications where high-data-rate transmission is not critical, such as in sensor networks, smart lighting systems, or basic indoor positioning systems with lower polling rates.

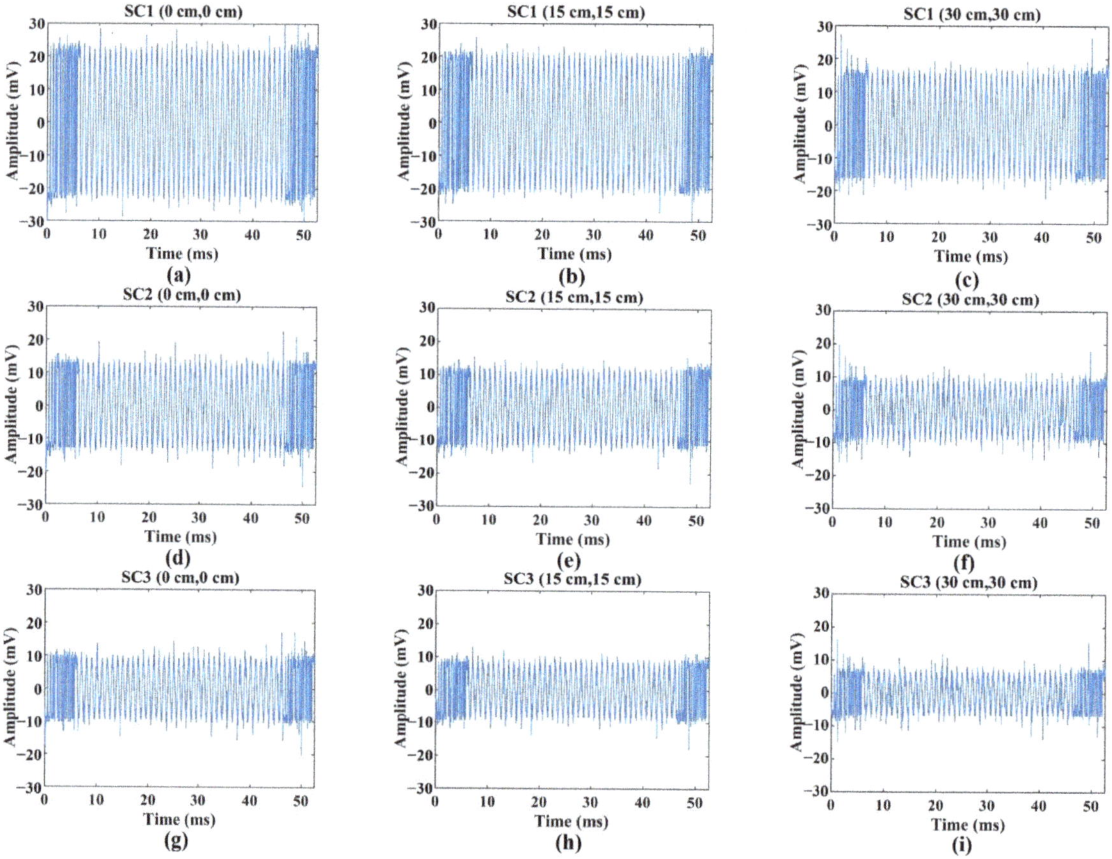

Figure 4. Downsampled waveforms received by SC1 (**a**) at point A, (**b**) at point B, and (**c**) at point C; downsampled waveforms received by SC2 (**d**) at point A, (**e**) at point B, and (**f**) at point C; downsampled waveforms received by SC3 (**g**) at point A, (**h**) at point B, and (**i**) at point C.

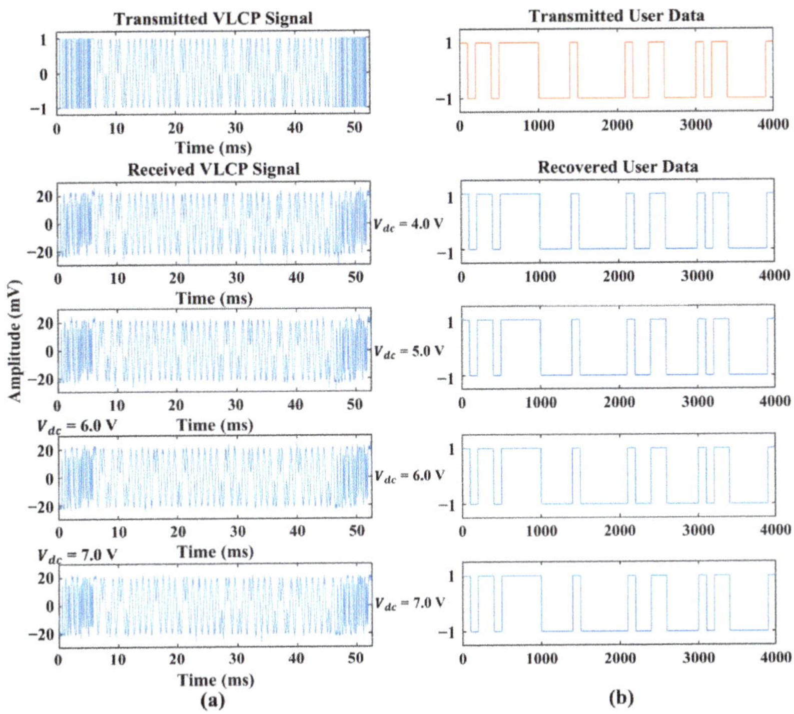

Figure 5. VLCP signal (**a**) waveforms and (**b**) demodulation results received by SC1 at point A under different LED DC drive voltages.

By performing a Fourier transform on the extracted signals collected from various points on the receiving surface, the corresponding signal strength was calculated. This method facilitated the estimation of the RSS distributions across the receiving plane for the three SCs, as illustrated in Figure 6. The RSS distributions for all three SCs exhibited a clear pattern, with lower values around the periphery than at the center. Additionally, the maximum RSS values for SC2 and SC3 were observed to shift toward the negative x-axis and negative y-axis, respectively. These shifts can be attributed to the L-shaped arrangement of the SCs, with SC2 positioned along the positive x-axis and SC3 along the positive y-axis. The experimental results were consistent with expectations and correctly reflected the spatial orientation of the SCs within the receiver.

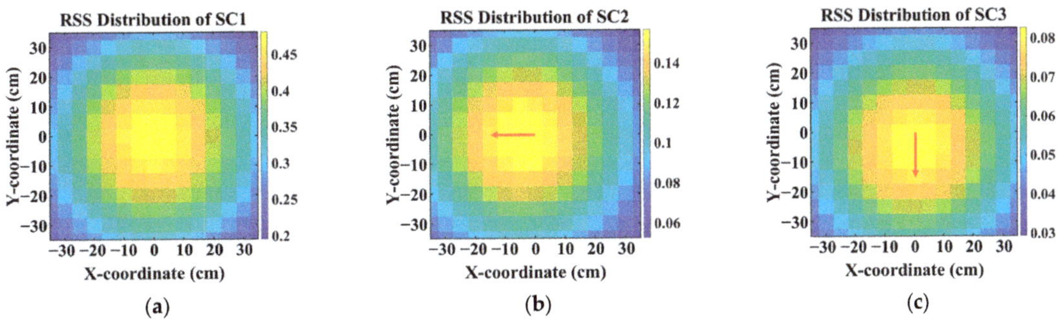

Figure 6. The RSS distributions of (**a**) SC1, (**b**) SC2, and (**c**) SC3.

After estimating the RSS for each SC channel, we applied a fitting method to derive a set of fitting functions based on Equation (9). The regression curves, representing the transmission distances between the LED and each corresponding SC, exhibited clear exponential decay. The derived fitting functions for estimating the transmission distance, d_i, from the i-th SC's RSS, $P_{r,i}$, are as follows:

$$\begin{cases} \tilde{d}_1 = 84.65 + 39.43\exp(-3.458 \cdot P_{r,1}) \\ \tilde{d}_2 = 79.59 + 35.49\exp(-12.63 \cdot P_{r,2}), \\ \tilde{d}_3 = 85.35 + 34.58\exp(-10.30 \cdot P_{r,3}) \end{cases} \quad (12)$$

To accurately quantify the precision of the transmission distance estimates within our system, we utilized the transmission distance error (TDE) metric, which measures the absolute difference between the estimated and theoretical transmission distances. The distributions of TDE are shown in Figure 7, illustrating a clear trend: the TDE increases as the distance from the center increases. This observed pattern indicates that the accuracy of the VLP system diminishes as the distance increases, likely due to factors such as signal attenuation and environmental interference.

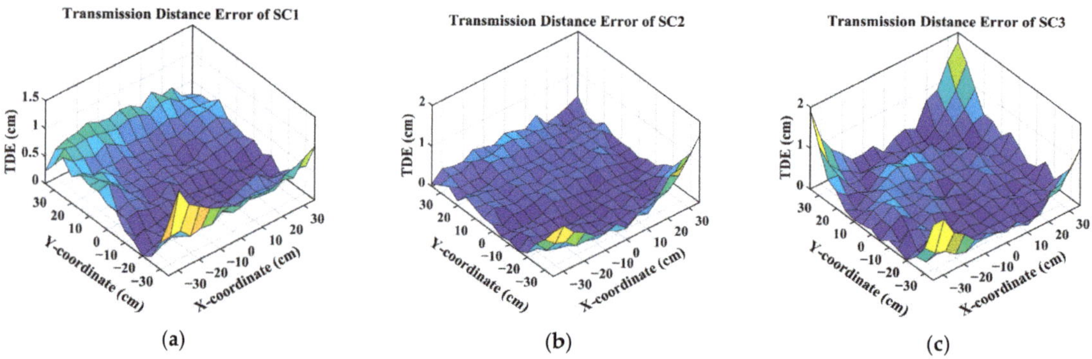

Figure 7. The transmission distance error of (**a**) SC1, (**b**) SC2, and (**c**) SC3.

By using Equation (11), we calculated the estimated positions of the test points, and the positioning results are displayed in Figure 8a. To enhance clarity in the diagram, only the central 11 × 11 points are retained, as more significant deviations were observed at the periphery points. In the figure, blue dots represent the theoretical positions of the test points, while red stars indicate the estimated positions. Lines connecting these points help visually illustrate the degree of positioning deviation.

From the figure, it is evident that within the central 7 × 7 points, covering a 30 cm × 30 cm area, the deviation between the estimated positions and the theoretical positions is relatively small. We conducted a statistical analysis of the positioning errors for the central 7 × 7 points, the central 11 × 11 points, and all 15 × 15 points, with the results shown in Figure 8b. For the 7 × 7 points, the positioning errors are well concentrated within 5 cm, with an average error of 3.40 cm. For the larger area of 11 × 11 points, most positioning errors are less than 10 cm. However, for all test points, the average error is 10.49 cm. The box plot shows that the median lines of the three groups of statistical data are all lower than the mean values. Additionally, as more peripheral points are included, this skewness gradually increases. This indicates that the positioning errors of the peripheral points significantly raise the overall positioning error, while the errors of the central points remain low. Figure 8c shows the cumulative number of points that are less than the corresponding positioning error; out of the total 225 test points in the experiment, 82 points had positioning errors of less than 5 cm, and over 140 points had errors of less than 10 cm. As the positioning error increases, the increment in the number of points gradually decreases, supporting the above observations. Overall, the VLCP system demonstrated good positioning accuracy.

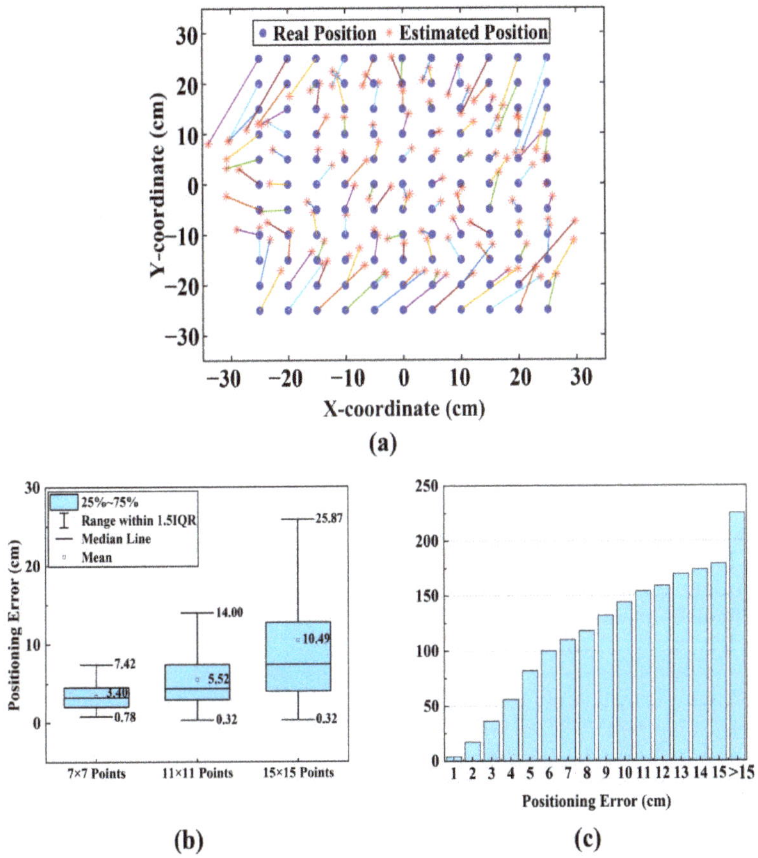

Figure 8. (**a**) The positioning results, including estimated position and real position; (**b**) box plots of positioning error within central 7 × 7 points, 11 × 11 points, and 11 × 11 points; (**c**) cumulative number of points below respective positioning errors.

As the positioning area increases, the discrepancy between the theoretical and estimated positions begins to increase significantly. Given the comprehensive consideration of previous results, this could be attributed to several factors. Firstly, the SCs used may have a poor low-light response, making it challenging to distinguish RSS accurately at the periphery, where the light intensity is relatively lower than the central area. Secondly, the undefined far-field light intensity distribution of the LED used in the experiments might contribute to the errors, especially if the LED has a narrow viewing half-angle, leading to a rapid decline in RSS measurements beyond the 50 cm × 50 cm area. Additionally, noise introduced by the receiver's peripheral circuitry could also be a contributing factor to these errors.

Lastly, we compared the results of this work with previous PD/SC-based VLC/VLP methods, as shown in Table 1. As the first single-source SC-VLCP, our work is unique in simultaneously achieving visible light communication and positioning using a solar cell. The VLC system in [12] is based on an α-Si SC panel, achieving a rate of 11.84 Mbps by applying OFDM, but it does not have VLP capability. This work can provide a reference for us to apply more advanced modulation techniques in the future. Refs. [10,24] present common multi-source VLP systems based on PD and APD, respectively. Compared with these, our work still lags slightly behind in overall positioning accuracy compared with PD-based methods, though it achieves comparable accuracy in narrow-range positioning. This also provides data for future optimization efforts.

Table 1. Comparison of this work with previous PD/SC-based VLC/VLP methods.

Ref	Illumination	Receiver Type	Rate	Positioning Error	Note
This work	Single white LED	Monocrystalline Si SC Array	1.21 kbps	Narrow-range 3.40 cm; wide-range 10.49 cm	First single-source SC-VLCP
[12]	Single white LED	α-Si SC Panel	11.84 Mbps (OFDM)	N/A	VLC
[24]	4 White LEDs	Si PD	N/A	4 cm	VLP
[10]	3 White LEDs	APD	N/A	6.08 cm	VLP

4. Future Work

Throughout this study, the feasibility of using SC as a VLCP receiver has been validated, achieving an initiative simultaneous SC-VLCP system with vast potential for applications in light-source-limited scenarios. Based on our work, it is believed that various derivative directions merit further investigation, including but not limited to channel model investigation; hardware optimization of SC receivers; and the integration of visible light communication, positioning, and energy harvesting within the SLIPT framework.

For channel model investigation, future research could include the further simulation of channel models that include NLoS components, particularly in SC-based VLCP systems. Advanced simulation techniques could be employed to model channel behavior under various reflection conditions, and practical experiments could be conducted that include complex indoor environmental factors [25]. This will help evaluate and quantify the overall impact of NLoS propagation on system communication performance and positioning accuracy. These steps are crucial for optimizing system design, enhancing the robustness of positioning algorithms, and advancing the application of SC-VLCP technology in real-world complex environments.

For hardware optimization, a well-designed receiver circuit could help obtain high-quality VLCP signals. Future work could involve optimizing the circuit design of the SC array receiving module to reduce the impact of circuit noise on the received signal, including implementing stricter power supply noise shielding and applying advanced filtering circuits. Applying nonlinear compensation circuits for SC under low illumination may also help improve the overall performance of the system [26].

For energy harvesting, the efficiency of solar cells and optimization strategies within the SLIPT framework could be key focus areas for future research. The next steps of this work could involve discussing and verifying methods to improve the energy harvesting efficiency of SC receivers and achieve self-powered receivers, enabling the simultaneous implementation of visible light communication, positioning, and energy harvesting. Emerging solar cells with novel structures and materials could be applied to achieve significant improvements in rate and power conversion efficiency [27]. Signal modulation and demodulation techniques could also be optimized, and efficient resource allocation strategies are worth studying to enhance the applicability of SC-VLCP systems. These advancements will help fully optimize and extend the application of SLIPT in VLCP systems.

5. Conclusions

This paper proposed a solar-cell-array-based indoor VLCP system with a single LED transmitter, coupled with the RATT-VLP algorithm based on multiple SC receivers. The LED emits VLCP signals that pass through the VLC channel and are captured by the SC array, serving dual purposes for communication and positioning. The alignment of the received BPSK signals is achieved through autocorrelation with an MLS, and the data are demodulated using threshold judgment. The RSS calculations employ the Fourier transform method, and the receiver position is determined using RATT-VLP. Experimental validation confirmed the feasibility of the proposed VLCP system and the RATT-VLP algorithm. For BPSK-modulated signals, the highest measured VLC communication data rate for this VLCP system was 1.21 kbps, demonstrating its ability to support low-speed VLC applications. Furthermore, the RATT-VLP algorithm applied within this VLCP system achieved an average positioning accuracy of 3.40 cm within a 30 cm $\times$ 30 cm area, with an overall average accuracy of 10.49 cm. The results showcase the SC array-based indoor VLCP system's potential for implementing communication and positioning functions simultaneously. Finally, future research directions based on this work are provided, including but not limited to channel model investigation, SC receiver hardware and circuitry optimization, and energy harvesting efficiency improvement for achieving SC-VLCP systems under the SLIPT concept.

Author Contributions: Conceptualization, D.X. and Z.L.; methodology, D.X. and Z.L.; software, D.X.; validation, D.X.; formal analysis, D.X.; investigation, D.X.; resources, C.Y.; data curation, D.X.; writing—original draft preparation, D.X.; writing—review and editing, D.X. and Z.L.; visualization, D.X.; supervision, C.Y.; project administration, C.Y.; funding acquisition, C.Y. All authors have read and agreed to the published version of the manuscript.

Funding: This research was funded by the Hong Kong Research Grants Council, grant number GRF15212720.

Institutional Review Board Statement: Not applicable.

Informed Consent Statement: Not applicable.

Data Availability Statement: The data underlying the results presented in this paper are not publicly available at this time but may be obtained from the authors upon reasonable request.

Conflicts of Interest: The authors declare no conflicts of interest.

References

1. Pathak, P.H.; Feng, X.; Hu, P.; Mohapatra, P. Visible light communication, networking, and sensing: A survey, potential and challenges. *IEEE Commun. Surv. Tutor.* **2015**, *17*, 2047–2077. [CrossRef]
2. Karunatilaka, D.; Zafar, F.; Kalavally, V.; Parthiban, R. LED based indoor visible light communications: State of the art. *IEEE Commun. Surv. Tutor.* **2015**, *17*, 1649–1678. [CrossRef]
3. Lee, S.H.; Jung, S.Y.; Kwon, J.K. Modulation and coding for dimmable visible light communication. *IEEE Commun. Mag.* **2015**, *53*, 136–143. [CrossRef]
4. Lou, P.; Zhang, H.; Zhang, X.; Yao, M.; Xu, Z. Fundamental analysis for indoor visible light positioning system. In Proceedings of the 2012 1st IEEE International Conference on Communications in China Workshops (ICCC), Beijing, China, 15–17 August 2012; pp. 59–63. [CrossRef]
5. Komine, T.; Nakagawa, M. Integrated system of white LED visible-light communication and power-line communication. *IEEE Trans. Consum. Electron.* **2003**, *49*, 71–79. [CrossRef]
6. Liu, Z.; Yu, C. Multi-user visible light communication and positioning system based on dual-domain multiplexing scheme. *Photonics* **2023**, *10*, 306. [CrossRef]
7. Yu, C.; You, X.; Liu, Z.; Chen, J. Conceptual design for indoor visible light communication and positioning cooperative systems. In Proceedings of the 22nd International Conference on Transparent Optical Networks (ICTON) 2020, Bari, Italy, 19–23 July 2020. [CrossRef]
8. Zhang, Y.; Wei, Z.; Liu, Z.; Cheng, C.; Wang, Z.; Tang, X.; Yang, Y.; Fu, H.Y. Optical communication and positioning convergence for flexible underwater wireless sensor network. *J. Light. Technol.* **2023**, *41*, 5321–5327. [CrossRef]
9. Shao, S.; Salustri, A.; Khreishah, A.; Xu, C.; Ma, S. R-VLCP: Channel Modeling and Simulation in Retroreflective Visible Light Communication and Positioning Systems. *IEEE Internet Things J.* **2023**, *10*, 11429–11439. [CrossRef]
10. Yang, H.; Chen, C.; Zhong, W.D.; Alphones, A.; Zhang, S.; Du, P. Demonstration of a Quasi-Gapless Integrated Visible Light Communication and Positioning System. *IEEE Photonics Technol. Lett.* **2018**, *30*, 2001–2004. [CrossRef]
11. Kim, S.M.; Won, J.S.; Nahm, S.H. Simultaneous reception of solar power and visible light communication using a solar cell. *Opt. Eng.* **2014**, *53*, 046103. [CrossRef]
12. Wang, Z.; Tsonev, D.; Videv, S.; Haas, H. On the design of a solar-panel receiver for optical wireless communications with simultaneous energy harvesting. *IEEE J. Sel. Areas Commun.* **2015**, *33*, 1612–1623. [CrossRef]

13. Shin, W.H.; Yang, S.H.; Kwon, D.H.; Han, S.K. Self-reverse-biased solar panel optical receiver for simultaneous visible light communication and energy harvesting. *Opt. Express* **2016**, *24*, A1300–A1305. [CrossRef] [PubMed]
14. Diamantoulakis, P.D.; Karagiannidis, G.K.; Ding, Z. Simultaneous lightwave information and power transfer (SLIPT). *IEEE Trans. Green Commun. Netw.* **2018**, *2*, 764–773. [CrossRef]
15. Wang, Z.; Tsonev, D.; Videv, S.; Haas, H. Towards self-powered solar panel receiver for optical wireless communication. In Proceedings of the 2014 IEEE International Conference on Communications (ICC), Sydney, NSW, Australia, 10–14 June 2014; pp. 3348–3353. [CrossRef]
16. Lorrière, N.; Bialic, E.; Pasquinelli, M.; Chabriel, G.; Barrere, J.; Escoubas, L.; Simon, J.J. An OFDM testbed for LiFi performance characterization of photovoltaic modules. In Proceedings of the 2018 Global LIFI Congress (GLC), Paris, France, 8–9 February 2018; pp. 1–5. [CrossRef]
17. Sepehrvand, S.; Theagarajan, L.N.; Hranilovic, S. Rate-power trade-off in simultaneous lightwave information and power transfer systems. *IEEE Commun. Lett.* **2020**, *25*, 1249–1253. [CrossRef]
18. Fakidis, J.; Helmers, H.; Haas, H. Simultaneous wireless data and power transfer for a 1-Gb/s GaAs VCSEL and photovoltaic link. *IEEE Photonics Technol. Lett.* **2020**, *32*, 1277–1280. [CrossRef]
19. Das, S.; Sparks, A.; Poves, E.; Videv, S.; Fakidis, J.; Haas, H. Effect of sunlight on photovoltaics as optical wireless communication receivers. *J. Light. Technol.* **2021**, *39*, 6182–6190. [CrossRef]
20. Hsu, C.W.; Wu, J.T.; Wang, H.Y.; Chow, C.W.; Lee, C.H.; Chu, M.T.; Yeh, C.H. Visible light positioning and lighting based on identity positioning and RF carrier allocation technique using a solar cell receiver. *IEEE Photonics J.* **2016**, *8*, 7905507. [CrossRef]
21. Wang, Z.; Zhong, W.-D.; Yu, C.; Chen, J.; Francois, C.P.S.; Chen, W. Performance of dimming control scheme in visible light communication system. *Opt. Express* **2012**, *20*, 18861–18868. [CrossRef] [PubMed]
22. Neokosmidis, I.; Kamalakis, T.; Walewski, J.W.; Inan, B.; Sphicopoulos, T. Impact of Nonlinear LED Transfer Function on Discrete Multitone Modulation: Analytical Approach. *J. Light. Technol.* **2009**, *27*, 4970–4978. [CrossRef]
23. Komine, T.; Nakagawa, M. Fundamental analysis for visible-light communication system using LED lights. *IEEE Trans. Consum. Electron.* **2004**, *50*, 100–107. [CrossRef]
24. Raes, W.; Knudde, N.; De Bruycker, J.; Dhaene, T.; Stevens, N. Experimental Evaluation of Machine Learning Methods for Robust Received Signal Strength-Based Visible Light Positioning. *Sensors* **2020**, *20*, 6109. [CrossRef]
25. Dos Santos, D.R.; Julien-Vergonjanne, A.; Dkhil, S.B.; Parmentier, M.; Combeau, P.; Sahuguède, S.; Bouclé, J. Toward Indoor Simulations of OPV Cells for Visible Light Communication and Energy Harvesting. *IEEE Access* **2024**, *12*, 41027–41041. [CrossRef]
26. Chen, S.; Liu, L.; Chen, L.K. Passive Nonlinear Compensation Circuits for Photovoltaic Visible Light Communications under Low Illuminance. In Proceedings of the Optical Fiber Communication Conference (OFC) 2022, San Diego, CA, USA, 6–10 March 2022; p. W2A.32.
27. Fu, Y.; Zhang, L.; Wangzhou, Y.; Zheng, S.; Sun, Y.; Liu, J.; Zhang, Y.; Li, S.; Hu, Z.; Yu, B.; et al. Wavelength division multiplexing visible light communication with wide incident angle enabled by perovskite quantum dots. *Opt. Commun.* **2024**, *559*, 130427. [CrossRef]

photonics

MDPI

Article

Optimisation of the Transmitter Layout in a VLP System Using an Aperture-Based Receiver

José Miguel Menéndez [1,2,*] and Heidi Steendam [1]

1 TELIN/IMEC, Ghent University, 9000 Ghent, Belgium; heidi.steendam@ugent.be
2 Department of Electrical and Computer Engineering, ESPOL University, Campus Gustavo Galindo, Guayaquil P.O. Box 09-01-5863, Ecuador
* Correspondence: josemiguel.menendezsanchez@ugent.be or jmmenend@espol.edu.ec

Abstract: In this paper, we consider a visible light positioning (VLP) system, where an array of photo diodes combined with apertures is used as a directional receiver and a set of inexpensive and energy-efficient light-emitting diodes (LEDs) is used as transmitters. The paper focuses on the optimisation of the layout of the transmitter, i.e., the number and placement of the LEDs, to meet the wanted position estimation accuracy levels. To this end, we evaluate the Cramer–Rao bound (CRB), which is a lower bound on the mean-squared error (MSE) of the position estimate, to analyse the influence of the LEDs' placement. In contrast to other works, where only the location of the LEDs was considered and/or the optimisation was carried out through simulations, in this work, the optimisation is carried out analytically and considers all the parameters involved in the VLP system as well as the illumination. Based on our results, we formulate simple rules of thumb with which we can determine the spacing between LEDs and the minimum number of LEDs, as well as their position on the ceiling, while also taking into account the requirements for the illumination.

Keywords: VLP; Cramer–Rao bound; horizontal illuminance; indoor navigation

1. Introduction

In the last decade, LEDs have become the main illumination source due to their excellent energy efficiency and their long lifetime. In contrast to traditional light sources, such as incandescent and fluorescent lights, LEDs can be modulated up to several MHz. Given that LEDs are typically mounted on the ceiling, an optical receiver pointing upward will experience a line-of-sight link with the LED. Due to this, LEDs are considered transmitters in visible light communication systems [1,2]. Moreover, considering that the channel gain of the LOS component is a simple function of the distance and the angle between the LED and the optical receiver, we can obtain position-related information from the received light. Therefore, visible-light LEDs are considered for indoor positioning. Several works have already investigated the accuracy of visible light positioning (VLP) systems [3–5], and reported an accuracy in the order of centimetres. As only a few changes need to be made to the illumination infrastructure to be able to modulate the LEDs, VLP systems can be considered as an accurate, low-cost solution for indoor positioning.

The type and amount of position-related information that can be extracted from the visible-light LEDs depends on the optical receiver that is considered. While charge-coupled device (CCD) cameras contain a large number of pixels with which not only the received signal strength (RSS) can be obtained, but also the direction from which the light is coming, a single photo diode (PD) will only be able to detect the RSS. However, due to practical problems with the CCD camera, including higher energy consumption compared to a PD and the shutter speed of the camera, which limits the modulation bandwidth of the LED, recent work has focused on VLP systems using PDs in the receiver [6,7]. As a single PD is not capable of extracting the direction of the light, several papers investigated other optical receiver structures that offer angular diversity, i.e., that are able to detect the direction of the light [8–10]. To make the receiver directional, in [11], a receiver with several tilted PDs was considered, while [12,13] considered PDs combined with apertures. In this paper, we consider the receiver structure from [12], consisting of an array of PDs and apertures, as this receiver offers a wide field of view (FOV) with high angular diversity.

To obtain a low-cost solution, a visible light positioning system is preferably combined with an existing illumination system. Most works on VLP focus on positioning accuracy and optimise the placement of the LEDs to improve the accuracy. However, in a VLP system that coexists with the illumination system, there are additional factors that require further analysis. Illuminance, for example, has a major effect on how people perceive and perform their visual tasks. Because of this, the illuminance levels and the illumination uniformity must be maintained in a range that ensures eye safety and comfort. In our paper [14], we analysed the placement of illumination LEDs to optimise the number of LEDs and their spacings. In this paper, we will employ our results from [14] to analyse the positioning performance of an aperture-based VLP system and to evaluate the coexistence of the VLP system with the illumination system. Our analytical approach provides a deeper insight into the impact of each of the parameters that makes up the system. This contrasts with [15], where a few scenarios were considered with a limited number of LEDs, and results were obtained after extensive experiments. To measure the positioning performance, we use the Cramer–Rao bound (CRB), which is a lower bound on the mean-squared error (MSE) of the positioning error. The results will provide useful information about the installation of LEDs used for illumination and VLP and will be shown to be applicable for rooms of various sizes.

The remainder of the paper is structured as follows. In Section 2, the system is described. Section 3 considers the derivation of the CRB and the illumination level, and the numerical results are shown in Section 4. Finally, the conclusions are given in Section 5. Furthermore, we provide in Table 1 a list of the most important parameters examined in Section 4.

Table 1. Summary of the main parameters of the system.

Parameter	Description
$X_{\max}, Y_{\max}, Z_{\max}$	Room size on the x, y and z coordinate axes. For rooms with square areas on the xy plane, $X_{\max} = Y_{\max} = \chi_{\max}$.
ρ_a	Scaling factor that sets the VLP-LED grid spread in the a direction. For rooms with square areas on the xy plane, $\rho_x = \rho_y = \rho$.
$\rho_{a,I}^{\min}$ $\rho_{a,I}^{\max}$ $\rho_{a,V}^{opt}$	Minimum ρ_a from the perspective of illumination. Maximum ρ_a from the perspective of illumination. Optimal ρ_a from the perspective of VLP.
$\delta_{L,a}$	Distance between VLP-LEDs in the a direction. For rooms with square areas on the xy plane, $\delta_{L,x} = \delta_{L,y} = \delta_L$.
$K_{L,a}$	Number of VLP-LEDs placed in the a direction. For rooms with square areas on the xy plane, $K_{L,x} = K_{L,y} = K_L$.
$h_{j,i}$	Vertical distance between LED i and RE j.
$\phi_{j,i}$	Incident angle between LED i and RE j.
$m_{S,i}$	Lambertian order of the LED i.
$\bar{\Phi}_{e,i}$	Average optical power of the LED i.
$\Phi_{v,i}$	Luminous flux of the LED i.
$\mathcal{U}_t$ $\mathcal{U}_s$	Uniformity level of illumination of the task area. Uniformity level of illumination of the surrounding area.
ξ^t	Task area delimiter (central sub-area of the area where the receiver is assessed). This is determined during the illumination design.
ζ	Maximum desired accuracy level of receiver position estimation.

2. System Description

2.1. Transmitter

This paper considers a VLP system that uses K_L VLP-LEDs attached to the ceiling. Although, for illumination purposes, LEDs are often grouped in arrays of LEDs, using multiple LEDs within one array for positioning purposes is not recommended for the following reason. Taking into account that the dimensions of an array are rather small compared to the dimensions of a room, this implies that two LEDs belonging to the same array can be considered as virtually co-located. Assuming that all LEDs within an array have the same orientation, it follows that position information that can be obtained from these two LEDs is strongly correlated, and will lead to a reduced dilution of precision (DOP) (the dilution of precision is a term used in GPS navigation to express the positional measurement precision) for a given number of VLP-LEDs. To obtain a good DOP, the LEDs need to have good spatial spreading. For this reason, when LED arrays are available for illumination, we will only use one LED per array for positioning purposes. We assume that all LEDs follow a Lambertian radiation pattern, where the Lambertian order of the VLP-LED i is expressed as $m_{S,i} = -\frac{\ln 2}{\ln\left(\cos\left(\Phi_{1/2,i}\right)\right)}$, with $i = 1, \ldots, K_L$ and $\Phi_{1/2,i}$ as the semi-angle at which half optical power is reached.

In order to estimate the position of a receiver within the room, we define a global coordinate system (GCS), (X, Y, Z), whose origin is fixed at the centre of the ceiling. Within this coordinate system, the ith VLP-LED has coordinates $v_{L,i}$ with $v_{L,i} = (x_{L,i}, y_{L,i}, z_{L,i})^T$. We assume that the coordinates of all LEDs are known by the receiver, and all LEDs point straight downward, i.e., their normal is $N_{L,i} = (0, 0, -1)^T$.

The receiver captures the modulated light emitted by the VLP-LEDs, the nonmodulated light from I-LEDs and background light. To separate the light from the different VLP-LEDs and the other light sources, we assume the VLP-LEDs are modulated using different orthogonal frequencies that can be separated by applying a discrete Fourier transform at the receiver.

2.2. Receiver

The receiver consists of an array of M receiving elements (REs). Each RE contains a bare photo diode (PD) and an aperture in an opaque screen [16]. We assume that the aperture and the PD have the same size and shape, and that the plane of the apertures is parallel to and at a distance h_A above the plane of the PDs. We restrict our attention to the case where the PDs are circular with radius R_D and assume that the only light that reaches the PD is the light that enters through its aperture. Taking into account that R_D is much larger than the wavelength of light, the incident light will introduce a circular light spot with radius R_D in the plane of the PDs.

To determine the receiver's position, we select a reference point, $v_U = (x_U, y_U, z_U)^T$, in the plane of the apertures. The centre $v_{AP,j} = \left(x_{AP,j}, y_{AP,j}, z_{AP,j}\right)^T$ of the aperture j is located in a relative position $v_{\delta_{AP,j}} = \left(x_{\delta_{AP},j}, y_{\delta_{AP},j}, z_{\delta_{AP},j}\right)^T$ from this reference point v_U, i.e., $v_{AP,j} = v_U + v_{\delta_{AP},j}$, $j = 1, \ldots, M$ (Figure 1). For simplicity, we assume that the receiver array points straight upward, that is, the normal of the receiver is $N_R = (0, 0, 1)^T$, implying that $z_{AP,j} = z_U$ and $z_{\delta_{AP},j} = 0$. However, the extension to the case where the receiver array or the REs within the array are tilted is straightforward.

PDs are relatively displaced compared to their apertures in order to create angular diversity [12] (see Figure 2). The centre $v_{PD,j} = \left(x_{PD,j}, y_{PD,j}, z_{PD,j}\right)^T$ of PD j has coordinates $v_{PD,j} = v_{AP,j} + \Delta v_{PD,j}$ with $\Delta v_{PD,j} = \left(d_{AP,j}\cos\alpha_{AP,j}, d_{AP,j}\sin\alpha_{AP,j}, z_{PD,j}\right)^T$, where $z_{PD,j} = -h_A$. The angles $\alpha_{AP,j} = j\frac{2\pi}{M}$ are chosen to obtain a receiver array with angular diversity, while the distance $d_{AP,j}$ and height h_A can be selected to change the FOV of the receiver array [17].

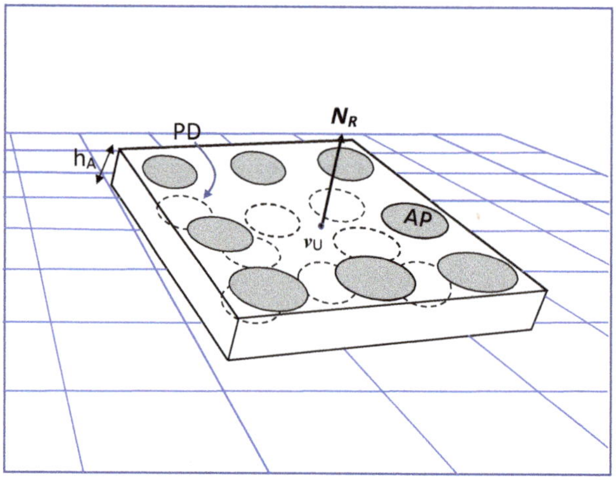

Figure 1. Receiver description: a 3D representation of the REs.

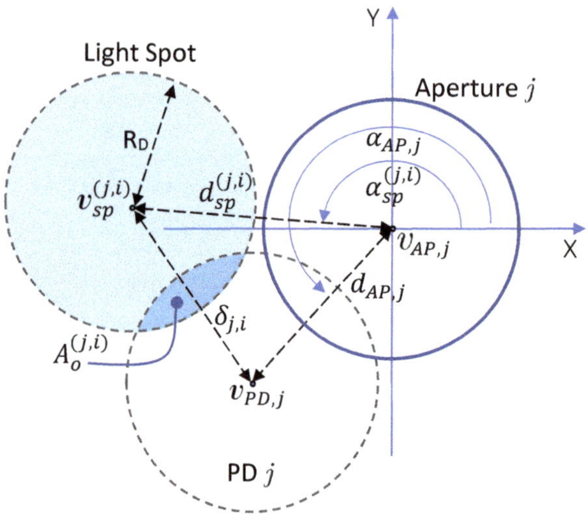

Figure 2. Receiver description: light spot introduced by the aperture and the overlap of the light spot with the PD.

2.3. *Received Signal Strength*

We will determine the position of the receiver by comparing the received signal strengths (RSS) in the different REs. The RSS level corresponding to the contribution from VLP-LED i detected at RE j mainly depends on the radiation angle at the source and the angle at which the light ray impinges on the PD, that is, the incident angle. Taking into account the assumption that the LED array and receiver array are parallel to the ceiling, this incident angle is equal to the radiation angle (see Figure 3) and can be expressed as

$$\phi_{j,i} = \arctan\left(\frac{z_{L,i} - z_{AP,j}}{\| \boldsymbol{v}_{L,i} - \boldsymbol{v}_{AP,j} \|}\right). \quad (1)$$

where, we recall, $\boldsymbol{v}_{AP,j} = \boldsymbol{v}_U + \boldsymbol{v}_{\delta_{AP},j}$, i.e., the RSS levels at the different REs depend on the unknown position $\boldsymbol{v}_U$ of the receiver and the known layout of the receiver and the known positions of the VLP-LEDs.

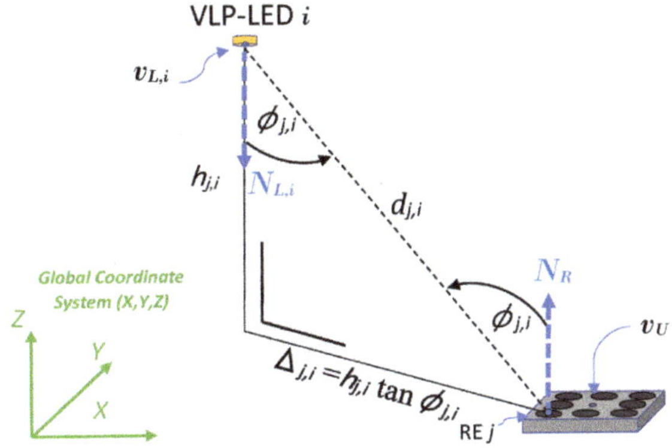

Figure 3. Global coordinate system and positions of the transmitter and receiver array within this coordinate system.

In each RE, we focus only on the contributions from the VLP-LEDs by filtering out the contributions originating from sources different from those of the VLP-LEDs. Collecting in RE j the contributions of the K VLP-LEDs, we obtain the vector $\mathbf{r}_j$, $j = 1, \ldots, M$, of observations:

$$\mathbf{r}_j[i] = R_{p,j}\, h_c^{(j,i)} \bar{\Phi}_{e,i} + \tilde{N}_j, \quad i = 1, \cdots, K, \tag{2}$$

where $R_{p,j}$ is the responsivity of PD j, $h_c^{(j,i)}$ is the optical channel gain between LED i and PD j, $\bar{\Phi}_{e,i}$ is the average transmitted optical power of LED i and $\tilde{N}_j$ is the shot noise in RE j. This shot noise is modelled as a zero-mean Gaussian random variable with variance $N_{0,j} = 2q\, R_{p,j} p_{n,j} A_{D,j} \Delta\lambda$ [18], with q being the charge of an electron, $p_{n,j}$ the optical power obtained at PD j from light sources different from the VLP-LEDs, $A_{D,j}$ the area of PD j and $\Delta\lambda$ the optical bandwidth. Furthermore, following [18], the channel gain $h_c^{(j,i)}$ can be written as

$$h_c^{(j,i)} = \frac{(m_{S,i}+1)}{2\pi d_{j,i}^2} A_0^{(j,i)} \cos^{m_{S,i}+1} \phi_{j,i} \tag{3}$$

where $d_{j,i}$ is the Euclidean distance between VLP-LED i and the centre $\boldsymbol{v}_{sp}^{(j,i)}$ of the light spot introduced by aperture j and coming from VLP-LED i, i.e., $d_{j,i} = \| \boldsymbol{v}_{L,i} - \boldsymbol{v}_{sp}^{(j,i)} \|$. In (3), $A_0^{(j,i)}$ is the surface of the overlap area between PD j and the light spot with the centre $\boldsymbol{v}_{sp}^{(j,i)}$. These coordinates of the centres of the light spots are given by $\boldsymbol{v}_{sp}^{(j,i)} = \boldsymbol{v}_{AP,j} + \boldsymbol{v}_{\delta_{SP}}^{(j,i)}$, where $\boldsymbol{v}_{\delta_{SP}}^{(j,i)} = (d_{sp}^{(j,i)} \cos \alpha_{sp}^{(j,i)}, d_{sp}^{(j,i)} \sin \alpha_{sp}^{(j,i)}, -h_A)^T$, with $d_{sp}^{(j,i)} = h_A \tan \phi_{j,i}$ being the horizontal distance between the centres of the light spot and aperture of RE j, and $\alpha_{sp}^{(j,i)} = \pi + \alpha_{j,i}$ (see Figure 2), where $\alpha_{j,i}$ is the azimuth angle of the incoming light. Defining the horizontal distance $\Delta_{j,i} = h_{j,i} \tan \phi_{j,i}$ between VLP-LED i and the centre of aperture j, with $h_{j,i} = z_{L,i} - z_{AP,j}$ as the vertical distance between the VLP-LED and the receiver (see Figure 3), the azimuth angle $\alpha_{j,i}$ can be written as

$$\cos \alpha_{j,i} = \frac{x_{L,i} - x_{AP,j}}{\Delta_{j,i}}. \tag{4}$$

Using the angles $\phi_{j,i}$ and $\alpha_{j,i}$, we establish the following relationship between the RE positions and the VLP-LEDs:

$$\begin{aligned} x_{L,i} &= x_{AP,j} + (z_{L,i} - z_{AP,j}) \tan \phi_{j,i} \cos \alpha_{j,i} \\ y_{L,i} &= y_{AP,j} + (z_{L,i} - z_{AP,j}) \tan \phi_{j,i} \sin \alpha_{j,i} \end{aligned} . \tag{5}$$

Taking into account the definition of the centre $v_{SP}^{(j,i)}$ of the light spot, the overlap area $A_0^{(j,i)}$ is given by

$$A_0^{(j,i)} = \begin{cases} 2R_D^2 \arccos\left(\frac{\delta_{j,i}}{2R_D}\right) & 0 \leq \delta_{j,i} \leq 2R_D \\ \quad - \frac{\delta_{j,i}}{2}\sqrt{4R_D^2 - \delta_{j,i}^2} & \\ 0 & \delta_{j,i} > 2R_D \end{cases}, \tag{6}$$

where the distance $\delta_{j,i}$ between the centre of the light spot and the centre of PD j equals

$$\delta_{j,i} = \left\| v_{PD,j} - v_{SP}^{(j,i)} \right\|. \tag{7}$$

2.4. Average Transmitted Optical Power

The objective of this paper is to relate the positioning accuracy of the VLP system to the illumination level of the area. As will be shown in the next section, the positioning accuracy is a function of the average transmitted optical power $\bar{\Phi}_{e,i}$, $i = 1, \ldots, K$ of the LEDs. This transmitted optical power is, in turn, a function of the forward current signal applied to the LED. Taking into account that the receiver must be able to distinguish the signals from the different LEDs, we assume that the different LEDs are modulated using different orthogonal frequencies. A standard technique for combining signals modulated with different carrier frequencies is orthogonal frequency division multiplexing (OFDM). However, standard OFDM is not real-valued or positive. Therefore, several real-valued positive variants of OFDM have been developed for LED modulation, e.g., DC-biased optical OFDM (DCO-OFDM), asymmetrically clipped OFDM (ACO-OFDM) and asymmetrically clipped DC-biased optical OFDM (ADO-OFDM) [19–21]. In these techniques, the orthogonality of the signals is affected by non-linear distortions. Therefore, the LED typically operates within the linear dynamic range of the current versus voltage (I–V) curve of the LED (see Figure 4). Assuming that each LED uses only a single subcarrier, for all OFDM variants mentioned above, the signal for LED i can be written as

$$s_i(t) = K_c A_i (1 + \cos(2\pi f_{c,i} t)). \tag{8}$$

where the frequencies $f_{c,i}$ of the LEDs are selected to be orthogonal over a time interval T_c, the amplitude A_i is selected so that the LED operates within the linear dynamic range of the current versus voltage (I–V) curve of the LED and K_c is the slope of the linear part of the I–V curve of the LED. The OFDM system also allows us to filter out the contributions of the nonmodulated I-LEDs, implying that the presence of the I-LEDs will have no effect on the observed signals (the presence of the I-LEDs will have an influence, though, on the operation of the PD as it will generate a DC bias current that depends on the optical power of the I-LEDs. Provided that the presence of the I-LEDs does not saturate the PDs, we assume that this constant bias current is filtered out before the amplifier).

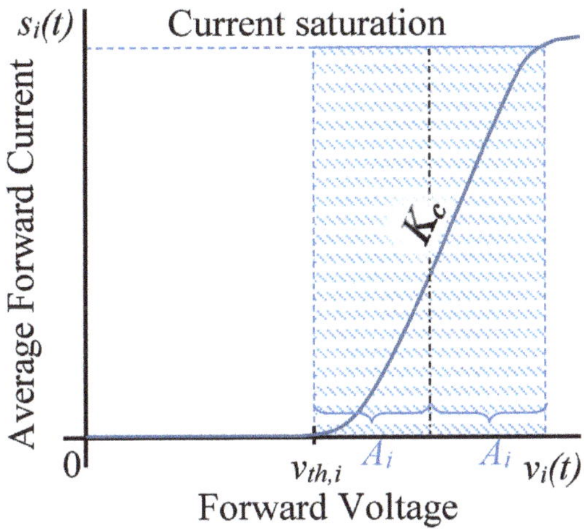

Figure 4. I–V curve for an LED. The blue region represents the linear operation range and $v_{th,i}$ the threshold voltage of the LED.

To relate the average optical power $\bar{\Phi}_{e,i}$ from (2) to the forward current $s_i(t)$, we need the luminous flux Φ_v, which is also used to determine the illumination level in an area, as will be explained in the next section. This luminous flux depends not only on the electrical current $s_i(t)$, but also on factors such as thickness, combination concentration and the geometrical distribution of the material of which it is composed. This relationship is summarised through $\Phi_{v,i} = K_\Phi \Phi_v^{typ} s_i(t)$, obtained from the forward current versus normalised relative luminous flux curve contained in the LED data sheets [22], where K_Φ is the slope of the linear part of the curve, and Φ_v^{typ} is the typical value of the luminous flux tested by the manufacturer. The instantaneous optical power $\Phi_{e,i}(t)$, radiated by the LED i, is proportional to this luminous flux:

$$\Phi_{e,i}(t) = \frac{\Phi_{v,i}}{\eta_{LER}} = \frac{K_\Phi \Phi_v^{typ}}{\eta_{LER}} s_i(t), \tag{9}$$

where η_{LER} is the luminous efficacy of optical radiation (η_{LER} (lm/Watt) relates to the optical power demanded by the emitting device and should not be confused with the luminous efficiency η_{LE} (lm/Watt) of a light source, which is a term widely found in the technical specifications of LEDs but expresses the electrical power demanded by the emitting device), which is obtained by considering the LED operational electrical power and the normalized relative luminous flux specified in [22]. Consequently, the average optical power $\bar{\Phi}_{e,i}$ from (2) is given by

$$\bar{\Phi}_{e,i} = \kappa \lim_{T\to\infty} \frac{1}{T} \int_0^T s_i(t)dt = \kappa A_i \tag{10}$$

where $\kappa = \dfrac{K_\Phi \Phi_v^{typ} K_c}{\eta_{LER}}$.

3. System Evaluation Criteria

The evaluation criterion that will be used to assess the accuracy of positioning in our VLP system is the Cramer–Rao bound [23,24], which is a lower bound on the MSE of the position estimate for unbiased estimators. In [25], it is shown that, in most cases, the bias in the position estimate is sufficiently small to be neglected, implying that the CRB is a suitable lower bound to assess the performance of practical estimators and to analyse the VLP system.

In this paper, we evaluate the CRB on the position estimate $\hat{\boldsymbol{v}}_U = (\hat{x}_U, \hat{y}_U, \hat{z}_U)$ of the receiver position $\boldsymbol{v}_U = (x_U, y_U, z_U)$ to bound the MSE [24]:

$$\begin{aligned} MSE &= E[(x_U - \hat{x}_U)^2 + (y_U - \hat{y}_U)^2 + (z_U - \hat{z}_U)^2] \\ &\geq \mathrm{tr}(\boldsymbol{F}_U^{-1}). \end{aligned} \tag{11}$$

In (11), $\mathrm{tr}(\cdot)$ is the trace operator, and $\boldsymbol{F}_U$ is the Fisher information matrix (FIM). This FIM measures the amount of information about $\boldsymbol{v}_U$ that can be extracted from the observed vector $\boldsymbol{r}$ (2) of RSS values and is given by

$$\boldsymbol{F}_U = E[(\boldsymbol{\nabla}_{\boldsymbol{v}_U} \ln p(\boldsymbol{r}|\boldsymbol{v}_U))(\boldsymbol{\nabla}_{\boldsymbol{v}_U} \ln p(\boldsymbol{r}|\boldsymbol{v}_U))^T], \tag{12}$$

where $\boldsymbol{r} = [\boldsymbol{r}_1^T \ldots \boldsymbol{r}_M^T]^T$. From (2), it follows that $\boldsymbol{r}_j[i]|\boldsymbol{v}_U$ is Gaussian distributed with average $R_{p,j} h_c^{(j,i)} \bar{\Phi}_{e,i}$ and variance $N_{0,j}$. Substituting in (12), we find, after straightforward derivations, that $\boldsymbol{F}_U = 2T_c(\boldsymbol{F}_U)_{a,b}$, with $a, b \in \{x_U, y_U, z_U\}$, where T_c is the time interval over which the signal is observed and

$$(\boldsymbol{F}_U)_{a,b} = \sum_{i,i'=1}^{K} \bar{\Phi}_{e,i} \left(\sum_{j=1}^{M} \frac{R_{p,j}^2}{N_{0,j}} \left(\frac{\partial}{\partial a} h_c^{(j,i)} \frac{\partial}{\partial b} h_c^{(j,i')} \right) \right)_{i,i'} \bar{\Phi}_{e,i'}. \tag{13}$$

By substituting (3) in (13), the CRB can be obtained in a straightforward way (more details regarding the derivation of the CRB-related equations can be found in the appendix in [16] and of the illuminance-related equations in [14]).

4. Numerical Results

In this section, we analyse the performance of the VLP system. Unless specified otherwise, we consider that both the area where the receiver is evaluated, i.e., the receiver area, and the ceiling area have dimensions $X_{\max} \times Y_{\max}$, with the latter located parallel and at a distance $Z_{\max}$ above the first, with $Z_{\max} = z_{L,i} = h_{j,i}$. We assume that the receiver contains $M = 8$ REs. The relative coordinates $\boldsymbol{v}_{AP,j}$ of the aperture centres with respect to the reference point $\boldsymbol{v}_U$ of the receiver are given by $\boldsymbol{v}_{AP,j} = (x_{AP,j}, y_{AP,j}, z_{AP,j})$, where $\boldsymbol{x}_{AP} = \epsilon h_A (\,-1 \;\; 0 \;\; 1 \;\; 1 \;\; 1 \;\; 0 \;\; -1 \;\; -1\,), \boldsymbol{y}_{AP} = \epsilon h_A (\,-1 \;\; -1 \;\; -1 \;\; 0 \;\; 1 \;\; 1 \;\; 1 \;\; 0)$ and $\boldsymbol{z}_{AP} = \boldsymbol{0}$, with $\epsilon = 5$ and $h_A = 1$ mm. The horizontal distance ϵh_A between the apertures, i.e., five times the vertical distance between the apertures and the PD, ensures that the only light that reaches the PD is the light coming through its own aperture. Furthermore, we assume that the coordinates $\Delta\boldsymbol{v}_{AP,j}$ of the PD centres, relative to the aperture centres, are given by $\Delta\boldsymbol{v}_{AP,j} = (d_{AP,j} \cos\alpha_{AP,j}, d_{AP,j} \sin\alpha_{AP,j}, -h_A)$, where $d_{AP,j} = 0.5R_D$, with $R_D = h_A$ and $\alpha_{AP,j} = j \cdot \pi/4$. The apertures and PD are positioned in such a way that they provide a large FOV to collect the light from a large number of different VLP-LEDs. To determine the level of the noise spectral density, we assume a background spectral irradiance $p_n = 5.8 \times 10^{-6}$ W/cm^2·nm [18] and PD responsivity $R_{p_j} = 0.4$ mA/mW for all $j = 1, \ldots, M$ [26]. The optical filter only passes visible light in the range 380 to 740 nm, resulting in an optical bandwidth of $\Delta\lambda = 360$ nm. We assume that the signals transmitted by the LEDs are confined to the bandwidth $B = 20$ MHz, and we observe the signals during an interval $T_c = 1$ ms.

We determine the average optical transmitted power and the luminous flux based on the characteristics of the LED GW CSSRMU.CM manufactured by OSRAM Opto Semiconductors Inc. [22]. This LED has a Lambertian order $m_{S,i} = 1$, which corresponds to a semi-angle $\Phi_{1/2,i} = 60°$, a typical luminous flux output $\Phi_v^{typ} = 270$ lm at forward voltage of 2.7 V and $\Phi_{v,i} \approx 300$ lm in the modulation zone. The luminous efficacy η_{LER}, which depends on the operation current and voltage, for this type of LED is $\eta_{LER} = 110$ lm/W [27]. This allows an average optical power of $\bar{\Phi}_e = \Phi_{v,i}/\eta_{LER} = 2.7$ W.

In our analysis, we investigate the impact on the VLP system from relevant aspects such as the number and position of LEDs used as VLP transmitters.

4.1. Visible Light Positioning Light-Emitting Diodes and Receiver Area

To evaluate the influence of the placement of VLP-LEDs on the positioning performance, we assume that the VLP-LEDs are attached in a grid of $K_L = K_{L,x} \times K_{L,y}$ LEDs attached to the ceiling. We assume that the receiver is located at a distance $Z_{\max} = h_{j,i} = 2$ m below the ceiling in an area of size $X_{\max} \times Y_{\max}$ m^2. Further, we assume that the xy coordinates of the centre of the grid correspond to the xy coordinates of the centre of the receiver area. We select the positions of the VLP-LEDs to be

$$\begin{aligned} x_L(\ell_x) &= \left(\tfrac{1}{K_{L,x}-1}(\ell_x - 1) - \tfrac{1}{2}\right)\rho_x\, X_{\max} \\ y_L(\ell_y) &= \left(\tfrac{1}{K_{L,y}-1}(\ell_y - 1) - \tfrac{1}{2}\right)\rho_y\, Y_{\max} \end{aligned}, \tag{14}$$

where $\ell_x = 1, \ldots, K_{L,x}$ and $\ell_y = 1, \ldots, K_{L,y}$, which corresponds to $i = (\ell_y - 1)K_{L,x} + \ell_x$, $i = 1, \ldots, K_L$.

The scaling factors ρ_x and ρ_y determine how far the grid is extended in the x direction and y direction. To ensure that all VLP-LEDs are contained in the $X_{\max} \times Y_{\max}$ m^2 area, the scaling factors ρ_x and ρ_y must satisfy $0 \leq \rho_x, \rho_y \leq 1$.

4.2. Optimal Ratios

We first analyse the impact of the choice of $K_{L,x}$ and $K_{L,y}$ on the positioning performance. To this end, we assume the total number of VLP-LEDs is constant, that is, $K_L = K_{L,x}K_{L,y} = 144$, and consider different combinations of $K_{L,x}$ and $K_{L,y}$: $K_{L,i_K} = (\mathbf{K}_{L,x}(i_K), \mathbf{K}_{L,y}(i_K))$ with $\mathbf{K}_{L,x} = [48, 36, 24, 18, 12]$ and $\mathbf{K}_{L,y} = [3, 4, 6, 8, 12]$, $i_K = 1, \ldots, 5$. We assume that the area has fixed size $X_{\max} \times Y_{\max} = 256$ m^2, but we consider rectangular areas with different x and y dimensions, i.e., $\mathcal{S}_c(i_s) = \left(\mathbf{X}_c^T(i_s), \mathbf{Y}_c^T(i_s)\right)$ with $\mathbf{X}_c = 16\left[\sqrt{\frac{47}{2}}, \sqrt{\frac{35}{3}}, \sqrt{\frac{23}{5}}, \sqrt{\frac{17}{7}},\right]$ m and $\mathbf{Y}_c = 16\left[\sqrt{\frac{2}{47}}, \sqrt{\frac{3}{35}}, \sqrt{\frac{5}{23}}, \sqrt{\frac{7}{17}},\right]$ m, $i_s = 1, \ldots, 5$.

To evaluate the positioning accuracy, we consider the spatial average of the square root of the CRB, i.e., $\overline{\text{rCRB}}$, and the normalised standard deviation $\hat{\sigma}_{\text{rCRB}} = \sigma_{\overline{\text{rCRB}}}/\overline{\text{rCRB}}$. While the first performance measure, $\overline{\text{rCRB}}$, serves to find the configuration resulting in the most accurate average positioning performance, the second performance measure, $\hat{\sigma}_{\text{rCRB}}$, is an indication of the spatial uniformity of the result. For every combination of the above-mentioned area dimensions and VLP-LED layouts, we determine the optimal values ρ_x and ρ_y that result in the minimum value of $\overline{\text{rCRB}}$. The resulting $\overline{\text{rCRB}}$ is shown in Figure 5a. Similarly, we determine the values ρ_x and ρ_y that result in the lowest $\hat{\sigma}_{\text{rCRB}}$. The resulting $\hat{\sigma}_{\text{rCRB}}$ is shown in Figure 5b.

Although the optimal values of ρ_x and ρ_y differ for the two figures, because they optimise different objective functions, we can observe in Figure 5a,b that, for a given area dimension $\mathbf{X}_c(i_s) \times \mathbf{Y}_c(i_s)$, the VLP-LED layout with

$$\frac{\mathbf{X}_c(i_s)}{\mathbf{Y}_c(i_s)} = \frac{\mathbf{K}_{L,x}(i_K) - 1}{\mathbf{K}_{L,y}(i_K) - 1} \tag{15}$$

results in the smallest $\overline{\text{rCRB}}$ and $\hat{\sigma}_{\text{rCRB}}$. In other words, ignoring the presence of the scaling factors ρ_x and ρ_y, this corresponds to an equal spacing between the VLP-LEDs in the x and y direction, respectively: $\frac{X_{\max}}{K_{L,x}-1} = \frac{Y_{\max}}{K_{L,y}-1}$. This brings us to a first rule of thumb. Equation (15) shows us that *the number of VLP-LEDs in the x and y direction should follow the dimensions of the receiver area.*

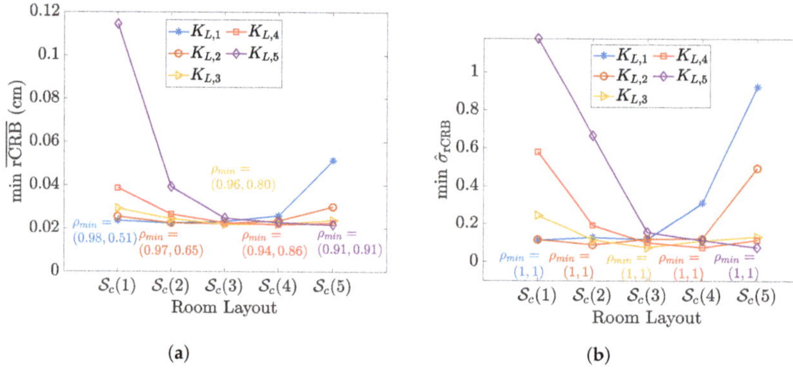

Figure 5. Minimum (over (ρ_x, ρ_y)) of (**a**) $\overline{\text{rCRB}}$ and (**b**) $\hat{\sigma}_{\text{rCRB}}$ for different combinations of VLP-LED layout and receiver area dimensions for a fixed number $K = 144$ of VLP-LEDs (with $m_{S,i} = 1$) and a fixed area of $X_{\text{max}} \times Y_{\text{max}} = 256\ \text{m}^2$.

4.3. Optimal Visible Light Positioning Light-Emitting Diode Spacing

In the previous discussion, we determined, for a given number of VLP-LEDs, the number of VLP-LEDs in each direction, i.e., $K_{L,x}$ and $K_{L,y}$. Now, we will further investigate the optimal spacing between VLP-LEDs, where the distance between VLP-LEDs in the x direction, $\delta_{L,x}$, and in the y direction, $\delta_{L,y}$, are derived from (14), resulting in $\delta_{L,x} = \frac{\rho_x X_{\text{max}}}{K_{L,x}-1}$ and $\delta_{L,y} = \frac{\rho_y Y_{\text{max}}}{K_{L,y}-1}$. To this end, we first restrict our attention to the case where the receiver area is square and the VLP-LEDs are placed in a square grid: $\rho_x = \rho_y = \rho$, $\delta_{L,x} = \delta_{L,y} = \delta_L$, $X_{\text{max}} = Y_{\text{max}} = \chi_{\text{max}}$ and $K_{L,x} = K_{L,y} = K = \sqrt{K_L}$. We will generalise later in this section the results to the rectangular layout. We consider three receiver areas with different sizes, i.e., with $\chi_{\text{max}} = 1, 4$ and 10 m. The scaling factor ρ with $\rho \in [0, 1]$, influencing the distance between neighbouring VLP-LEDs, will be optimised.

In this optimisation, we use as an objective function the spatial average $\overline{\text{rCRB}}$ over the $\chi_{\text{max}} \times \chi_{\text{max}}\ \text{m}^2$ area of a receiver placed at a vertical distance of $h_{j,i} = 2$ m below the ceiling. First, we plot $\overline{\text{rCRB}}$ as a function of the scaling factor ρ for different values of K_L. As expected, the results in Figure 6a show that increasing the number $K_L = K^2$ of VLP-LEDs leads to a lower rCRB, and thus a better positioning accuracy, as more information is available to estimate the user's position. Further, Figure 6a reveals that the positioning accuracy first improves when ρ increases but deteriorates after reaching a minimum. This can be explained as follows. When ρ is small, all VLP-LEDs are concentrated above the centre of the receiver area, implying that the centre is well lit, but the edges of the area receive less light due to distance-dependent attenuation and relatively large radiation and incident angle. Hence, although the position of a receiver can be determined accurately if it is located in the centre of the receiver area, the positioning accuracy will be strongly reduced when the receiver is near the boundary of the area because of the larger distance and angle between the receiver and the VLP-LEDs. Obviously, this will also have a negative impact on the uniformity $\hat{\sigma}_{\text{rCRB}}$ of the positioning performance, see Figure 6b. By increasing ρ, the positioning accuracy near the edges will improve due to the better channel response for the closest VLP-LEDs, improving both the average positioning accuracy and the uniformity. However, when the distance becomes too large, the VLP-LEDs furthest away from the receiver start to drop out of the FOV of the receiver, implying that less information is available to determine the position, resulting in a degradation of the accuracy. The effect of ρ on the position accuracy is greater when χ_{max} increases as the incident and radiation angles, as well as the distance between the VLP-LEDs and the receiver, increase when the area becomes larger. In Figure 6, we can observe that, in most cases, the optimal value of the spacing ρ for $\overline{\text{rCRB}}$ corresponds more or less to the optimal spacing for $\hat{\sigma}_{\text{rCRB}}$, and, in the cases where there is a discrepancy between the optimal spacing for optimising $\overline{\text{rCRB}}$ and $\hat{\sigma}_{\text{rCRB}}$, the normalised standard deviation $\hat{\sigma}_{\text{rCRB}}$ is relatively small. Therefore, in the remainder of this section, we concentrate on the optimisation of $\overline{\text{rCRB}}$.

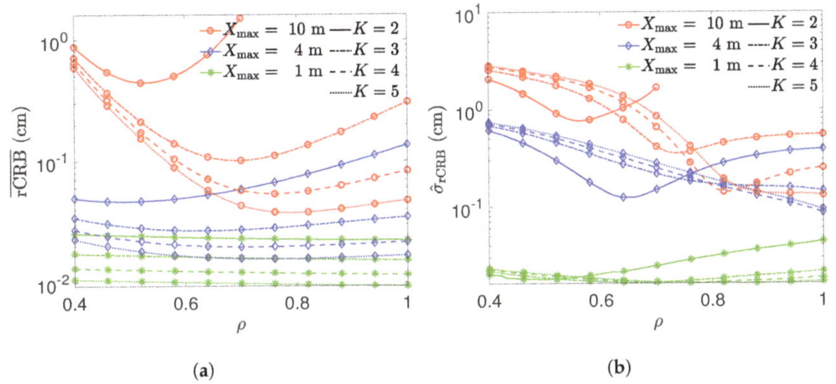

Figure 6. (**a**) $\overline{\text{rCRB}}$ and (**b**) $\overline{\text{rCRB}}$ uniformity, $\hat{\sigma}_{\text{rCRB}}$, for K^2 VLP-LEDs with $m_S = 1$ placed in a square grid with side $\rho\chi_{\text{max}}$.

Let us look more closely at the optimal scaling factor ρ and the corresponding optimal distance between the VLP-LEDs. In Figure 7a, we determine the optimal scaling factor ρ that minimises $\overline{\text{rCRB}}$ as a function of χ_{max}. We observe that, when $\chi_{\text{max}} < 2$ m, the scaling factor $\rho \approx 1$, i.e., the VLP-LEDs must be placed near the edges of the area. Due to the small area, the distance between the receiver and the VLP-LEDs is limited, as well as the radiation and incident angle, implying that the channel gain will be good for all positions of VLP-LEDs and the receiver. As the Fisher information matrix becomes closer to singular when ρ is small (as the measurements from the different VLP-LEDs will be correlated strongly because the VLP-LEDs are approximately co-located), the VLP-LEDs must be placed apart as far as possible. On the other hand, when $\chi_{\text{max}} \gg 2$ m, we observe that ρ becomes approximately independent of χ_{max}, and the optimal distance increases with the increasing number of VLP-LEDs. In Figure 7b, we show the ratio between the $\frac{\delta_L}{\chi_{\text{max}}}$ for the cases considered in Figure 7a, and we compare the results with the level $\frac{1}{K}$. From these results, we find that the optimal spacing $\delta_L^{opt}(K)$ can be approximated by

$$\delta_L^{opt}(K) \approx \frac{\chi_{\text{max}}}{K} \triangleq \tilde{\delta}_L^{opt}(K), \tag{16}$$

which corresponds to

$$\tilde{\rho}_{opt} = \frac{K-1}{K}. \tag{17}$$

To evaluate the positioning performance for this approximation of the optimal spacing, we compare the $\overline{\text{rCRB}}$ for the spacing $\tilde{\delta}_L^{opt}(K)$ (16) with the performance for the true optimal spacing $\delta_L^{opt}(K)$ from Figure 7. The results, shown in Figure 8, reveal that the approximated spacing (16) provides a close-to-optimal positioning performance, even for small values of χ_{max}. This can be explained by evaluating Figure 6a; for small areas, $\overline{\text{rCRB}}$ is essentially independent of ρ, implying that, for the spacing (16), the resulting $\overline{\text{rCRB}}$ will be close to the minimum.

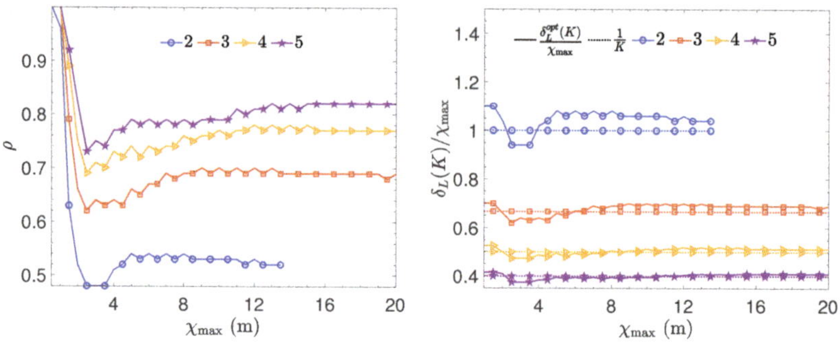

(**a**) (**b**)

Figure 7. Optimal VLP-LED spacing minimizing $\overline{\text{rCRB}}$ for $K_L \in \{2, 3, 4, 5\}$. (**a**) optimal scaling factor $\rho_{\min}$, (**b**) normalised optimal spacing $\delta_L^{opt}(K)/\chi_{\max}$ between VLP-LEDs and the approximation $1/K$. For all the VLP-LEDs, $\Phi_{v,i} = 57.6$ lm and $\eta_{LER} = 423$ lm/W are used.

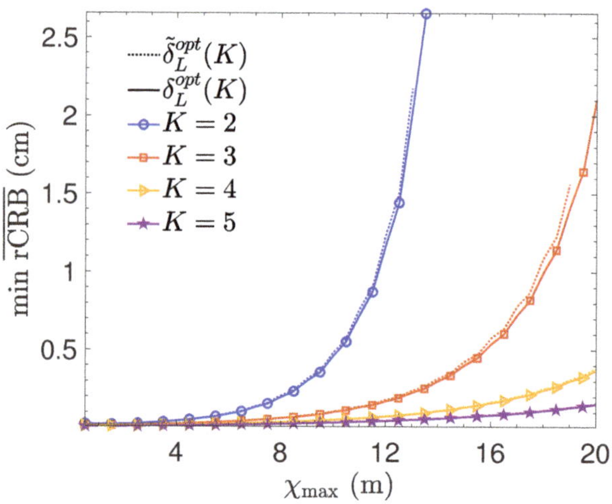

Figure 8. Performance of the approximation $\tilde{\delta}_L^{opt}(K)$ in order to obtain the minimum value of $\overline{\text{rCRB}}$ considering different $\chi_{\max}$.

In the above analysis, we restricted our attention to the case of a square receiver area. We now extend the results to the rectangular areas discussed earlier in this section. Using Equation (17), we extrapolate the approximation to

$$\begin{aligned} \tilde{\rho}_x^{opt} &\approx \frac{K_{L,x} - 1}{K_{L,x}} \\ \tilde{\rho}_y^{opt} &\approx \frac{K_{L,y} - 1}{K_{L,y}} \end{aligned} \tag{18}$$

and

$$\frac{\delta_{L,x}^{opt}}{\delta_{L,y}^{opt}} = \frac{\frac{\rho_x^{opt} X_{\max}}{K_{L,x}-1}}{\frac{\rho_y^{opt} Y_{\max}}{K_{L,y}-1}} \approx \frac{\frac{\tilde{\rho}_x^{opt} X_{\max}}{K_{L,x}-1}}{\frac{\tilde{\rho}_y^{opt} Y_{\max}}{K_{L,y}-1}} \approx \frac{C_S}{C_L}, \tag{19}$$

where $C_S = \frac{X_{\max}}{Y_{\max}}$ and $C_L = \frac{K_{L,x}}{K_{L,y}}$. Comparing the approximation (18) with the optimal values of (ρ_x, ρ_y) for the different scenarios considered in Figure 5a, we notice that $\tilde{\rho}_x^{opt}$ closely approximates all the values obtained for $X_{\max}$. However, for the y dimension, a larger deviation occurs between the optimal scaling factor ρ_y and the approximation $\tilde{\rho}_y^{opt}$ for areas with a small $Y_{\max}$. However, this does not imply that the approximation is not useful, as, similarly to the square area, when $\chi_{\max}$, $X_{\max}$ or $Y_{\max}$ is small, the $\overline{\text{rCRB}}$ becomes largely independent of ρ, ρ_x or ρ_y, respectively. Hence, *the approximations* (17) *and* (18) *can be used as a general rule of thumb to determine the optimal spacing between the VLP-LEDs.*

4.4. Average Accuracy Level

Now that the optimal placement of the VLP-LEDs has been determined, we are interested in finding the minimum number of VLP-LEDs required to obtain a given average positioning accuracy. To this end, we show in Figure 9 the positioning accuracy that can be obtained with $K_L = K^2$ VLP-LEDs in square areas with dimensions $\chi_{\max} \times \chi_{\max}$ for different Lambertian orders and transmitted optical power. From these figures, we can determine the minimum number of VLP-LEDs that results in $\overline{\text{rCRB}} < \zeta$ for areas up to 30×30 m^2. The figures reveal a linear relationship between the minimum K to obtain $\overline{\text{rCRB}} < \zeta$ and $\chi_{\max}$. This linear relationship can be expressed as

$$K \geq \mathcal{S}(m_{S,i}, h_{j,i}, \zeta, \bar{\Phi}_e)\chi_{\max} + \mathcal{B}(m_{S,i}, h_{j,i}, \zeta, \bar{\Phi}_e), \tag{20}$$

where the slope $\mathcal{S}$ and bias $\mathcal{B}$ depend on the system parameters $m_{S,i}$, $h_{j,i}$ and $\bar{\Phi}_e$ and on the wanted accuracy level ζ. Our simulations reveal that the bias $\mathcal{B}$ is very close to zero for all considered scenarios, implying that we can simplify (20) to

$$K \geq \mathcal{S}(m_{S,i}, h_{j,i}, \zeta, \bar{\Phi}_e)\chi_{\max}. \tag{21}$$

This behaviour is expected because the positioning performance is mainly determined by the distance between the receiver and the nearest VLP-LEDs, indicating that, if the size of the area increases, the number of VLP-LEDs should increase accordingly to keep this distance between the nearest VLP-LEDs and the receiver constant. Combining this linear behaviour between $\chi_{\max}$ and K with (16), it follows that the optimal spacing δ_L^{opt} is constant for a given accuracy level and equal to the inverse of this slope. In the following, we investigate the dependency of the slope $\mathcal{S}$ on the system parameters. Let us first evaluate the slopes for the cases shown in Figure 9. The slopes of the curves and the resulting optimal spacing are listed in Table 2.

Table 2. Definition of the ranges and the slopes and the resulting δ_L^{opt} for Figure 9.

Legend	**Average Accuracy Range (in cm)**	**$\bar{\Phi}_e = 2.7$ W, $m_{S,i} = 1$, $h_{j,i} = 2$ m**		**$\bar{\Phi}_e = 1.3$ W, $m_{S,i} = 1$, $h_{j,i} = 2$ m**		**$\bar{\Phi}_e = 2.7$ W, $m_{S,i} = 3$, $h_{j,i} = 2$ m**	
		Slope	**δ_L^{opt} (m)**	**Slope**	**δ_L^{opt} (m)**	**Slope**	**δ_L^{opt} (m)**
0.025 cm	$0 < \overline{\text{rCRB}} \leq 0.025$	0.65	1.5	1.25	0.8	0.46	2.2
0.05 cm	$0.025 < \overline{\text{rCRB}} \leq 0.05$	0.40	2.5	0.65	1.5	0.34	2.9
0.1 cm	$0.05 < \overline{\text{rCRB}} \leq 0.1$	0.29	3.4	0.40	2.5	0.28	3.5
0.5 cm	$0.1 < \overline{\text{rCRB}} \leq 0.5$	0.19	5.2	0.26	3.8	0.20	5

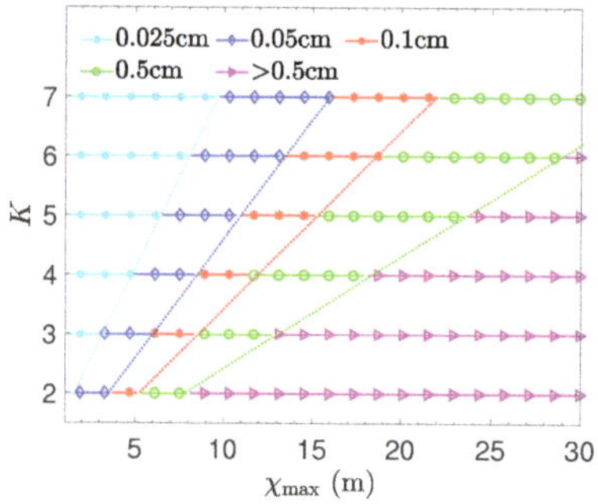

(a)

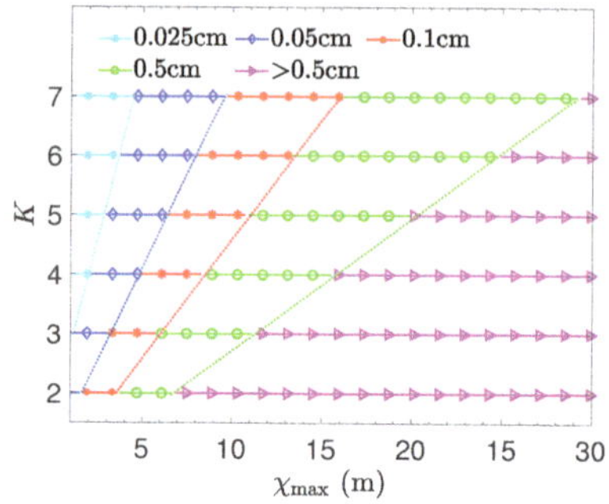

(b)

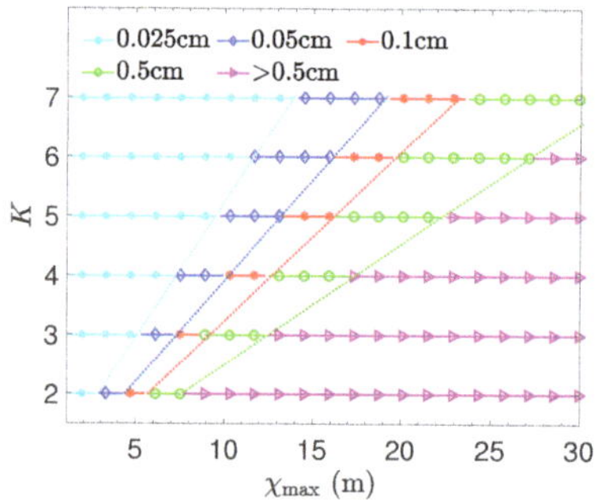

(c)

Figure 9. Minimum K required to obtain $\overline{\text{rCRB}} = \zeta$ cm as function of χ_{max}, with $\zeta \in \{0.025, 0.05, 0.1, 0.5, > 0.5\}$ cm for **(a)** $m_{S,i} = 1$, $\bar{\Phi}_e = 2.7$ W and $h_{j,i} = 2$ m, **(b)** $m_{S,i} = 1$, $\bar{\Phi}_e = 2.7/2$ W and $h_{j,i} = 2$ m and **(c)** $m_{S,i} = 3$, $\bar{\Phi}_e = 2.7$ W and $h_{j,i} = 2$ m.

Comparing Figure 9a ($m_{S,i} = 1$ and $\bar{\Phi}_e = 2.7$ W) with Figure 9b ($m_{S,i} = 1$ and $\bar{\Phi}_e = 2.7/2$ W) and the slopes of the curves, given in Table 2, we observe that, if we halve the transmitted optical power, this results in the curve $\overline{\text{rCRB}} = 0.05$ cm in Figure 9a being equivalent to the $\overline{\text{rCRB}} = 0.1$ cm in Figure 9b. In general, if we increase the transmitted optical power with a factor β, the side χ_{max} of the square area may increase with the same factor β to obtain the same average positioning accuracy (of course, this is provided that the scaling factor β is small enough that the VLP-LEDs do not disappear out of the field of view of the receiver and the receiver does not fall out of the field of view of the VLP-LEDs) with the same number of VLP-LEDs. The explanation can be found by verifying Equations (11) and (13); the rCRB is proportional to $1/\bar{\Phi}_e$, so increasing the optical power $\bar{\Phi}_e$ will reduce the rCRB with the same factor. In other words, the slope $\mathcal{S}$ in (20) will be a function of the product $\zeta\bar{\Phi}_e$.

In Figure 9c ($m_{S,i} = 3$ and $\bar{\Phi}_e = 2.7$ W), we change the Lambertian order so that $m_{S,i} + 1$ is doubled compared to Figure 9a ($m_{S,i} = 1$ and $\bar{\Phi}_e = 2.7$ W). The channel gain (3) depends on the Lambertian order through the prefactor $m_{S,i} + 1$, but also through the exponent of the cosine of the incident angle $\phi_{j,i}$. This incident angle depends on the position of the receiver, implying that the dependency of the rCRB on the Lambertian order is more complex than on the optical power. However, it is expected that, when this incident angle is sufficiently small, we can approximate $\cos\phi_{j,i} \approx 1$. In that case, the channel gain will depend on $m_{S,i}$ through the prefactor $m_{S,i} + 1$ only, implying that the rCRB will behave approximately proportionally to $1/(m_{S,i} + 1)$. This will be the case when the height $h_{j,i}$ is sufficiently large as, when the height increases, the incidence angle will become smaller. In Figure 9c, the rCRB only roughly reduces by factor $m_{S,i}+1$, indicating that the incident angle is not small enough for this approximation to hold. In general, the slope $\mathcal{S}$ in (20) will be a complicated function of the Lambertian order $m_{S,i}$.

In Figure 9a–c, we keep the same height of VLP-LEDs, i.e., $h_{j,i} = 2$ m. Changing this height will have an impact on $\overline{\text{rCRB}}$ that is more complex than when changing the optical power or Lambertian order as this height is directly related to the z coordinate z_U of the receiver position, that is, $h_{j,i} = z_{L,i} - z_U + h_A$. This can be observed in Figure 10, where we show the impact of the receiver height $h_{j,i}$ on the $\overline{\text{rCRB}}$ for $K^2 = 49$ VLP-LEDs with Lambertian order $m_{S,i} \in \{1,2,3\}$ and optical power $\Phi_e \in \{2.7, 1.4\}$ W and for two different receiver areas $\chi_{\max} \in \{8,16\}$ m. As predicted by our analysis of Figure 9a–c, we observe in Figure 10 that, for a given $\chi_{\max}$, the blue and red curves are parallel; doubling the transmitted optical power results in a reduction in the rCRB by a factor of two. In addition, we stated that the rCRB is approximately proportional to $1/(m_{S,i}+1)$ when the incident angle is small. Hence, for a large $h_{j,i}$, we expect that the yellow curve will become parallel to the blue curve. This effect indeed can be observed in the figure. This is further analysed in detail in Figure 11, where we show the ratio

$$\varrho(m_{S,1}, m_{S,2}) = \frac{\overline{\text{rCRB}}\big|_{m_{S,1}}}{\overline{\text{rCRB}}\big|_{m_{S,2}}}. \tag{22}$$

as a function of height for $m_{S,i} \in \{1,2,3\}$ and $\chi_{\max} \in \{8,16\}$ m. It can be observed that the curves converge to a constant for a sufficiently large $h_{j,i}$. This constant is slightly larger than $\frac{m_{S,2}+1}{m_{S,1}+1}$ as $\cos^{m_{S,2}+1}\phi_{j,i} \leq \cos^{m_{S,1}+1}\phi_{j,i}$ for $m_{S,2} > m_{S,1}$, implying that the overestimation of the channel gain (underestimation of the rCRB) when approximating $\cos\phi_{j,i} \approx 1$ is larger for $m_{S,2}$ than for $m_{S,1}$. For a given $\chi_{\max}$, the height $h^{\dagger}_{j,i}$ at which $\varrho(m_{S,1}, m_{S,2})$ converges is approximately independent of the value $m_{S,i}$ of the Lambertian order. Comparing the height $h^{\dagger}_{j,i}$ for $\chi_{\max} = 8$ m and $\chi_{\max} = 16$ m, we find that the height $h^{\dagger}_{j,i}$ for $\chi_{\max} = 16$ m is twice the height $h^{\dagger}_{j,i}$ for $\chi_{\max} = 8$ m, that is, the height of convergence scales with $\chi_{\max}$. To explain this, let us consider the following scenario. Consider the four LEDs surrounding the receiver and assume that the receiver is located in the centre of these four LEDs. Further, suppose that the LEDs are optimally spaced, that is, their spacing is $\delta_L^{opt}(K) = \frac{\chi_{\max}}{K}$ (16). Then, the incidence angle $\phi_{j,i}$ is for the four considered LEDs $h_{j,i}\tan\phi_{j,i} = \frac{\chi_{\max}}{\sqrt{2}K}$. Inserting $h^{\dagger}_{j,i} = 3.75$ m for $\chi_{\max} = 8$ m ($h^{\dagger}_{j,i} = 7.5$ m for $\chi_{\max} = 16$ m), the resulting incidence angle for $K = 7$ approximately equals $\phi^{\dagger}_{j,i} = 12^\circ$. We evaluated the convergence height for various values of K and $\chi_{\max}$ and found, for all cases, that convergence occurs when the incidence angle is in the interval $\phi^{\dagger}_{j,i} \in [11^\circ, 14^\circ]$. This corresponds to a height $h^{\dagger}_{j,i} = \alpha_h \frac{\chi_{\max}}{K}$, where $\alpha_h \in [2.8, 3.6]$, i.e., when the height is greater than α_h times the optimal spacing $\frac{\chi_{\max}}{K}$ between the VLP-LEDs, the rCRB will be approximately proportional to $1/(m_{S,i}+1)$.

Until now, we have concentrated on the relative differences between the curves in Figure 10. Now, we will discuss the behaviour of the $\overline{\text{rCRB}}$ as a function of height $h_{j,i}$. The figure shows that, for a small $h_{j,i}$, the $\overline{\text{rCRB}}$ first decreases as a function of $h_{j,i}$, and, after reaching a minimum, it increases again. This is explained as follows. For a small $h_{j,i}$, the receiver sees only a limited number of LEDs because it falls out of the FOV of most LEDs. By increasing the height, the receiver will gradually see more LEDs, so more information will be available to determine its position, resulting in an improvement of the positioning accuracy. However, at some point, the additional information that is obtained by detecting more LEDs will be outweighed by the reduction in the channel gain because the receiver is further away from the LEDs. As the signal-to-noise ratio reduces as a result of the smaller channel gain, the position accuracy will degrade. For a large $h_{j,i}$, the rCRB increases approximately proportionally to $h^2_{j,i}$. This follows from Equation (3); when $h_{j,i}$

is large, $\cos\phi_{j,1} \approx 1$, and, for the nearest LEDs, $d_{j,i} \approx h_{j,i}$, implying that the channel gain approximately is proportional to $1/h_{j,i}^2$, which leads to the rCRB being proportional to $h_{j,i}^2$.

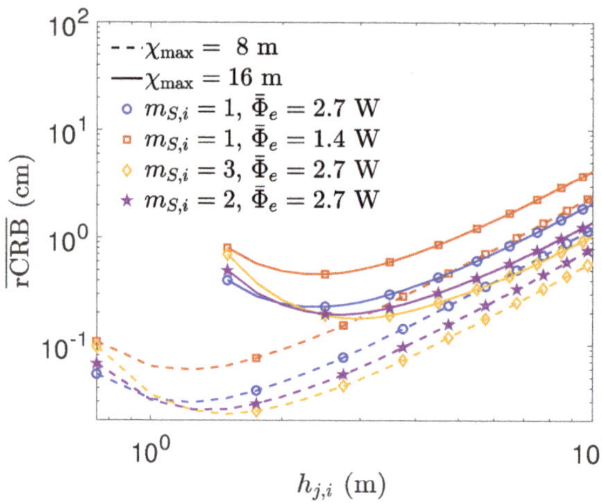

Figure 10. Impact of the vertical distance $h_{j,i}$ between VLP-LEDs and the receiver on $\overline{\text{rCRB}}$ for $K^2 = 49$ VLP-LEDs with Lambertian order $m_{S,i} \in \{1,3\}$ and transmitted optical power $\bar{\Phi}_e \in \{1.4, 2.7\}$ W for two receiver area sizes, i.e., $\chi_{\max} \in \{8, 16\}$ m.

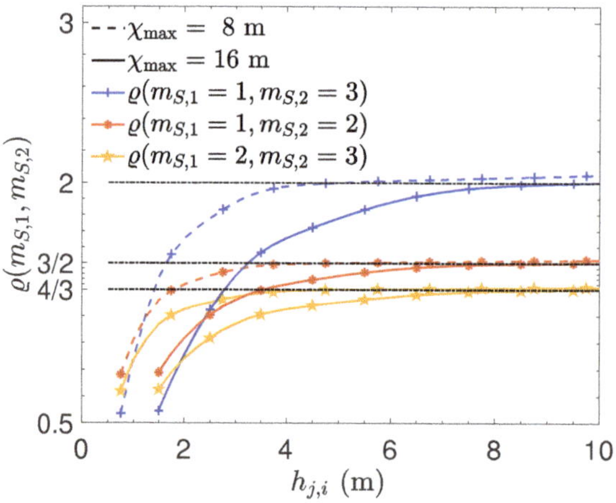

Figure 11. The ratio $\varrho(m_{S,1}, m_{S,2})$ (22) for $K = 7$, $\bar{\Phi}_e = 2.7$ W, $\chi_{\max} \in \{8, 16\}$ m and $m_{S,i} \in \{1, 2, 3\}$.

In conclusion, the rCRB scales with the transmitted optical power $\bar{\Phi}_e$, and, when the incidence angle (for the four closest LEDs) is sufficiently small, the rCRB is proportional to $1/(m_{S,i}+1)$ and $h_{j,i}^2$. This will occur when $h_{j,i} \geq h_{j,i}^{\dagger}$, where $h_{j,i}^{\dagger} = \alpha_h \frac{\chi_{\max}}{K}$ with $\alpha_h \in [2.8, 3.6]$ and $\frac{\chi_{\max}}{K}$ is the optimal spacing between the LEDs.

Although the above analysis gives us insight into the behaviour of the rCRB as a function of the system parameters, we are not able to find an accurate closed-form expression for the slope $\mathcal{S}$ in (21) as a function of the system parameters $\bar{\Phi}_e$, $m_{S,i}$ and $h_{j,i}$ and accuracy level ζ. Therefore, we will derive a lower and upper bound on the minimum number K of VLP-LEDs.

To find a lower bound $K_{\min}$ in K, i.e., $K \geq K_{\min}$, we analyse the FOV of the receiver. To estimate the receiver position, at least one VLP-LED must be in this FOV, otherwise the Fisher information matrix will become singular. Let us focus on the receiver position furthest from the centre of the area, i.e., in a corner, and consider the VLP-LED closest to the receiver. This VLP-LED will be in the FOV when the incident angle $\phi_{j,i}$ is smaller than the FOV angle Φ_{rec}, i.e.,

$$\tan(\Phi_{\text{rec}}) > \tan(\phi_{j,i}). \tag{23}$$

Assuming the LEDs are optimally spaced, i.e., $\delta_L^{opt} = \frac{\chi_{\text{max}}}{K}$, the horizontal distance between the LED and the receiver in this scenario equals $\Delta = \sqrt{2}\frac{\chi_{\text{max}}}{2K}$, which is related to the incident angle through $\Delta = h_{j,i}\tan\phi_{j,i}$, leading to the lower bound

$$K_{\text{min}} = \frac{\sqrt{2}}{4\tan(\Phi_{\text{rec}})h_{j,i}}\chi_{\text{max}} \triangleq \mathcal{S}_{\text{min}}\chi_{\text{max}}. \tag{24}$$

For the receiver structure specified at the beginning of the numerical results section, $\tan(\Phi_{\text{rec}}) = 2.5$. The slope $\mathcal{S}_{\text{min}}$ for this lower bound is independent of m_S and $\zeta\bar{\Phi}_e$. As (24) is a strict lower bound on K, which implies that the slope $\mathcal{S}_{\text{min}}$ is a lower bound on the true slope $\mathcal{S}$, it is expected that the lower bound will become less tight when χ_{max} increases. This is observed in Figure 12, which shows a histogram where, as a function of K, the difference $\Delta K_{\text{min}} = K - K_{\text{min}}$ is depicted for 1000 scenarios, where we randomly choose for each parameter in Table 3 a value within the range specified in the table. As can be observed, when $K = 1$, corresponding to a small χ_{max}, the lower bound is tight, that is, $\Delta K_{\text{min}} = 0$ for all scenarios. The difference ΔK_{min} increases for increasing K and, thus, χ_{max}. For $K > 2$, the difference $\Delta K_{\text{min}} \geq 1$ for all scenarios.

As the above lower bound becomes less tight for a larger χ_{max}, we designed an approximation for the minimum number of VLP-LEDs in an empirical way. To this end, we added to the lower bound (24), a factor that depends on the parameters m_S, $h_{j,i}$ and $\zeta\bar{\Phi}_e$, and tuned the factor based on the simulations. This led to the following approximation:

$$K_{\text{max}} = \left\lceil \frac{1}{4\tan(\Phi_{\text{rec}})}\sqrt{\frac{2}{h_{j,i}}}\left(\frac{\zeta\bar{\Phi}_e}{0.175\zeta\bar{\Phi}_e - 0.12}\right)^{0.3} \right\rceil \chi_{\text{max}}. \tag{25}$$

The difference $\Delta K_{\text{max}} = K_{\text{max}} - K$ between this approximation K_{max} and the true minimum value K is also visualised in Figure 12 for the 1000 simulated scenarios. As can be observed in the figure, for all scenarios, $\Delta K_{\text{max}} = 0$ or $\Delta K_{\text{max}} = 1$, implying that the approximation can serve as an upper bound on the true minimum K, i.e., $K_{\text{min}} \leq K \leq K_{\text{max}}$. We also observe in the histogram that the upper bound (25) becomes more tight when K increases. This will occur when χ_{max} increases (for a given $h_{j,i}$) or $h_{j,i}$ reduces (for a given χ_{max}).

Table 3. Parameter ranges for the simulations of the lower and upper bound (24) and (25).

	Min.	Max.	Units
$h_{j,i}$	1.5	4	m
χ_{max}	5	25	m
FoV	40	70	Deg
$m_{S,i}$	0.65	4.82	
$\bar{\Phi}_e$	1	5	W
ζ	1	10	cm

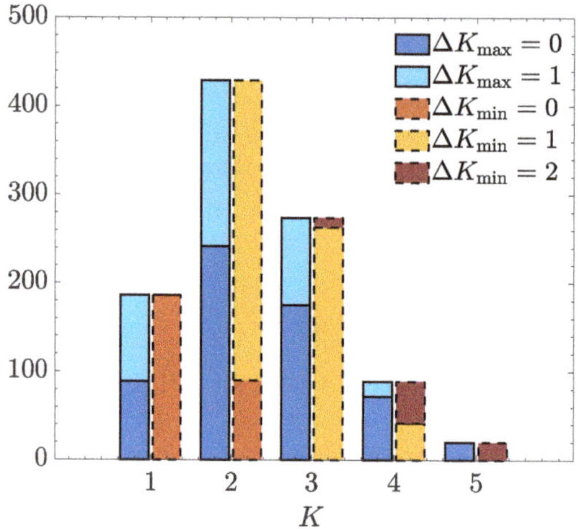

Figure 12. Tightness analysis of the lower and upper bound (24) and (25).

In this analysis, we restricted our attention to square areas. However, the results can be extended to rectangular areas where $C_S > 1$. We found that the lower and upper bounds (24) and (25) can also be used to find a lower and upper bound on $K_{L,x}$ and $K_{L,y}$ for the rectangular case by replacing in (24) and (25) $\chi_{\max}$ with $X_{\max}$ and $Y_{\max}$, respectively, for the ranges indicated in Table 3 provided that $\frac{C_S}{C_L} \approx 1$ and $C_S \leq 2$, i.e., $X_{\max} \leq 2Y_{\max}$.

4.5. Illumination- versus Visible-Light-Positioning-Based Optimisation

Until now, we have focused on the placement of the VLP-LEDs in order to optimise the positioning performance without considering the primary purpose of the LEDs, i.e., illumination. In [14], we analysed the placement of LEDs to satisfy the constraints on the horizontal illuminance and uniformity of illumination according to the DIN EN 12464-1 standard [28]. Assuming that the LEDs are grouped into luminaries, in [14], we showed that the average illuminance depends mainly on the total number of LEDs contained in the luminaries, while the uniformity of the illuminance depends primarily on the placement of the luminaries. That analysis also revealed that the number and spacing of the luminaries to meet the required illuminance and uniformity of illumination depend in a complex manner on the system parameters such as the size of the room, the layout of luminaries to be used and the desired level of lighting uniformity in the lit area.

In this section, we will use the results of [14] to evaluate the feasibility of combining illumination with VLP on the same infrastructure. To this end, we assume that the illumination LEDs are grouped in luminaries and distributed on the ceiling in a rectangular grid, similarly to in [14]. Further, we assume that each luminary contains one VLP-LED placed in the centre of the luminary. Taking into account that the average illuminance depends mainly on the number of LEDs and is essentially independent of the placement of the LEDs, in our analysis, we focus on the illuminance uniformity.

We first compare the results of the optimal luminary placement obtained from [14] with the optimal VLP-LED spacing from this paper. To this end, we determine the spacing between the luminaries in order to maintain the illumination uniformity levels as prescribed in the standard [28]. Similarly to in Section 4.3, this spacing is expressed in terms of the fraction $\rho_{x,I} \in [0,1]$ of $X_{\max}$. As the standard defines a range for the illumination level and uniformity, the value of $\rho_{x,I}$ for which the illumination conditions are satisfied will also belong to a range, i.e., $[\rho_{x,I}^{\min}, \rho_{x,I}^{\max}]$. In Figure 13, we show this range of $\rho_{x,I}$ as a function of $X_{\max}$ for different values of $K_{L,x}$ and $C_S = \frac{X_{\max}}{Y_{\max}}$, assuming $K_{L,y} = 5$. For the parameters used in the figure, the upper bound is equal to $\rho_{x,I}^{\max} = 1$ for all values of $X_{\max}$, while the lower bound $\rho_{x,I}^{\min}$ varies as a function of $X_{\max}$. In this figure, we also show the optimal spacing for the VLP-LEDs in terms of the fraction $\rho_{x,V}^{opt}$ of $X_{\max}$. As can be observed in the figure, the optimal value $\rho_{x,V}^{opt}$ for the spacing of the VLP-LEDs lies below the curve of $\rho_{x,I}^{\min}$. In other words, the minimum spacing between the luminaries according to the illumination constraints is larger than the optimal spacing between the LEDs for VLP. Hence, it is not possible to satisfy the illumination constraints and obtain optimal positioning accuracy at the same time. This implies that we need to find a compromise between the illumination and positioning constraints.

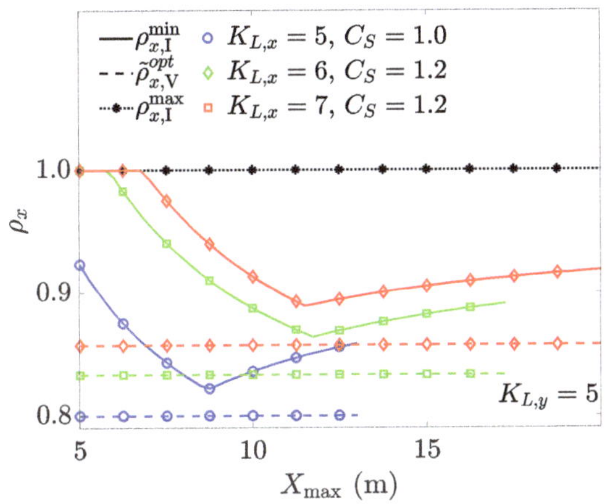

Figure 13. Visualisation of the optimal ρ_x for both the illumination and the VLP approach for different $K_{L,x}$ values, assuming $K_{L,y} = 5$.

For this reason, we evaluate how the spacing resulting from the illumination conditions degrades the positioning performance, and how the spacing obtained from the VLP constraints analysed in this paper impacts the illumination levels and uniformity. In Figure 14a, we show the rCRB for the optimal spacing for VLP, i.e., corresponding to the fraction $\rho_{x,V}^{opt}$, and for the lower bound $\rho_{x,I}^{\min}$ as, for all spacings satisfying the illumination constraints, this spacing is closest to the optimal spacing for VLP, implying that the degradation of the rRCB will be the smallest. As can be seen in the figure, the degradation is negligibly small for areas up to $X_{\max} = 15$ m. For $X_{\max} > 15$ m, the degradation increases but is still small. On the other hand, in Figure 14b,c, we show the uniformity for the task area (the task area is defined as the central part of the area where the illuminance uniformity should be higher, i.e., more uniform, than in the surrounding area, which is defined as the part of the area near its boundaries. For the task area, this uniformity should be $\mathcal{U}_t \geq 0.7$, and, for the surrounding area, $\mathcal{U}_s \geq 0.5$) $\mathcal{U}_t$ and the surrounding area $\mathcal{U}_s$ for different values of $K_{L,x}$ and C_S, assuming $K_{L,y} = 5$, $m_{S,i} = 1$ and $\xi^t = 0.8$ (ξ^t defines the central sub-area of the area where the receiver is assessed, and $\xi^t = 0.8$ means that 80% is taken in both the x and y directions). As can be observed in Figure 14b, for the task area, the uniformity for $\rho_{x,I}^{\min}$ exceeds the required level of uniformity for all considered area sizes (except for $K_{L,x} = 5$, where it is only met for $X_{\max} < 18$). For the optimum spacing for VLP, the figure shows that, for a larger $X_{\max}$, the uniformity level for $\rho_{x,V}^{opt}$ lies above that for $\rho_{x,I}^{\min}$. Hence, we still meet the uniformity condition. However, for smaller area sizes, i.e., $X_{\max} < 9$–11 m, we cannot satisfy the uniformity requirement for the task area when using the optimal spacing for VLP. On the other hand, as can be seen in Figure 14c, in the surrounding area, the minimum uniformity level for $\rho_{x,I}^{\min}$ is reached for all area sizes, but this is not the case for the optimum spacing $\rho_{x,V}^{opt}$ for VLP. Here, only for a smaller $X_{\max}$, that is, $X_{\max} < 8$–9 m, is the uniformity sufficiently large. Hence, when the optimal spacing for VLP is used, these results show that it is not possible to satisfy the illumination constraints for both the task and surrounding area.

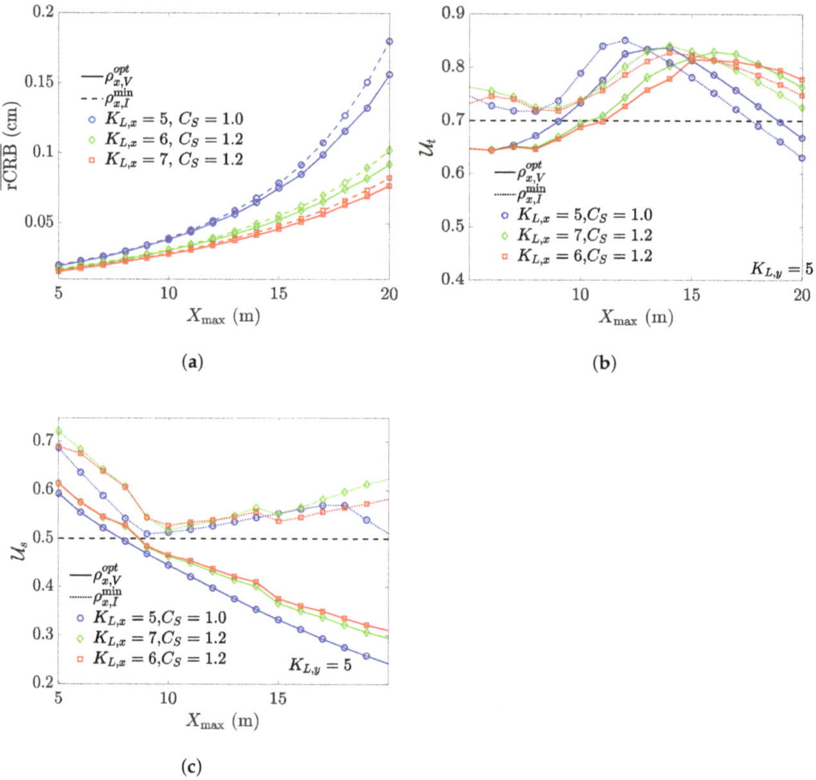

Figure 14. Performance comparison of optimal LED spacing for illumination and positioning: (**a**) $\overline{\mathrm{rCRB}}$, (**b**) uniformity $\mathcal{U}_t$ in the task area, (**c**) uniformity $\mathcal{U}_s$ in the surrounding area for different $K_{L,x}$, C_S, $\mathcal{U}_t = 0.7$, $\xi^t = 0.8$, $K_{L,y} = 5$ and $m_{S,i} = 1$.

To summarise this subsection, our results show that the optimal spacing for the VLP does not conform to the spacing range in which the illumination constraints are satisfied, implying that a compromise needs to be established between VLP and illumination. Based on our results, if one wants to design a combined illumination and positioning system, our recommendation is to retain the (minimum) spacing for illumination because, when using this spacing, the degradation of the positioning accuracy is limited, while, in the other case, when we use the optimal spacing for VLP, the uniformity of illumination will be seriously affected.

5. Conclusions

We have presented an analysis of a VLP system comprising a set of non-directional white LEDs used as transmitters and, as receiver, an aperture-based receiver with a wide FOV and high angular diversity. The receiver was evaluated in an area of $X_{max} \times Y_{max}$ m^2 at a vertical distance of Z_{max} m below the ceiling where the VLP-LEDs were distributed in a rectangular grid. Moreover, we assumed that the LEDs were perfectly pointing downward and that the receiver was orientated parallel to the ceiling.

The performance of the VLP system was evaluated by obtaining a lower bound on the mean-squared error of the receiver position estimation, the CRB. More specifically, both the highest accuracy and the uniformity of receiver position estimation were considered in the analysis. Unlike in other studies, the simulations carried out in the optimisation process involved parameters such as the room dimension, the number of VLP-LEDs and the Lambertian order of LEDs, as well as the uniformity in the illuminance.

The results obtained allowed us to establish rules of thumb for the design of VLP systems for the optimal distance at which the VLP-LEDs should be spaced, depending mainly on the room dimensions and the number of VLP-LEDs, as well as a simple approximation for calculating the overall average receiver position estimation accuracy.

Furthermore, we further contrasted the results of the optimal spacing of VLP-LEDs with the recommendation on the optimal spacing distance between luminaries made in [14], where only the primary function of LEDs, illumination, was considered. It was found that, although the optimal spacing distance between the VLP-LEDs is below the optimal range determined in the analysis based on adequate illumination, the degradation in average positioning accuracy is negligible for small areas and of greater impact in large areas, which could be mitigated by increasing the number of VLP-LEDs.

Author Contributions: Research, J.M.M.; Writing, J.M.M.; Editing, H.S.; Supervising, H.S.; Funding, H.S. All authors have read and agreed to the published version of the manuscript.

Funding: José Miguel Menéndez acknowledges the National Secretariat of Higher Education, Science, Technology and Innovation of Ecuador (SENESCYT) for their financial support. This work is partially funded by the EOS grant 30452698 from the Belgian Research Councils FWO and FNRS. Further, it is funded by the Flemish Government (AI Research Program).

Institutional Review Board Statement: Not applicable.

Informed Consent Statement: Not applicable.

Data Availability Statement: The data presented in this study are available on request from the corresponding author.

Conflicts of Interest: The authors declare no conflict of interest.

References

1. Pathak, P.H.; Feng, X.; Hu, P.; Mohapatra, P. Visible light communication, networking, and sensing: A survey, potential and challenges. *IEEE Commun. Surv. Tutor.* **2015**, *17*, 2047–2077. [CrossRef]
2. Burchardt, H.; Serafimovski, N.; Tsonev, D.; Videv, S.; Haas, H. VLC: Beyond point-to-point communication. *IEEE Commun. Mag.* **2014**, *52*, 98–105. [CrossRef]
3. Yang, S.; Jeong, E.; Kim, D.; Kim, H.; Son, Y.; Han, S. Indoor three-dimensional location estimation based on LED visible light communication. *Electron. Lett.* **2013**, *49*, 54–56. [CrossRef]
4. Wang, T.Q.; Sekercioglu, Y.A.; Neild, A.; Armstrong, J. Position Accuracy of Time-of-Arrival Based Ranging Using Visible Light With Application in Indoor Localization Systems. *J. Light. Technol.* **2013**, *31*, 3302–3308. [CrossRef]
5. Hou, Y.; Xiao, S.; Bi, M.; Xue, Y.; Pan, W.; Hu, W. Single LED Beacon-Based 3-D Indoor Positioning Using Off-the-Shelf Devices. *IEEE Photonics J.* **2016**, *8*, 6806211. [CrossRef]
6. Hijikata, S.; Terabayashi, K.; Umeda, K. A simple indoor self-localization system using infrared LEDs. In Proceedings of the 2009 Sixth International Conference on Networked Sensing Systems (INSS), Pittsburgh, PA, USA, 17–19 June 2009; pp. 1–7.
7. Nadeem, U.; Hassan, N.U.; Pasha, M.A.; Yuen, C. Highly accurate 3D wireless indoor positioning system using white LED lights. *Electron. Lett.* **2014**, *50*, 828–830. [CrossRef]
8. Yin, L.; Wu, X.; Haas, H. Indoor Visible Light Positioning with Angle Diversity Transmitter. In Proceedings of the 2015 IEEE 82nd Vehicular Technology Conference (VTC2015-Fall), Boston, MA, USA, 6–9 September 2015; pp. 1–5.
9. Taylor, M.T.; Hranilovic, S. Angular Diversity Approach to Indoor Positioning Using Visible Light. In Proceedings of the 2013 IEEE Globecom Workshops (GC Wkshps), Atlanta, GA, USA, 9–13 December 2013; pp. 1093–1098.
10. Cincotta, S.; He, C.; Neild, A.; Armstrong, J. High angular resolution visible light positioning using a quadrant photodiode angular diversity aperture receiver (QADA). *Opt. Express* **2018**, *26*, 9230–9242. [CrossRef]
11. Yang, S.; Kim, H.; Son, Y.; Han, S. Three-Dimensional Visible Light Indoor Localization Using AOA and RSS With Multiple Optical Receivers. *J. Light. Technol.* **2014**, *32*, 2480–2485. [CrossRef]
12. Wang, T.Q.; He, C.; Armstrong, J. Angular Diversity for Indoor MIMO Optical Wireless Communications. In Proceedings of the 2015 IEEE International Conference on Communications (ICC), London, UK, 8–12 June 2015; pp. 5066–5071.

13. Cincotta, S.; He, C.; Neild, A.; Armstrong, J. QADA-PLUS: A novel two-stage receiver for visible light positioning. In Proceedings of the 2018 International Conference on Indoor Positioning and Indoor Navigation (IPIN), Nantes, France, 24–27 September 2018; pp. 1–5.
14. Menéndez, J.M.; Steendam, H. On the Optimisation of Illumination LEDs for VLP Systems. *Photonics* **2022**, *9*, 750. [CrossRef]
15. Bastiaens, S.; Goudos, S.; Joseph, W.; Plets, D. Metaheuristic Optimization of LED Locations for Visible Light Positioning Network Planning. *IEEE Trans. Broadcast.* **2021**, *67*, 894–908. [CrossRef]
16. Steendam, H.; Wang, T.; Armstrong, J. Theoretical lower bound for indoor visible light positioning using received signal strength measurements and an aperture-based receiver. *J. Light. Technol.* **2016**, *35*, 309–319. [CrossRef]
17. Steendam, H.; Wang, T.; Armstrong, J. Cramer-Rao Bound for Indoor Visible Light Positioning Using an Aperture-Based Angular-Diversity Receiver. In Proceedings of the 2016 IEEE International Conference on Communications (ICC), Kuala Lumpur, Malaysia, 22–27 May 2016.
18. Kahn, J.M.; Barry, J.R. Wireless infrared communications. *Proc. IEEE* **1997**, *85*, 265–298. [CrossRef]
19. Armstrong, J. OFDM for Optical Communications. *J. Light. Technol.* **2009**, *27*, 189–204. [CrossRef]
20. Armstrong, J.; Lowery, A. Power efficient optical OFDM. *Electron. Lett.* **2006**, *42*, 370–372. [CrossRef]
21. Dissanayake, S.D.; Armstrong, J. Comparison of ACO-OFDM, DCO-OFDM and ADO-OFDM in IM/DD Systems. *J. Light. Technol.* **2013**, *31*, 1063–1072. [CrossRef]
22. Seoul Semiconductor Inc. Manufacturer Part Number: WZ10150-02-U2-AA, Product: XZ10150. 2018. Available online: http://www.seoulsemicon.com/upload2/XZ10150.pdf (accessed on 8 April 2019).
23. Kay, S.M. *Fundamentals of Statistical Signal Processing: Estimation Theory*; Prentice Hall PTR: Upper Saddle River, NJ, USA, 1993.
24. Trees, H.L.V. *Detection, Estimation, and Modulation Theory, Part III*; John Wiley & Sons: Hoboken, NJ, USA, 2001.
25. Stevens, N.; Steendam, H. Magnitude of the Distance Estimation Bias in Received Signal Strength Visible Light Positioning. *IEEE Commun. Lett.* **2018**, *22*, 2250–2253. [CrossRef]
26. Zeng, L.; O'Brien, D.; Minh, H.; Faulkner, G.; Lee, K.; Jung, D.; Oh, Y.; Wong, E. High Data Rate Multiple Input Multiple Output (MIMO) Optical Communications Using White LED Lighting. *IEEE J. Sel. Areas Commun.* **2009**, *27*, 1654–1662. [CrossRef]
27. LED Lighting-White. Available online: https://www.digikey.com/en/products/filter/led-lighting-white/124 (accessed on 8 October 2012).
28. Normung, D. *DIN EN 12464-1, Licht und Beleuchtung—Beleuchtung von Arbeitsstätten. Teil 1, Arbeitsstätten in Innenräumen: Light and Lighting—Lighting of Work Places*; Part 1, Indoor Work Places; Beuth Verlag GmbH: Berlin, Germany, 2019. Available online: https://books.google.be/books?id=azKIzgEACAAJ (accessed on 8 April 2019).

hv photonics

MDPI

Article

Avalanche Photodiode-Based Deep Space Optical Uplink Communication in the Presence of Channel Impairments

Wenjng Guo [1,2], Xiaowei Wu [1] and Lei Yang [1,*]

1 Technology and Engineering Center for Space Utilization, Chinese Academy of Sciences, Beijing 100094, China; gwjsmail@163.com (W.G.); wuxiaowei@csu.ac.cn (X.W.)
2 University of Chinese Academy of Sciences, Beijing 101400, China
* Correspondence: yang.lei@csu.ac.cn

Abstract: Optical communication is a critical technology for future deep space exploration, offering substantial advantages in transmission capacity and spectrum utilization. This paper establishes a comprehensive theoretical framework for avalanche photodiode (APD)-based deep space optical uplink communication under combined channel impairments, including atmospheric and coronal turbulence induced beam scintillation, pointing errors, angle-of-arrival (AOA) fluctuations, link attenuation, and background noise. A closed-form analytical channel model unifying these effects is derived and validated through Monte Carlo simulations. Webb and Gaussian approximations are employed to characterize APD output statistics, with theoretical symbol error rate (SER) expressions for pulse position modulation (PPM) derived under diverse impairment scenarios. Numerical results demonstrate that the Webb model achieves higher accuracy by capturing APD gain dynamics, while the Gaussian approximation remains viable when APD gain exceeds a channel fading-dependent gain threshold. Key system parameters such as APD gain and field-of-view (FOV) angle are analyzed. The optimal APD gain significantly influences the achievement of optimal SER performance, and angle of FOV design balances AOA fluctuations tolerance against noise suppression. These findings enable hardware optimization under size, weight, power, and cost (SWaP-C) constraints without compromising performance. Our work provides critical guidelines for designing robust APD-based deep space optical uplink communication systems.

Keywords: deep space optical communication; uplink channel model; avalanche photodiode; Webb and Gaussian models; symbol error rate

1. Introduction

In deep space wireless communication, optical communication is a well-appreciated method due to numerous advantages, such as large transmission capacity, license-free frequency spectrum, etc., [1]. Deep space optical communication has been utilized in the National Aeronautics and Space Administration (NASA) Lunar Laser Communication Demonstration (LLCD) project [2]. Current research in deep space optical communication has concentrated predominantly on achieving high-rate downlink transmission capabilities [3], while uplink systems remain inadequately addressed. Nevertheless, studying optical uplink communication is essential for several key reasons, summarized as follows. First, channel models for uplink and downlink are different, so receiver-focused approaches optimized for downlink prove inadequate to address cascading impairments in uplink scenarios [4]. Uplink transmissions suffer from atmospheric disturbances in the initial propagation stage, which leads to cumulative beam wandering and intensity fading [5]. Data from European Space Agency (ESA) experiments reveal that transmit-end atmospheric turbulence introduces much higher wavefront aberration variance compared to downlink [6]. Second, deep space optical uplink systems face critical limitations in both hardware miniaturization and signal processing capabilities [7]. Achieving both high photon collection and compact device design requires careful trade-offs due to size, weight, power, and cost (SWaP-C) constraints. For example, NASA's launch standards limit optical receivers to apertures below 0.5 m and total masses below 50 kg [8]. Third, uplink communication proves indispensable in deep space missions by ensuring mission-critical command transmission (e.g., orbital maneuvers) where even 0.1% data loss risks total mission failure. Therefore, this paper focuses on addressing these critical challenges for deep space optical uplink communication.

For deep space optical channel modeling, Ivanov et al. in [3] investigated the Poisson channel in a controlled laboratory environment. Works in [9–11] took into account the coronal turbulence-induced beam scintillation effect. However, the uplink of deep space optical communication suffers from more impairments. As shown in Figure 1, a schematic of a deep space optical communication system is demonstrated. First, atmosphere and solar wind will result in beam scintillation [10,12]. For modeling turbulence-induced beam scintillation, various distributions were proposed, such as Log-normal [13], Gamma-Gamma [14], K-distribution [15], Exponentiated Weibull [16], Fisher–Snedecor [17], and Málaga distribution [18]. Gamma-Gamma has higher accuracy than both Log-normal and K-distribution [14]. In addition, the two-parameter structure of the Gamma-Gamma model enables higher computational efficiency than multi-parameter alternatives like Exponentiated Weibull, Fisher–Snedecor, and Málaga distribution [17,18]. This efficiency benefits real-time optical systems requiring rapid turbulence compensation with minimal latency. Second, beam misalignment is one of the inescapable impairments [19]. Due to the movement of the satellite, it is difficult for an optical beam to aim perfectly at the receiving aperture [20]. Beam misalignments include two aspects, pointing errors and angle of arrival (AOA) fluctuations. In [21], pointing errors were assumed to be identical in both vertical and horizontal axes, and are modeled by Rayleigh distribution. For AOA fluctuations, it was modeled by two-valued (1 and 0) distribution in [22,23]. Factor 1 shows that no signal power is lost and 0 indicates that complete optical signal dropout occurs. Third, link attenuation is also a serious impairment in the optical uplink [24]. The aerosol particles and molecules in the atmosphere cause the scattering and absorption of the optical beam [25,26]. Geometric diffusion-induced attenuation is related to link length, beam divergence angle, and optical receiving aperture size [24]. Last, background noise is a non-negligible damaging factor of deep space optical communication, which depends on parameters of noise sources, FOV, optical filter bandwidth, and receiving aperture area [27]. In conclusion, in the uplink of deep space optical communication, atmospheric and coronal turbulence, pointing errors, AOA fluctuations, link attenuation, and background noise are important impairments and need to be considered in the channel model.

To address SWaP-C constraints while maintaining high communication performance in miniaturized hardware platforms, pulse position modulation (PPM) and on-off keying (OOK) are widely used [2,3,5,9,28]. LLCD utilized the PPM [2], and works in [3,5,7,28] also considered the PPM in deep space optical communication. Xu et al. in [9] highlighted the use of OOK for deep space exploration. For the photodetection of PPM and OOK signals, several optical detectors are commonly deployed, including P-i-N photodiodes (PINs), avalanche photodiodes (APDs), single-photon avalanche diodes (SPADs), and superconducting nanowire single-photon detectors (SNSPDs) [29–32]. APDs exhibit a compact footprint and lower power consumption compared to PINs [30], and have higher detection efficiency compared to SPADs. In addition, SPADs operate within a temperature range of 0–25 °C [31]. SNSPDs require cryogenic cooling below −268 °C, resulting in prohibitively high power demands, prolonged startup times, and bulky system designs [32]. Thus, APDs are a preferred choice for deep space optical uplink communication. Webb [33] and Gaussian [34] can be used to approximate APD output. Webb distribution offers superior accuracy but is more complex [35]. In previous studies, the Gaussian distribution was widely used due to its simplicity [36–41]. However, when the system performance evaluated by the Gaussian model is better than the actual communication performance, the actual link budget will be lower than the analytical link budget analyzed during system design, which significantly increases the probability of link outage.

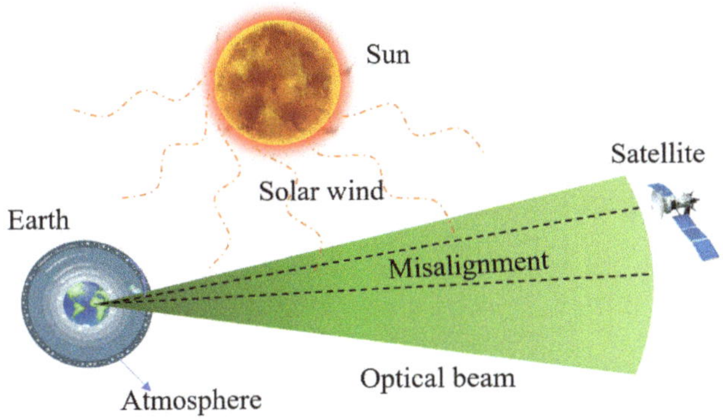

Figure 1. Schematic of deep space optical uplink communication over various channel impairments.

Motivated by the above analysis, this paper studies an APD-based deep space optical uplink communication in the presence of various channel impairments. The block diagram of our studied system is shown in Figure 2. Specifically, our contributions are as follows:

- We study the combined effects of critical channel impairments in deep space optical uplink systems, including atmospheric and coronal turbulence, pointing errors, AOA fluctuations, link attenuation, and background noise. A closed-form analytical uplink channel model integrating these factors is derived, and its accuracy is rigorously validated through Monte Carlo simulations.
- We use Webb and Gaussian distributions to approximate APD output. For both models, theoretical expressions of PPM communication SER performance are derived. A comparative analysis is performed for the Webb and Gaussian models' evaluated SER performance.
- We discuss various numerical results, including characteristics of channel destructive factors, Webb and Gaussian modeled APD output and SER performance. While the Gaussian model exhibits inherent limitations in capturing the dynamics of avalanche noise, its operational constraints are analyzed. The results reveal non-monotonic dependencies of SER on the APD gain and FOV angle. Hence, these parameters are strategically optimized to improve the performance of our considered system.

The rest of this article is organized as follows. In Section 2, we analyze various channel impairments and a closed-form expression of the uplink model is derived. Section 3 first introduces the APD output and noise model. In Section 4, theoretical expressions of PPM SER performance based on Webb and Gaussian models are derived. The simulation results are shown and discussed in Section 5. Finally, Section 6 concludes this article.

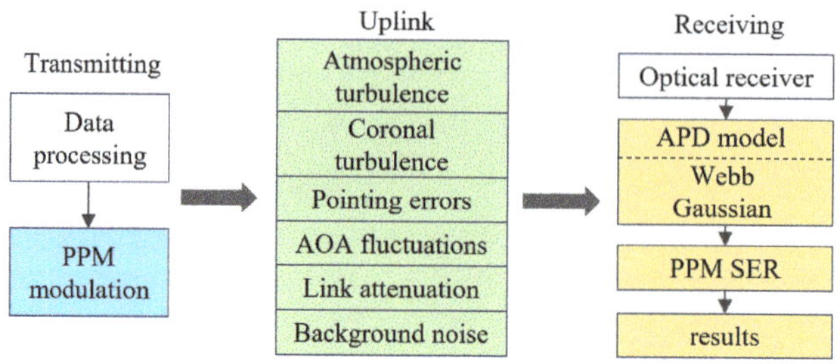

Figure 2. Block diagram of studied deep space optical communication system.

2. Optical Uplink Channel Model

Considering PPM modulation, after beam propagating from ground station to deep space satellite, received optical power of a PPM slot is expressed as

$$P_r = hP_t + P_b, \tag{1}$$

where h, P_t, and P_b are the channel fading coefficient, transmitted optical power, and power of background noise, respectively. In a high speed optical communication scenario [42], h is presumed to be constant over a PPM symbol duration. Various crucial factors impair optical power. We denote the power fading coefficient h as

$$h = h_t h_s h_p h_a h_c, \tag{2}$$

where h_t and h_s are caused by atmospheric turbulence and coronal turbulence-induced beam scintillation, respectively. Moreover, h_p, h_a, and h_c denote the fading coefficients that result from pointing errors, AOA fluctuations, and link attenuation, respectively.

In the following, the statistical properties of the above channel fading factors are first introduced. Then, a closed-form expression of the probability density function (PDF) of h is derived.

2.1. Statistical Properties of Channel Fading Factors

2.1.1. Atmospheric and Coronal Turbulence-Induced Beam Scintillation

Turbulence will cause optical intensity fluctuations, i.e., beam scintillation. Numerous statistical models for beam scintillation have been proposed, such as Log-normal, Gamma-Gamma, K-distribution, Exponentiated Weibull, Fisher–Snedecor, and Málaga distribution. We adopt the commonly used Gamma-Gamma model, whose PDF is defined as [14]

$$f_g(x|\alpha,\beta) = \frac{2(\alpha\beta)^{\frac{\alpha+\beta}{2}}}{\Gamma(\alpha)\Gamma(\beta)} x^{\frac{\alpha+\beta}{2}-1} \mathrm{K}_{\alpha-\beta}\left(2\sqrt{\alpha\beta x}\right), \tag{3}$$

where α and β represent the small-scale and large-scale turbulent eddies, respectively. $\Gamma(\cdot)$, and $\mathrm{K}_{\alpha-\beta}(\cdot)$ denote the standard gamma function and the modified Bessel function of the second kind of order $(\alpha-\beta)$, respectively. In addition, both α and β are closely related to the Rytov variance σ_g^2 of turbulence.

For atmospheric turbulence-caused optical power fading coefficient, it is denoted as h_t. The distribution of h_t is as (3), which is rewritten as $f_g(h_t|\alpha_t,\beta_t)$. Rytov variance σ_g^2 for atmospheric turbulence is rewritten as σ_{ht}^2, which is given by [43]

$$\sigma_{ht}^2 \simeq 2.25 k_n^{7/6} \sec^{11/6}(\zeta_g) \int_{H_g}^{H_s} C_n^2(H)(H_s - H)^{\frac{5}{6}} dH, \tag{4}$$

where ζ_g and $k_n = 2\pi/\lambda$ are the zenith angle and the optical wave number with λ being the wavelength, respectively. In addition, H, H_s, and H_g denote the height parameter, satellite height, and ground station altitude, respectively. $C_n^2(H)$ is the structure parameter for refractive index fluctuation.

Coronal turbulence-caused fading coefficient is expressed as h_s. Similarly, the PDF of h_s is as (3) and rewritten as $f_g(h_s|\alpha_s, \beta_s)$. Rytov variance σ_g^2 for coronal turbulence is rewritten as σ_{hs}^2, which takes the form [9]

$$\sigma_{hs}^2 \simeq \frac{1}{2} r_e^2 (2\pi)^{\frac{p_l-3}{2}} \pi(3-p_l)\Gamma\left(\frac{p_l}{2}\right)\Gamma\left(\frac{p_l-1}{2}\right)\Gamma\left(1+\frac{p_l}{2}\right)\zeta_s^2 N_e^2 L_o^{3-p_l}(L_l+\lambda)^{1+\frac{p_l}{2}}\sec\left(\frac{\pi p_l}{4}\right), \quad (5)$$

where r_e, p_l, and ζ_s represent the classical electron radius, spectral index, and relative solar wind density fluctuations ratio, respectively. In addition, N_e, L_o, and L_l represent the solar wind density, outer scale of coronal turbulence, and communication distance, respectively.

The parameters of σ_{ht}^2 and σ_{hs}^2 are significant for the intensity of atmospheric and coronal turbulence, respectively. The typical values of σ_{ht}^2 and σ_{hs}^2 are analyzed in Appendix A.1.

2.1.2. Pointing Errors and AOA Fluctuations

The continuous motion of satellites introduces significant challenges in maintaining the precise alignment of the optical beam with the receiving aperture. Beam wander effects further exacerbate pointing inaccuracies and AOA fluctuations. As established in prior research [37], the instantaneous pointing errors can be statistically modeled using Rayleigh distribution. In particular, the optical power fading coefficient h_p, induced by these pointing errors, demonstrates a parametric dependence on system characteristics including beam divergence angle and receiver aperture diameter. This relationship underscores the necessity for comprehensive system parameter optimization to mitigate power degradation in space optical communications. Considering a Gaussian beam, PDF of h_p is [44]

$$f_{h_p}(h_p) = \frac{w_b^2 \cos\theta_a}{4\sigma_p^2}\left(\mathrm{erf}\left(\frac{\sqrt{\pi}d_r}{2\sqrt{2}w_d}\right)\right)^{-\frac{w_b^2}{4\sigma_p^2}} h_p^{\frac{w_b^2}{4\sigma_p^2}-1}, \quad (6)$$

where w_d, σ_p^2, and θ_a are the beam width, normalized variance of pointing errors, and instantaneous AOA, respectively. In addition, erf is the Gaussian error function, and the equivalent beam width w_b is related to w_d.

For AOA fluctuations induced fading coefficient h_a, a two-valued distribution can be used, which is [23]

$$h_a \approx \begin{cases} 1, & \theta_a < \Omega_F, \\ 0, & \theta_a \geq \Omega_F, \end{cases} \quad (7)$$

where Ω_F denotes the angle of FOV. The rationality of this two-valued distribution is discussed here. The intensity of light spot on the APD photon sensing aperture is expressed by the Airy pattern. Let d_{Ai} be the distance between the center and any point in the airy light spot. When $d_{Ai} \geq 25\lambda$, the power of circle spot with radius d_{Ai} accounts for 99% of the total optical power [45]. Taking $\lambda = 1550$ nm as an example, $25\lambda = 3.87\ \mu$m is much smaller than the APD photon sensing aperture. Therefore, when the light beam converges at the APD photon sensing aperture, the total received power can be detected; that is, $h_a = 1$. When $\theta_a > \Omega_F$, almost no optical power is detected, i.e., $h_a = 0$. Figure 3 shows an example for both cases. Instantaneous AOA θ_a in (7) follows the Rayleigh distribution [44], which is

$$f_{\theta_a}(\theta_a) = \frac{\theta_a}{\sigma_\theta^2} \exp\left(-\frac{\theta_a^2}{2\sigma_\theta^2}\right), \tag{8}$$

where σ_θ^2 is the variance of AOA.

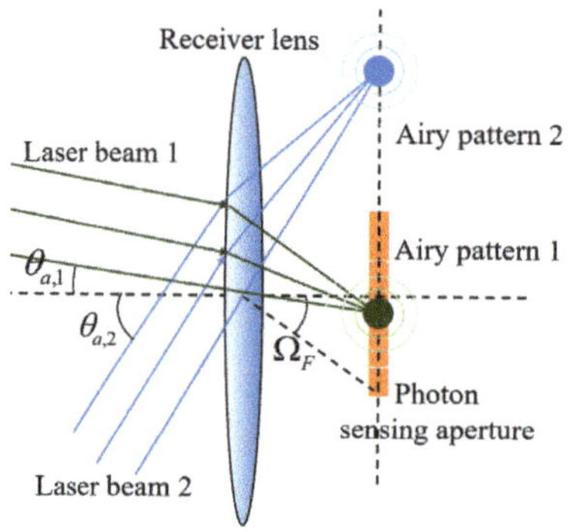

Figure 3. Schematic of AOA fluctuations and Airy pattern.

2.1.3. Link Attenuation

In deep space optical uplink, two essential factors cause the optical power attenuation, geometric diffusion and atmosphere attenuation. Due to the optical receiving aperture size being limited and the optical divergence angle being nonzero, the light spot at the receiver is much larger than the receiving aperture. Consequently, most of the optical power is lost. This is the geometric diffusion attenuation, which is denoted as $\varsigma_g = \left(\pi L_l \tan^2\theta_d\right)^{-1}$, where θ_d is the optical divergence angle. Moreover, the atmosphere contains numerous molecules and aerosol particles, which cause absorption and scattering of optical power. Specific attenuation effects vary with weather conditions, such as clear weather, foggy, hazy, rainy, and snowy. Attenuation in the atmosphere is represented by the Beers–Lambert law, which is given by $\varsigma_a = \exp(-L_a\xi)$ [46]. Here, L_a stands for the length of the atmospheric link and ξ is the atmospheric attenuation coefficient related to weather conditions. In addition, interplanetary dust is not a critical impairment for the scenarios analyzed in this work. Consequently, the link attenuation coefficient h_c is derived as

$$h_c = \varsigma_g \varsigma_a = \frac{1}{\pi L_l \tan^2\theta_d \exp(L_l\xi)}. \tag{9}$$

2.1.4. Background Noise

In deep space optical communication, usually, the optical power received is very small because of the ultra-long link distance. Therefore, background noise can easily impair communication performance. The power of received background noise P_b is [27]

$$P_b = 0.25\pi d_r^2 B_o \Omega_F N_b(\lambda), \tag{10}$$

where B_o is the bandwidth of optical filter, and $N_b(\lambda)$ is the background radiation spectral radiance. Moreover, $N_b(\lambda)$ is expressed as [27]

$$N_b(\lambda) = \frac{2\hbar c R_f}{\lambda^3 \exp(\hbar c/\kappa_c T_b \lambda) - 1}, \tag{11}$$

where $\hbar$, c, and κ_c are the Planck's constant, light speed, and Boltzmann's constant, respectively. The parameter R_f is the spectral albedo of a celestial body, defined as the ratio of radiation reflected by the planet (including its atmosphere) to total incident light. In the solar system, the Sun is the luminous body; therefore, the Sun's spectral albedo value is set to 1. The average spectral albedo of the Earth is $R_f = 0.25$. The values of spectral albedo for other important celestial bodies can be found in [27]. It is critical to recognize that other celestial bodies inherently exhibit blackbody radiation even in the absence of solar reflection. Consequently, when such bodies act as independent radiation sources, R_f is equivalently set to unity. In addition, the parameter T_b denotes the average blackbody temperature of a celestial body; for example, the Sun's $T_b = 5776$ K and Earth's $T_b = 288$ K. When the specifics of a deep space mission are known (e.g., communication range, satellite position), the variables in (11) are nearly fixed, thereby determining the value of background radiation spectral radiance.

2.2. Closed-Form Expression for the Channel Model

The comprehensive channel impairments in deep space optical uplinks encompass turbulence-induced beam scintillation (atmospheric h_t, coronal h_s), pointing errors h_p, AOA fluctuations h_a, link attenuation h_c, and background noise P_b. While these parameters are critical for channel modeling, their synergistic impacts remain scarcely addressed in the existing literature. To bridge this gap, we present a novel closed-form analytical model (Equation (2)) that integrates all degradation mechanisms. The methodological framework for deriving the PDF of the composite channel coefficient h is as follows.

First, let $h_2 = h_t h_s$ represent the beam scintillation caused by atmospheric turbulence and coronal turbulence, the cumulative distribution function (CDF) of h_2 is

$$F_{h_2}(h_2) = \int_0^\infty \int_0^\infty f_g(h_t) f_g(h_s) dh_t dh_s. \tag{12}$$

The PDF of h_2 is written as $f_{h_2}(h_2)$, which is the derivative of the CDF of h_2. Hence, with $h_t = h_2/h_s$, $f_{h_2}(h_2)$ is derived as

$$f_{h_2}(h_2) = \frac{\partial}{\partial h_2} \int_0^\infty \int_0^\infty f_g\left(\frac{h_2}{h_s}\right) f_g(h_s) d\frac{h_2}{h_s} dh_s = \int_0^\infty \frac{1}{h_s} f_g\left(\frac{h_2}{h_s}\right) f_g(h_s) dh_s. \tag{13}$$

According to (3), (13) is rewritten as

$$f_{h_2}(h_2) = \frac{4(\alpha_t\beta_t)^{\frac{\alpha_t+\beta_t}{2}} (\alpha_s\beta_s)^{\frac{\alpha_s+\beta_s}{2}} h_2^{\frac{\alpha_t+\beta_t}{2}-1}}{\Gamma(\alpha_t)\Gamma(\beta_t)\Gamma(\alpha_s)\Gamma(\beta_s)} \int_0^\infty \frac{\mathrm{K}_{\alpha_t-\beta_t}\left(2\sqrt{\alpha_t\beta_t h_2/h_s}\right)\mathrm{K}_{\alpha_s-\beta_s}\left(2\sqrt{\alpha_s\beta_s h_s}\right)}{h_s^{1-\frac{\alpha_s+\beta_s-\alpha_t-\beta_t}{2}}} dh_s, \tag{14}$$

where $K_v(x) = \frac{1}{2}\hat{G}_{0,2}^{2,0}\left[\frac{x^2}{4}\Big|_{v/2,-v/2}\right]$, and $\hat{G}_{p,q}^{m,n}[\cdot]$ is the Meijer's G function. Employing (07.34.21.0011.01) in [47], the result of (14) is

$$f_{h_2}(h_2) = \frac{\alpha_t\beta_t\alpha_s\beta_s}{h_2\Gamma(\alpha_t)\Gamma(\beta_t)\Gamma(\alpha_s)\Gamma(\beta_s)} \hat{G}_{0,4}^{4,0}\left[\alpha_t\beta_t\alpha_s\beta_s h_2\Big|_{\alpha_t-1,\beta_t-1,\alpha_s-1,\beta_s-1}\right]. \tag{15}$$

The specific derivation can be found in Appendix A.2.

Next, let $h_3 = h_2 h_p$ represent the joint effect of turbulence-induced scintillation and pointing errors. Combining the PDF expressions in (6) and (15), the PDF of h_3 is derived as

$$f_{h_3}(h_3) = \frac{C_{\alpha\beta}C_{p1}C_{p2}^2}{\Gamma(\alpha_t)\Gamma(\beta_t)\Gamma(\alpha_s)\Gamma(\beta_s)} \hat{G}_{1,5}^{5,0}\left[C_{\alpha\beta}C_{p1}h_3\Bigg|\begin{matrix} C_{p2}^2 \\ C_{p2}^2-1, \alpha_t-1, \beta_t-1, \alpha_s-1, \beta_s-1\end{matrix}\right], \tag{16}$$

where $C_{\alpha\beta} = \alpha_t\beta_t\alpha_s\beta_s$, $C_{p1} = w_b^2/2d_r^2$, and $C_{p2} = w_b/2\sigma_p$.

Finally, we take into account AOA fluctuations and link attenuation, i.e., $h = h_3 h_a h_c$, where h_a is modeled as (7), and h_c is a constant when link length, divergence angle, and weather condition are given. Therefore, the PDF of h is derived as

$$f_h(h) = \frac{C_{\alpha\beta}C_{p1}C_{p2}^2}{h_c\Gamma(\alpha_t)\Gamma(\beta_t)\Gamma(\alpha_c)\Gamma(\beta_c)}\left(1-\exp\left(-\frac{\Omega_F^2}{2\sigma_\theta^2}\right)\right)\hat{G}_{1,5}^{5,0}\left[\frac{C_{\alpha\beta}C_{p1}}{h_c}h\,\middle|\,\begin{matrix} C_{p2}^2 \\ C_{p2}^2-1,\alpha_t-1,\beta_t-1,\alpha_c-1,\beta_c-1\end{matrix}\right] + \delta(h)\cdot\exp\left(-\frac{\Omega_F^2}{2\sigma_\theta^2}\right), \tag{17}$$

where $\delta(\cdot)$ denotes the Dirac delta function.

In addition, background noise constituting an additive noise component in the received optical power, it is fundamentally governed by the spectral radiance density $N_b(\lambda)$ through the radiometric relationship shown in (10) and (11). This spectral radiance is intrinsically related to environmental radiation sources, with three dominant scenarios considered in this work: direct solar radiation, Earth-reflected sunlight, and thermal emission from the Earth's night side, corresponding to maximum, moderate, and minimum noise conditions, respectively [27]. At an operational wavelength of $\lambda = 1550$ nm, under blackbody temperature for the Sun (5776 K) and Earth (288 K) with $R_f = 1$, $N_b(\lambda)$ exhibits extreme variability, ranging from 6.4×10^{-20} W/Hz in the solar-dominated worst-case scenario to 2.5×10^{-33} W/Hz under optimal Earth nighttime conditions. This thirteen-order-of-magnitude disparity highlights the critical role of orbital positioning and receiver pointing accuracy in mitigating background noise, particularly when operating near solar illumination zones, where adaptive filtering strategies become essential for maintaining communication fidelity.

3. APD Output Models

APD efficiently converts optical signals into electrical currents through photon-triggered carrier multiplication, making them compact and energy-efficient detectors for deep space optical uplinks. The McIntyre model [48] fundamentally describes the stochastic detection process of APD through exact conditional probabilities. Subsequent studies developed simplified Webb [33] and Gaussian [35] approximations to facilitate the analysis of communication systems, trading precision for computational tractability. Comparative studies [49] reveal that the Webb approximation better preserves detection statistics despite increased mathematical complexity. This section presents a comparative analysis of both approaches, structured as follows: First, we establish the foundational APD detection principles and noise characteristics. Second, we derive the Webb-approximated probability distribution for the APD output. Third, we formulate the Gaussian-approximated APD output model. Finally, we compare the two models and conclude that the Webb model better characterizes the gain dynamics of APD with higher precision, while the Gaussian approximation is primarily applicable in noise-dominated scenarios.

3.1. APD Detection Process and Noise Model

The photon absorption process in an APD during a PPM time slot follows Poisson statistics, where the absorbed photon count n has a mean value $\bar{n} = P_r\eta_a T_s/\hbar v$. Here, P_r denotes the optical power derived in (1). The parameters η_a, T_s, and v are the APD quantum efficiency, PPM time slot width, and optical frequency, respectively. The resultant electron count m, governed by the McIntyre distribution, arises from stochastic avalanche multiplication processes. Therefore, the PDF of electron generation is obtained by convolution of the Poisson and McIntyre PDFs, as established in [48].

The statistical fluctuation in m, arising from both the photon absorption randomness and the avalanche multiplication uncertainty, constitutes the fundamental mechanism of the shot noise in APD-based receivers. The shot noise magnitude depends intrinsically on the quantum efficiency and multiplication statistics governed by the McIntyre model. Beyond fundamental shot noise, the output current of APD exhibits two critical noise components: thermal noise and dark current noise [30]. The thermal noise, originating from electron thermal agitation in the front-end load resistor, follows additive white Gaussian statistics with power spectral density $\sigma_{th}^2 = 4\kappa_c T_e B_e T_s^2 / R_L$, where T_e, B_e, and R_L are the APD temperature, bandwidth, and load resistance, respectively. Dark current noise consists of surface leakage current and bulk current, obeying Gaussian statistics with mean $I_D T_s$ and variance $\sigma_D^2 = 2qI_D B_e T_s^2$. The composite noise current i_{td} comprising both dark current and thermal noise components follows a composite Gaussian distribution expressed as

$$\phi\left(i_{td} | \mu_{td}, \sigma_{td}^2\right) = \frac{1}{\sqrt{2\pi}\sigma_{td}} \exp\left(-\frac{(i_{td} - \mu_{td})^2}{2\sigma_{td}^2}\right), \tag{18}$$

where $\mu_{td} = I_D T_s$ and $\sigma_{td}^2 = \sigma_{th}^2 + \sigma_D^2$.

By integrating the McIntyre–Poisson convolution framework for photon-generated electron statistics [48] with the Gaussian noise component (18), the comprehensive APD output current model is derived as

$$p_m(i) = \sum_{m=0}^{\infty} \sum_{n=1}^{m} \left(\frac{(\bar{n})^n \Gamma(c_{ka} m / k_a + 1)(c_G - c_G k_a)^{m-n}}{\sqrt{2\pi}\sigma_m m(n-1)!(m-n)!} \frac{(1/G + k_a c_G)^{n + c_{ka} m}}{\Gamma(c_{ka} m + n + 1)} \exp\left(-\bar{n} - \frac{(i - \mu_m)^2}{2\sigma_m^2}\right)\right), \tag{19}$$

where G and k_a are the APD gain and the APD ionization ratio, respectively. In addition, $c_{ka} = k_a/(1 - k_a)$, $c_g = (G-1)/G$, $\sigma_m^2 = \sigma_{td}^2$, and $\mu_m = mq_e + \mu_{td}$ with q_e being the electron charge.

3.2. Webb and Gaussian-Approximated APD Output Model

To balance computational efficiency with model fidelity in APD characterization, Webb model [33] provides an optimal framework for approximating the stochastic response of APD output. Given the mean received photons $\bar{n}$, the conditional PDF of electron multiplication $f_W(m|\bar{n})$ approximated by the Webb distribution is

$$f_W(m|\bar{n}) = \frac{\left(1 + (m - G\bar{n}) / c_f \bar{n}\right)^{2/3}}{\sqrt{2\pi \bar{n} G^2 F_a}} \exp\left(-\frac{(m - G\bar{n})^2}{2\bar{n} G^2 F_a \left(1 + (m - G\bar{n}) / c_f \bar{n}\right)}\right), \tag{20}$$

where F_a is the APD noise factor, and $c_F = \frac{GF_a}{F_a - 1}$. Through statistical convolution of the Webb-distributed photon response (20) with Gaussian noise component (18), the composite APD output model preserving avalanche multiplication characteristics is formulated as

$$p_w(i) = \sum_{m=0}^{\infty} \frac{\left(1 + (m - G\bar{n}) / c_f \bar{n}\right)^{2/3}}{2\pi\sigma_m \sqrt{\bar{n} G^2 F_a}} \exp\left(-\frac{(m - G\bar{n})^2}{2\bar{n} G^2 F_a \left(1 + \frac{m - G\bar{n}}{c_f \bar{n}}\right)} - \frac{(i - \mu_m)^2}{2\sigma_m^2}\right). \tag{21}$$

To further simplify the Webb modeled APD output, Gaussian approximation is strategically implemented. That is, the conditional PDF of electron multiplication $f_W(m|\bar{n})$ under mean photon flux $\bar{n}$ transitions to Gaussian statistics. Hence, (20) is rewritten as

$$f_G(m|\bar{n}) = \frac{1}{\sqrt{2\pi\sigma_{sh}^2}} \exp\left(-\frac{(m - \mu_{sh})^2}{2\sigma_{sh}^2}\right), \tag{22}$$

where $\sigma_{sh}^2 = 2G^2 q_e^2 F_a \bar{n} B_e T_s$ and $\mu_{sh} = Gq_e\bar{n}$ are the variance and the mean of this Gaussian distribution, respectively. Incorporating APD noise characteristics, the APD output approximated by Gaussian model is

$$p_g(i) = \frac{1}{\sqrt{2\pi\sigma_g^2}} \exp\left(-\frac{(i-\mu_g)^2}{2\sigma_g^2}\right), \tag{23}$$

where $\mu_g = \mu_{sh} + \mu_m$ and $\sigma_g^2 = \sigma_{sh}^2 + \sigma_m^2$.

The Webb model [33] establishes a rigorous framework for APD avalanche gain dynamics through the statistical convolution of photon absorption statistics and nonlinear multiplication processes. This approach effectively captures multiplicative shot noise mechanisms (i.e., avalanche fluctuations) that dominate in gain-saturated operational regimes, as quantified in Section 5. In contrast, the Gaussian approximation model employs an additive Gaussian noise assumption to simplify APD output analysis, sacrificing fidelity in modeling nonlinear gain dynamics. While this abstraction disregards the inherent nonlinear gain dynamics captured by the Webb model, it offers significant computational advantages for noise-limited scenarios. Specifically, under high background radiation conditions as discussed in Section 5, where system performance becomes dominated by external noise sources rather than intrinsic APD gain characteristics, the Gaussian model achieves acceptable accuracy.

4. PPM SER Performance Analysis

In deep space optical uplink communication systems constrained by power budgets, PPM is strategically implemented due to its superior photon efficiency under average transmission power constraints. To quantify communication reliability, we adopt the SER as the critical performance metric, defined as the probability of erroneous identification of the PPM symbol. This section presents a comprehensive SER analysis based on both Webb and Gaussian-approximated APD output models, specifically incorporating channel impairments: atmospheric and coronal turbulence-induced fading, pointing error fluctuations, AOA fluctuations, link attenuation, and background noise. Comprehensive analysis provides critical insights for designing robust deep space optical uplink communication systems.

The conditional SER for PPM systems under channel fading coefficient h is analytically formulated as

$$E_{p|h} = 1 - \int_{-\infty}^{\infty} p_e(i_x|1)di_x \left(\int_{-\infty}^{i_x} p_e(i_y|0)di_y\right)^{Q-1}, \tag{24}$$

where Q denotes the PPM symbol size. Conditional probability densities $p_e(i_x|1)$ and $p_e(i_y|0)$ characterize the APD output under pulse transmission (bit 1) and silence states (bit 0), respectively. These distributions are formulated through both Webb (21) and Gaussian (23)-approximated stochastic APD output. The statistical characterization of the deep space optical uplink channel is analytically formulated as $f_h(h)$ in (17). Under the influence of various channel impairments, the unconditioned SER of the PPM system is obtained by

$$E_p = \int_0^{\infty} f_h(h) E_{p|h} dh. \tag{25}$$

Consequently, through Webb-modeled APD statistical output, the PPM SER performance is derived as

$$E_W = \int_0^{\infty} f_h(h) \left(1 - \mathrm{InS}\left(i_x, \frac{\frac{1}{c_2\sqrt{x}}\left(1+\frac{c_x}{c_f x}\right)^{2/3}}{\exp\left(\frac{c_x^2}{c_1\bar{n}+\frac{c_x c_2 2\bar{n}}{c_f x}} + \frac{(i_x-\mu_m)^2}{2\sigma_m^2}\right)} \right) \mathrm{InS}^{Q-1}\left(i_y, \frac{\frac{1}{c_2\sqrt{\bar{n}_b}}\left(1+\frac{c_y}{c_f \bar{n}_b}\right)^{2/3}}{\exp\left(\frac{c_y^2}{c_1\bar{n}_b+\frac{c_y c_2 2\bar{n}_b}{c_f \bar{n}_b}} + \frac{(i_y-\mu_m)^2}{2\sigma_m^2}\right)} \right) \right) dh. \tag{26}$$

where $\mathrm{InS}(i,\cdot)$ represents the operation of $\int_{-\infty}^{\infty}\sum_{m=0}^{\infty}(\cdot)di$. Parameters $x = h\bar{n}_s + \bar{n}_b$, $c_x = m - Gx$, $c_y = m - G\bar{n}_b$, $c_1 = 2G^2F_a$, $c_2 = \pi\sigma_m\sqrt{2c_1}$. The complete mathematical derivation of this SER formulation is provided in Appendix A.2. The Webb-modeled SER formulation (26) encounters analytical intractability in resolving this multidimensional integration complexity. We can use trapezoidal rules and simulation tools to obtain significant numerical computation results.

For the Gaussian-approximated APD output model operating under composite channel conditions, the SER of PPM framework is analytically formulated as

$$\begin{aligned} E_G = & \frac{Q-1}{2}\exp\left(-\frac{\Omega_F^2}{2\sigma_\theta^2}\right) + \frac{(Q-1)C_{\alpha\beta}C_{p1}C_{p2}^2}{2h_c\Gamma(\alpha_t)\Gamma(\beta_t)\Gamma(\alpha_s)\Gamma(\beta_s)}\left(1-\exp\left(-\frac{\Omega_F^2}{2\sigma_\theta^2}\right)\right) \\ & \times \int_0^{\infty} \hat{G}_{1,5}^{5,0}\left[\frac{C_{\alpha\beta}C_{p1}}{h_c}h \,\middle|\, \begin{matrix} C_{p2}^2 \\ C_{p2}^2-1, \alpha_t-1, \beta_t-1, \alpha_s-1, \beta_s-1 \end{matrix}\right] \mathrm{erfc}\left(\frac{Gq_e\bar{n}_s h}{\sqrt{c_{g1}(2\bar{n}_b + h\bar{n}_s) + c_{g2}}}\right) dh. \end{aligned} \tag{27}$$

where $c_{g1} = 2q_e^2G^2F_a$, $c_{g2} = 4q_eI_DT_s + \frac{8\kappa_c T_e T_s}{R_L}$, and $\mathrm{erfc}(\cdot)$ is the complementary error function. Correspondingly, the mathematical derivation process underlying these results is rigorously presented in Appendix A.2.

5. Numerical Results and Discussion

This section presents various numerical results. First, the effects of various fading factors are discussed. Then, we compare Webb and Gaussian-approximated APD output and SER performance. Finally, key system parameters are properly designed to optimize communication performance.

For a single transmission, where the link length, divergence angle, and atmosphere loss factor are fixed, the channel attenuation is constant. Therefore, for simplicity, we normalize the channel attenuation coefficient, i.e., $h_c = 1$. In addition, Table 1 summarizes other constant parameters used in simulations.

Table 1. Simulation parameters.

Parameter	Symbol	Value
Optical wavelength	λ	1550 nm
PPM symbol size	Q	16
Slot width	T_s	2×10^{-8} s
Optical filter bandwidth	B_o	12.5×10^{9} Hz
APD quantum efficiency	η_a	0.9
APD Bandwidth	B_e	2.5×10^{7} Hz
Dark current	I_D	1×10^{-9} A
Load resistance	R_L	146,650 Ω
Ionization ratio	k_a	0.007
Field of view	Ω_F	150 µrad
Spectral albedo of the Earth	R_f	0.25
Optical receiving aperture diameter	d_r	0.1 m
Background radiation spectral radiance	$N_b(\lambda)$	6.4×10^{-20} W/Hz

5.1. Results of the Channel Model

For atmospheric and coronal turbulence, values of their Rytov variances are shown in Appendix A.1. We use $\sigma_{ht}^2 = 0.4$ and $\sigma_{hc}^2 = 0.6$ for weak turbulence, and $\sigma_{ht}^2 = 2$ and $\sigma_{hc}^2 = 3$ for strong turbulence. Consider that the radius of light spot have been normalized at the receiver, we use $\sigma_p = 0.1$ and $\sigma_p = 0.3$ to represent small and large pointing errors, respectively. For simulation of AOA fluctuations, we use $\sigma_\theta = 20$ μrad for small fluctuations and use $\sigma_\theta = 100$ μrad for large fluctuations.

Our derived closed-form expression for the uplink channel model is shown as (17). The analytical results and the Monte Carlo simulation results under different σ_{ht}^2, σ_{hs}^2, σ_p, and σ_θ are shown in Figure 4. It can be seen that the analytical results agree well with Monte Carlo simulation results, which confirms the accuracy of our derived closed-form expression of the uplink model. Moreover, it is clear that the average channel fading coefficient, i.e., mean of h, decreases with the increase in σ_{ht}^2, σ_{hc}^2, σ_p, and σ_θ.

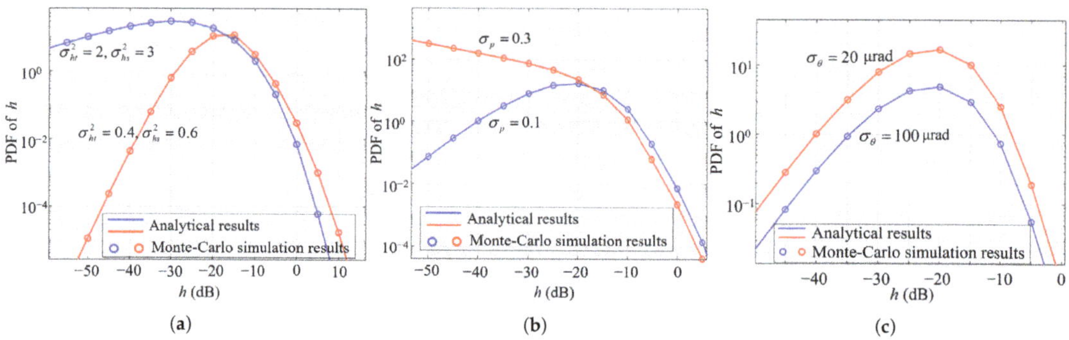

Figure 4. Channel distribution under various: (**a**) atmospheric and coronal turbulence, (**b**) pointing errors, and (**c**) AOA fluctuations.

In addition, as shown in Figure 4a,b, respectively, the distribution shapes of the PDF curves under various values of σ_{ht}^2, σ_{hs}^2, and σ_p are different. However, in Figure 4c, the distribution shapes of the PDF curves are similar for different AOA fluctuations. Specifically, curvilinear trends are the same, and only absolute PDF values are different when σ_θ is 20 µrad and 100 µrad. The reasons for these cases are that various atmospheric turbulence, coronal turbulence, and pointing errors will cause different fading in the channel. However, in the model of AOA fluctuations, h_a is reasonably assumed to be two values, 1 and 0, which correspond to the optical beam being captured successfully and unsuccessfully, respectively. That is, h_a does not influence the channel distribution when $h_a = 1$, and the probability of $h_a = 0$ increases with increasing σ_θ when FOV is given. Hence, in addition to the results shown in Figure 4c, it is reasonable to infer that the probability of $h_a = 0$ of the blue curve is bigger than that of the red curve.

5.2. Results of the Webb and Gaussian Modeled SER

For performance analysis of APD-based optical communication systems, many papers have shown that the Webb model is more precise than the Gaussian model [33,35,49]. Of course, the Gaussian model is also widely used for performance analysis due to the simplicity of expressions [36–41]. Here, we first investigate whether the Gaussian model is precise enough to characterize the considered deep space optical uplink system. In particular, we compare the SER estimated by Webb model and Gaussian model, with the Webb-estimated SER as the performance benchmark. SER performance is studied under various channel conditions. The applicable conditions where the Gaussian model can give a close fit to the Webb model, i.e., having a low approximation error, are discussed.

Figure 5a shows comparisons of SER using Webb and Gaussian models under different average background noise and over the lossless link, where APD gain is 40. Parameters $\bar{n}_b$ and $\bar{n}_s$ denote the average number of background noise photons and received optical photons in a PPM symbol, respectively. Under both Webb and Gaussian models, SER decreases monotonically with increasing received photons, i.e., signal-to-noise ratio (SNR), aligning with expectations. In addition, SER exhibits an inverse dependence on the background noise level. In particular, under $\bar{n}_b = 1$, the Gaussian approximation achieves remarkable congruence with the Webb model, validating the efficacy of the Gaussian model in noise-dominated regimes as analyzed in Section 3.2. At $\bar{n}_b = 0.01$, the Gaussian model systematically underestimates SER relative to the Webb framework, highlighting its inherent limitations in high-sensitivity regimes. However, in this scenario, the Gaussian model can still be utilized to evaluate the system performance. This is because the communication performance estimated by the Gaussian model is worse than the actual system performance. Consequently, during system design, to achieve specific communication performance targets, the designed system redundancy (referring to the performance margin, such as additional power or error correction capability) becomes higher than the actual system redundancy. For instance, the actual received signal power might be greater than pre-research design assumptions. Consequently, over the lossless link, Webb with high accuracy and Gaussian with low complexity are both applicable in optical communications.

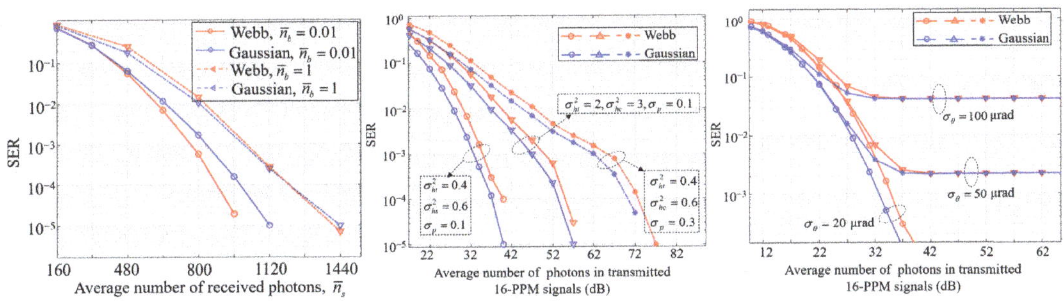

(**a**) (**b**) (**c**)

Figure 5. (**a**) Comparisons of SER using Webb and Gaussian models over lossless link with $G = 40$. (**b**) Comparisons of Webb and Gaussian models estimated SER under various σ_{ht}^2, σ_{hs}^2, and σ_p with $G = 40$ and $\bar{n}_b = 0.01$. (**c**) Under different AOA fluctuations, comparisons of Webb and Gaussian models estimated SER with $G = 40$ and $\bar{n}_b = 0.01$.

In the fading channel, comparisons of SER using Webb and Gaussian models with APD gain being 40 are shown in Figure 5b. It can be seen that, when average received photons are given, the SER becomes worse as pointing errors, AOA fluctuations, and atmospheric and coronal turbulence intensity increase. Moreover, in Figure 5b, SER decreases with the increase in the average received photons under various σ_{ht}^2, σ_{hs}^2, and σ_p, but in Figure 5c, the SER first decreases, and then remains constant, which means average received photons reaching saturation point. This is because σ_θ can cause an optical receiving failure, which further results in SER error floors. In other words, the error floor level is decided primarily by σ_θ. In particular, the error floor increases with increasing σ_θ.

Moreover, when the error floor is not present in Figure 5b,c, Gaussian model estimated SER is smaller than Webb model estimated SER under a given average received photons, which differs from Figure 5a. That is, communication redundancy estimated by the Gaussian model is higher than that by the Webb model. The actual redundancy of APD-based communication system can be evaluated by the Webb model, so the Gaussian model overestimates communication performance and no longer applies under the circumstances of Figure 5b,c.

As illustrated in Figure 6a, comparative analysis of SER dynamics under varying APD gain G reveals fundamentally distinct behaviors between the Webb and Gaussian models. For fixed turbulence parameters σ_{ht}^2 and σ_{hs}^2, the Gaussian-modeled SER exhibits a non-monotonic trend: initially decreasing sharply with G, reaching a minimum, then gradually increasing due to noise amplification. The Webb modeled SER rapidly decreases before stabilizing asymptotically. This divergence is due to the models' contrasting treatments of signal-to-shot noise ratio (SSNR) shown in Figure 6b. At low-to-moderate gains ($G < G_T$), both models predict SER reduction dominated by signal amplification, where photon-generated electrons scale quadratically with G while thermal and dark current noise remain constant. The critical threshold G_T, defined as the intersection point of the Webb and Gaussian SER curves, decreases logarithmically with increasing σ_{ht}^2 and σ_{hs}^2. Beyond G_T, Gaussian-modeled SER degrades due to its large noise amplification, whereas Webb-modeled SER stabilizes by incorporating McIntyre-distributed gain statistics that inherently bound excess noise growth. These dynamics underscore the necessity of the Webb model in gain-dominated regimes ($G < G_T$). In addition, the threshold gain G_T exhibits a positive correlation with channel fading, decreasing from 760 to 300 when σ_{ht}^2, σ_{hs}^2 decrease from (2, 3) to (0.4, 0.6), corresponding to a 60.5% mitigation. Similar trends appear under pointing errors and AOA fluctuations, as detailed in Appendix A.3. These significant results reveal that adaptive model switching based on real-time gain monitoring and channel state estimation optimizes receiver efficiency for deep space optical uplink communication under various channel impairments.

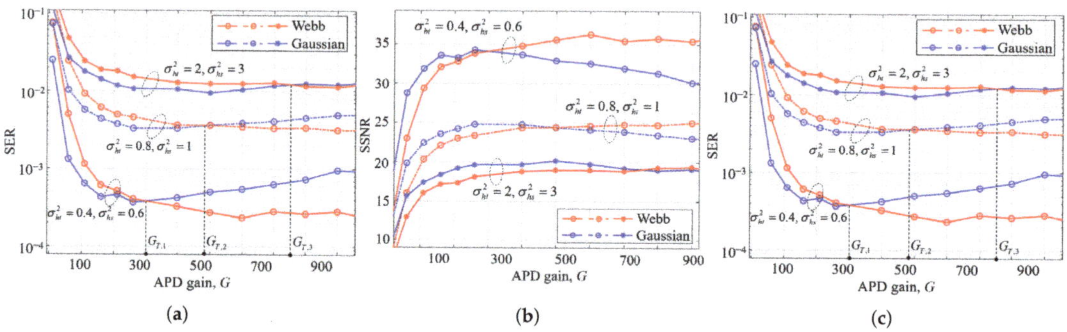

Figure 6. (**a**) SER versus APD gain under various atmospheric and coronal turbulence. ($\sigma_p = 0.1$, $\sigma_\theta = 20\ \mu$rad, $\bar{n}_s = 47$ dB, and $\bar{n}_b = 0.01$). (**b**) SSNR versus APD gain under various atmospheric and coronal turbulence. ($\sigma_p = 0.1$, $\sigma_\theta = 20\ \mu$rad, $\bar{n}_s = 47$ dB, and $\bar{n}_b = 0.01$). (**c**) SER versus APD gain under different background noise, where $\sigma_{ht}^2 = 0.8$, $\sigma_{hs}^2 = 1$, $\sigma_p = 0.1$, $\sigma_\theta = 20\ \mu$rad, $\bar{n}_s = 47$ dB.

In general, deep space optical uplink systems confront multi-factor channel impairments, including atmospheric and coronal turbulence, pointing errors, AOA fluctuations, and background noise. For modeling the APD output and estimating system SER performance, Webb model achieves high accuracy by greatly resolving the APD gain dynamics. Gaussian approximation becomes operationally viable when the APD gain exceeds the critical threshold G_T, where noise is dominated.

5.3. Parameters Design

The performance of deep space optical uplink communication can be optimized by designing key parameters of the system. First, the FOV angle Ω_F critically governs the performance through its dual influence on AOA fluctuations and background noise. As illustrated in (7) and (10), small Ω_F configurations mitigate background noise by spatially filtering extraneous photons but exacerbate signal loss from AOA fluctuations. In contrast, a large Ω_F improves beam capture robustness at the cost of amplified background noise. Figure 7 illustrates the relationship between the received background photon count and the field-of-view angle Ω_F. The results demonstrate that, when implementing 16-PPM modulation with a 10 MHz slot rate, the average background photon number $\bar{n}_b$ exceeds 10 photons/slot once Ω_F exceeds 0.5 mrad. Therefore, it is necessary to consider the trade-off between AOA fluctuations and background noise when designing the angle of FOV.

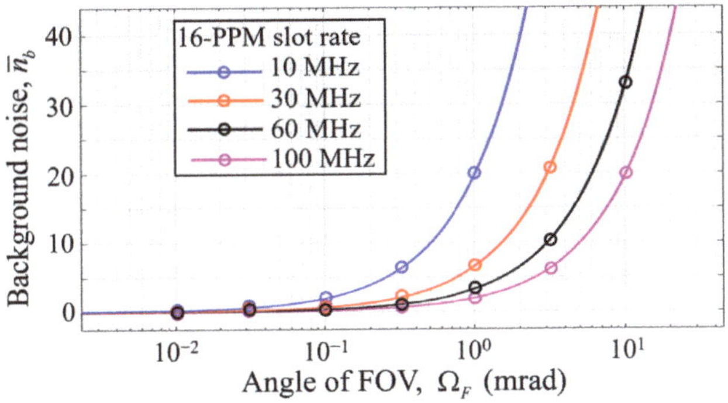

Figure 7. Photons number of received background noise versus angle of FOV under various PPM slot rate, where $d_r = 0.1$ m and $B_0 = 12.5 \times 10^9$ Hz.

From Figure 6, it is concluded that the APD gain is a significant parameter affecting system performance. As the APD gain G increases, the SER predicted by the Webb model initially decreases before stabilizing at a relatively constant level, while the Gaussian model exhibits a distinct behavior in which the SER first reduces and then subsequently increases. Given that the Webb model provides more accurate characterization of practical APD operation compared to the Gaussian approximation, the optimal APD gain determination should primarily rely on the SER trend observed through Webb modeling. Specifically, this optimal gain corresponds to the transition point where the Webb model's SER curve approaches its stable plateau region. For example, as shown in Figure 6c, the Webb model identifies approximately 150 as the optimal value of G. However, when employing the computationally simpler Gaussian model while requiring enhanced estimation accuracy, the optimal gain selection must satisfy dual criteria: achieving minimal SER according to the Webb model reference while simultaneously meeting the Gaussian model's operational precondition $G \geq G_T$ established in prior analysis. This explains why, under identical conditions in Figure 6c, despite the Gaussian model suggesting $G = 150$ as its minimum SER point, the practically viable value of G should be elevated to 500 to comply with both performance optimization requirements and model validity constraints.

6. Conclusions

This study established a comprehensive analytical framework for APD-based deep space optical uplink systems under combined channel impairments, including atmospheric and coronal turbulence-induced scintillation, pointing errors, AOA fluctuations, link attenuation, and background noise. A closed-form analytical channel model integrating these effects was derived and validated through Monte Carlo simulations, demonstrating strong alignment with the empirical results. The Webb and Gaussian models were used to characterize the APD output statistics. The Webb model exhibited higher accuracy in capturing nonlinear gain dynamics and SER trends. Gaussian approximation introduced significant deviations in the SER estimation unless the APD gain exceeded a critical threshold, which decreased under intensified channel fading. This highlighted the need for adaptive model selection based on real-time gain monitoring and channel conditions. Key system parameters, such as APD gain, angle of FOV, and o ptimal APD gain significantly influenced the achievement of optimal SER performance, balancing signal amplification and noise saturation.FOV design required a trade-off: small FOV angles reduced background noise but increased AOA-induced signal loss, while large FOV angles improved alignment robustness at the cost of noise intrusion. These findings enabled hardware optimization under SWaP-C constraints. Our work provided critical guidelines for designing robust APD-based deep space optical uplink communication systems.

Author Contributions: Conceptualization, W.G.; methodology, W.G.; software, W.G.; validation, W.G. and X.W.; formal analysis, W.G.and X.W.; investigation, W.G.; resources, W.G. and X.W.; data curation, X.W. and L.Y.; writing—original draft preparation, W.G.; writing—review and editing, X.W.; visualization, W.G., X.W. and L.Y.; supervision, X.W. and L.Y.; project administration, L.Y.; funding acquisition, L.Y. All authors have read and agreed to the published version of the manuscript.

Funding: This research was funded in part by the National Natural Science Foundation of China under Grants 62371439, 62301533, and U24A20214; in part by the Youth Innovation Promotion Association CAS under Grant 2022161; and in part by the Prospective Foundation of Technology and Engineering Center for Space Utilization, Chinese Academy of Sciences under Grant T40347110Y.

Institutional Review Board Statement: Not applicable.

Informed Consent Statement: Not applicable.

Data Availability Statement: Data underlying the results are all presented in this paper.

Conflicts of Interest: The authors declare no conflicts of interest.

Abbreviations

The following abbreviations are used in this manuscript:

APD	Avalanche photodiode
SER	Symbol error rate
FOV	Field of view
AOA	Angle-of-arrival
SWaP-C	Size, weight, power, and cost
PPM	Pulse position modulation
OOK	On-off keying
PDF	Probability density function
CDF	Cumulative distribution function
SNR	Signal-to-noise ratio
SSNR	Signal-to-shot noise ratio

Appendix A

Appendix A.1

Rytov variances σ_{ht}^2 and σ_{hs}^2 are essential for atmospheric and coronal turbulence, respectively. Therefore, here we analyze and calculate the typical values for σ_{ht}^2 and σ_{hs}^2. Structure parameter $C_n^2(H)$ in (4) is expressed as the Hufnagel–Valley model ([50], p. 481),

$$C_n^2(H) = \left(C_0^2 + 594H^{10}\left(\frac{V_d}{27}\right)^2\right)\exp\left(-\frac{H}{1000}\right) + 2.7\times10^{-16}\exp\left(-\frac{H}{1500}\right), \quad \text{(A1)}$$

where V_d denotes the root mean square of the wind speed, and C_0^2 is the turbulence structure constant at the ground level. Solar wind density in (5) is given by [9]

$$N_e = 2.21 \times 10^{14} \left(\frac{L_{sl}}{r_s} \right)^{-6} + 1.55 \times 10^{12} \left(\frac{L_{sl}}{r_s} \right)^{-2.3}, \tag{A2}$$

where r_s denotes the radius of the Sun, and L_{sl} is the vertical range between the Sun and the optical link. $L_{sl} = L_{se} \sin(\theta_{sep})$, where L_{se} is the distance between the Sun and the Earth, and θ_{sep} is the angle of Sun–Earth–probe (SEP).

Consequently, based on parameters shown in Table A1 [9,44], results of σ_{ht}^2 and σ_{hs}^2 are calculated and shown in Figure A1a,b, respectively. It can be seen that, for atmospheric turbulence, σ_{ht}^2 increases with the increase in wind speed v_d and structure constant at the ground level C_0^2. For coronal turbulence, when the angle of SEP θ_{sep} increases or the spectral index decreases, σ_{hs}^2 will decrease. In general, as the value of Rytov variance is much bigger than 1, turbulence is defined as strong turbulence.

Table A1. Parameters for atmospheric and coronal turbulence.

Parameter	Symbol	Value
Ground station altitude	H_g	0.1 km
Wave number	k_n	4.05×10^6
Zenith angle	ζ_g	30°
Satellite height	H_s	5×10^5 km
Classical electron radius	r_e	2.8×10^{-15} m
Relative solar wind density fluctuations ratio	ζ_s	0.2
Radius of the Sun	r_s	8.96×10^5 km
Coronal turbulence outer scale	L_o	1×10^6 km
Distance between the Sun and the Earth	L_{se}	1.5×10^8 km

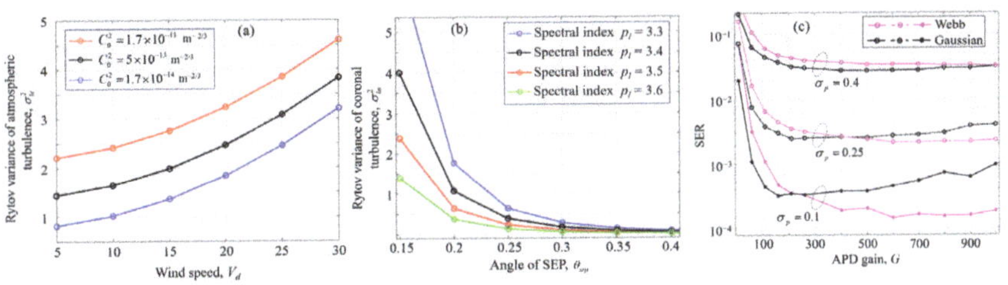

Figure A1. (**a**) Atmospheric turbulence Rytov variance σ_{ht}^2 versus wind speed V_d under different C_0^2. (**b**) Coronal turbulence Rytov variance σ_{hs}^2 versus angle of SEP θ_{sep} under different p_l. (**c**) SER versus APD gain under different pointing errors.

Appendix A.2

According to (07.34.21.0011.01) in [47], which is denoted as

$$\int_0^\infty \tau^{\alpha-1} \hat{G}_{u,v}^{s,t}\left(\tau w \middle| \begin{matrix} c_1,\ldots,c_t,c_{t+1},\ldots,c_u \\ d_1,\ldots,d_s,d_{s+1},\ldots,d_v \end{matrix}\right) \hat{G}_{p,q}^{m,n}\left(\tau z \middle| \begin{matrix} a_1,\ldots,a_n,a_{n+1},\ldots,a_p \\ b_1,\ldots,b_m,b_{m+1},\ldots,b_q \end{matrix}\right) d\tau ==$$
$$w^{-\alpha} \hat{G}_{v+p,u+q}^{m+t,n+s}\left(\frac{z}{w} \middle| \begin{matrix} a_1,\ldots,a_n,1-\alpha-d_1,\ldots,1-\alpha-d_s,1-\alpha-d_{s+1},\ldots,1-\alpha-d_v,a_{n+1},\ldots,a_p \\ b_1,\ldots,b_m,1-\alpha-c_1,\ldots,1-\alpha-c_t,1-\alpha-c_{t+1},\ldots,1-\alpha-c_u,b_{m+1},\ldots,b_q \end{matrix}\right). \tag{A3}$$

where parameters $s = 0$, $t = 2$, $u = 2$, $v = 0$, $m = 3$, $n = 0$, $p = 1$, and $q = 1$. Consequently, (14) is derived as (15).

According to (17) and (24), the Webb based PPM SER derivation process is as follows:

$$
\begin{aligned}
E_W &= \int_0^\infty f_h(h)\left(1-\int_{-\infty}^{\infty} p_w(i_x|h\bar{n}_s+\bar{n}_b)di_x\left(\int_{-\infty}^{i_x} p_w(i_y|\bar{n}_b)di_y\right)^{Q-1}\right)dh \\
&= \int_0^\infty f_h(h)\left\{\begin{array}{l} 1-\int_{-\infty}^{\infty}\sum_{m=0}^{\infty}\left(\frac{\left(1+\frac{m-G(h\bar{n}_s+\bar{n}_b)}{c_F(h\bar{n}_s+\bar{n}_b)}\right)^{2/3}}{2\pi\sigma_m\sqrt{(h\bar{n}_s+\bar{n}_b)G^2F_a}}\exp\left(-\frac{\frac{(m-G(h\bar{n}_s+\bar{n}_b))^2}{2(h\bar{n}_s+\bar{n}_b)G^2F_a}}{1+\frac{m-G(h\bar{n}_s+\bar{n}_b)}{c_F(h\bar{n}_s+\bar{n}_b)}}-\frac{(i_x-\mu_m)^2}{2\sigma_m^2}\right)\right)di_x \\ \cdot\left(\int_{-\infty}^{i_x}\sum_{m=0}^{\infty}\left(\frac{\left(1+\frac{m-G\bar{n}_b}{c_F\bar{n}_b}\right)^{2/3}}{2\pi\sigma_m\sqrt{\bar{n}_bG^2F_a}}\exp\left(-\frac{(m-G\bar{n}_b)^2}{2\bar{n}G^2F_a\left(1+\frac{m-G\bar{n}_b}{c_F\bar{n}_b}\right)}-\frac{(i_y-\mu_m)^2}{2\sigma_m^2}\right)\right)di_y\right)^{Q-1}\end{array}\right\}dh \\
&= \int_0^\infty f_h(h)\left(1-\mathrm{InS}\left(i_x,\frac{\frac{1}{c_2\sqrt{x}}\left(1+\frac{c_x}{c_fx}\right)^{2/3}}{\exp\left(\frac{c_x^2}{c_1\bar{n}+\frac{c_xc_{22}\bar{n}}{c_fx}}+\frac{(i_x-\mu_m)^2}{2\sigma_m^2}\right)}\right)\mathrm{InS}^{Q-1}\left(i_y,\frac{\frac{1}{c_2\sqrt{\bar{n}_b}}\left(1+\frac{c_y}{c_f\bar{n}_b}\right)^{2/3}}{\exp\left(\frac{c_y^2}{c_1\bar{n}_b+\frac{c_yc_{22}\bar{n}_b}{c_f\bar{n}_b}}+\frac{(i_y-\mu_m)^2}{2\sigma_m^2}\right)}\right)\right)dh.
\end{aligned}
\tag{A4}
$$

For Gaussian model estimated PPM SER, the deriving process is shown as

$$
\begin{aligned}
E_G &= \int_0^\infty f_h(h)\frac{Q-1}{2}\mathrm{erfc}\left(\sqrt{\frac{SNR}{2}}\right)dh \\
&= \int_0^\infty\left\{\begin{array}{l}\frac{Q-1}{2}\mathrm{erfc}\left(Gq_e\bar{n}_s\left(2q_e^2G^2F_a(2\bar{n}_b+\bar{n}_s)+4q_eI_DT_s+\frac{8\kappa_cT_eT_s}{R_L}\right)^{-1/2}\right) \\ \left\{\begin{array}{l}\frac{C_{\alpha\beta}C_{p1}C_{p2}^2}{h_c\Gamma(\alpha_t)\Gamma(\beta_t)\Gamma(\alpha_c)\Gamma(\beta_c)}\left(1-\exp\left(-\frac{\Omega_F^2}{2\sigma_\theta^2}\right)\right)\hat{G}_{1,5}^{5,0}\left[\frac{C_{\alpha\beta}C_{p1}}{h_c}h\left|\begin{array}{c}C_{p2}^2\\ C_{p2}^2-1,\alpha_t-1,\beta_t-1,\alpha_c-1,\beta_c-1\end{array}\right.\right] \\ +\delta(h)\cdot\exp\left(-\frac{\Omega_F^2}{2\sigma_\theta^2}\right)\end{array}\right\}\end{array}\right\}dh \\
&= \frac{Q-1}{2}\exp\left(-\frac{\Omega_F^2}{2\sigma_\theta^2}\right)+\frac{(Q-1)C_{\alpha\beta}C_{p1}C_{p2}^2}{2h_c\Gamma(\alpha_t)\Gamma(\beta_t)\Gamma(\alpha_s)\Gamma(\beta_s)}\left(1-\exp\left(-\frac{\Omega_F^2}{2\sigma_\theta^2}\right)\right) \\
&\quad\times\int_0^\infty\hat{G}_{1,5}^{5,0}\left[\frac{C_{\alpha\beta}C_{p1}}{h_c}h\left|\begin{array}{c}C_{p2}^2\\ C_{p2}^2-1,\alpha_t-1,\beta_t-1,\alpha_s-1,\beta_s-1\end{array}\right.\right]\mathrm{erfc}\left(\frac{Gq_e\bar{n}_sh}{\sqrt{c_{g1}(2\bar{n}_b+h\bar{n}_s)+c_{g2}}}\right)dh.
\end{aligned}
\tag{A5}
$$

Appendix A.3

Figures A1c and A2a show the SER and SSNR results under different σ_p, and Figure A2b,c show the SER and SSNR results under different σ_θ.

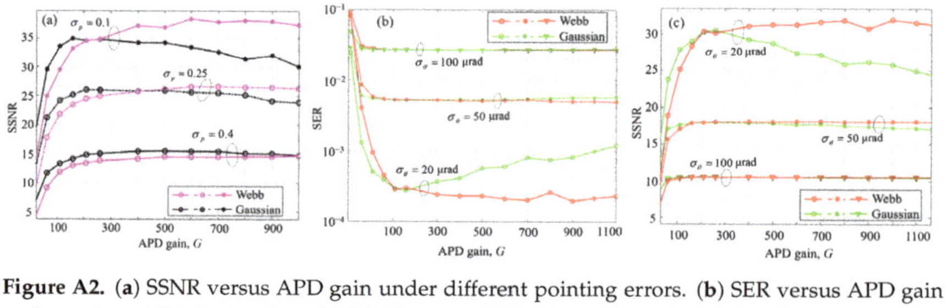

Figure A2. (**a**) SSNR versus APD gain under different pointing errors. (**b**) SER versus APD gain under various AOA fluctuations. (**c**) SSNR versus APD gain under various AOA fluctuations.

References

1. Biswas, A.; Srinivasan, M.; Piazzolla, S.; Hoppe, D. Deep space optical communications. In Proceedings of the Free-Space Laser Communication and Atmospheric Propagation XXX, San Francisco, CA, USA, 29–30 January 2018; SPIE: Bellingham, WA, USA, 2018; Volume 10524, p. 105240U. [CrossRef]
2. Boroson, D.M.; Robinson, B.S.; Murphy, D.V.; Burianek, D.A.; Khatri, F.; Kovalik, J.M.; Sodnik, Z.; Cornwell, D.M. Overview and results of the Lunar Laser Communication Demonstration. In Proceedings of the Free-Space Laser Communication and Atmospheric Propagation XXVI, San Francisco, CA, USA, 2–4 February 2014; SPIE: Bellingham, WA, USA, 2014; p. 89710S. [CrossRef]
3. Ivanov, H.; Leitgeb, E.; Freiberger, G. Characterization of poisson channel for deep space FSO based on SNSPD technology by experimental demonstration. In Proceedings of the 11th International Symposium on Communication Systems, Budapest, Hungary, 18–20 July 2018; IEEE: Piscataway, NJ, USA, 2018; pp. 1–5. [CrossRef]
4. Li, B.; Liu, Y.; Tong, S.; Zhang, L.; Yao, H. BER Analysis of a Deep Space Optical Communication System Based on SNSPD Over Double Generalized Gamma Channel. *IEEE Photon. J.* **2018**, *10*, 7907607. [CrossRef]
5. Zhang, Y.; Jiao, J.; Zhang, K.; Wang, Y.; Lu, R.; Zhang, Q. Synodic Period Channel Modeling and Coding Scheme for Deep Space Optical Communications. *IEEE Trans. Aerosp. Electron. Syst.* **2024**, *60*, 2364–2378. [CrossRef]
6. Kudielka, K.; Fischer, E.; Berkefeld, T.; Fischer, C. Optical feeder link demonstrations between the ESA Optical Ground Station and Alphasat. In Proceedings of the 2023 IEEE International Conference on Space Optical Systems and Applications (ICSOS), Vancouver, BC, Canada, 11–13 October 2023; IEEE: Piscataway, NJ, USA, 2024; pp. 64–68. [CrossRef]
7. Kaushal, H.; Kaddoum, G. Optical Communication in Space: Challenges and Mitigation Techniques. *IEEE Commun. Surv. Tutor.* **2017**, *19*, 57–96. [CrossRef]
8. Stahl, H.P.; Hopkins, R.C.; Schnell, A.; Smith, D.A.; Jackman, A.; Warfield, K.R. Designing astrophysics missions for NASA's Space Launch System. *J. Astron. Telesc. Instruments, Syst.* **2016**, *2*, 041213. [CrossRef]
9. Xu, G.; Zeng, M. Solar Scintillation Effect for Optical Waves Propagating Through Gamma–Gamma Coronal Turbulence Channels. *IEEE Photon. J.* **2019**, *11*, 1–15. [CrossRef]
10. Xu, G.; Zhang, Q. Mixed RF/FSO Deep Space Communication System Under Solar Scintillation Effect. *IEEE Trans. Aerosp. Electron. Syst.* **2021**, *57*, 3237–3251. [CrossRef]
11. Xu, G.; Song, Z. Effects of Solar Scintillation on Deep Space Communications: Challenges and Prediction Techniques. *IEEE Wirel. Commun.* **2019**, *26*, 10–16. [CrossRef]
12. Hennes, H.; Otakar, W. An Introduction to Free-space Optical Communications. *Radioengineering* **2010**, *19*, 203–212. [CrossRef]
13. Lee, E.J.; Chan, V.W. Part 1: Optical communication over the clear turbulent atmospheric channel using diversity. *IEEE J. Sel. Areas Commun.* **2004**, *22*, 1896–1906. [CrossRef]
14. Sandalidis, H. Performance Analysis of a Laser Ground-Station-to-Satellite Link With Modulated Gamma-Distributed Irradiance Fluctuations. *J. Opt. Commun. Netw.* **2010**, *2*, 938–943. [CrossRef]
15. Mishra, N.; Kumar, D.S. Outage analysis of relay assisted FSO systems over K-distribution turbulence channel. In Proceedings of the 2016 ICEEOT, Chennai, India, 3–5 March 2016; IEEE: Piscataway, NJ, USA, 2016; pp. 2965–2967. [CrossRef]
16. Barrios, R.; Dios, F. Exponentiated Weibull model for the irradiance probability density function of a laser beam propagating through atmospheric turbulence. *Opt. Laser Technol.* **2013**, *45*, 13–20. [CrossRef]
17. Badarneh, O.S.; Mesleh, R. Diversity analysis of simultaneous mmWave and free-space-optical transmission over F-distribution channel models. *J. Opt. Commun. Netw.* **2020**, *12*, 324–334. [CrossRef]
18. Shen, B.; Chen, J.; Xu, G.; Chen, Q.; Wang, J. Performance Analysis of a Drone-Assisted FSO Communication System over Málaga Turbulence under AOA Fluctuations. *Drones* **2023**, *7*, 374. [CrossRef]
19. Donmez, B.; Azam, I.; Kurt, G.K. Mitigation of Misalignment Errors Over Inter-Satellite FSO Energy Harvesting: (Invited Paper). In Proceedings of the PIMRC, Toronto, ON, Canada, 5–8 September 2023; IEEE: Piscataway, NJ, USA, 2023; pp. 1–5. [CrossRef]
20. Mao, B.; Zhou, X.; Liu, J.; Kato, N. On an Intelligent Hierarchical Routing Strategy for Ultra-Dense Free Space Optical Low Earth Orbit Satellite Networks. *IEEE J. Sel. Areas Commun.* **2024**, *42*, 1219–1230. [CrossRef]
21. Ninos, M.P.; Spirito, V.; Cossu, G.; Ciaramella, E. Outage Performance of Uplink Pre-Amplified FSO Links Over Turbulence, Beam Wander, and Pointing Errors. *IEEE Commun. Lett.* **2023**, *27*, 3275–3279. [CrossRef]
22. Safi, H.; Dargahi, A.; Cheng, J.; Safari, M. Analytical Channel Model and Link Design Optimization for Ground-to-HAP Free-Space Optical Communications. *J. Light. Technol.* **2020**, *38*, 5036–5047. [CrossRef]
23. Xu, G.; Zhang, N.; Xu, M.; Xu, Z.; Zhang, Q.; Song, Z. Outage Probability and Average BER of UAV-Assisted Dual-Hop FSO Communication With Amplify-and-Forward Relaying. *IEEE Trans. Veh. Technol.* **2023**, *72*, 8287–8302. [CrossRef]
24. Al-Gailani, S.A.; Mohd Salleh, M.F.; Salem, A.A.; Shaddad, R.Q.; Sheikh, U.U.; Algeelani, N.A.; Almohamad, T.A. A Survey of Free Space Optics (FSO) Communication Systems, Links, and Networks. *IEEE Access* **2021**, *9*, 7353–7373. [CrossRef]
25. Al Naboulsi, M.; Sizun, H.; de Fornel, F. Fog attenuation prediction for optical and infrared waves. *Opt. Eng.* **2004**, *43*, 319–329. [CrossRef]
26. Guo, W.; Zhan, Y.; Tsiftsis, T.A.; Yang, L. Performance and Channel Modeling Optimization for Hovering UAV-Assisted FSO Links. *J. Light. Technol.* **2022**, *40*, 4999–5012. [CrossRef]

27. Farrell, T.C. Sources of background light on space based laser communications links. In Proceedings of the Sensors and Systems for Space Applications XI, Anaheim, CA, USA, 10–11 April 2017; SPIE: Bellingham, WA, USA, 2018; Volume 10641, pp. 175–191. [CrossRef]
28. Xu, G. BER and channel capacity of a deep space FSO communication system using L-PPM-MSK-SIM scheme during superior solar conjunction. *Opt. Express* **2019**, *27*, 24610–24623. [CrossRef]
29. Gagliardi, R.M.; Karp, S. *Optical Communications*; Wiley Interscience: Hoboken, NJ, USA, 1976.
30. Xu, F.; Khalighi, M.A.; Bourennane, S. Impact of different noise sources on the performance of PIN-and APD-based FSO receivers. In Proceedings of the IEEE 11th ICTCE, Paphos, Cyprus, 31 August–2 September 2011; IEEE: Piscataway, NJ, USA, 2023; pp. 211–218.
31. Sarbazi, E.; Safari, M.; Haas, H. The Bit Error Performance and Information Transfer Rate of SPAD Array Optical Receivers. *IEEE T. Commmun.* **2020**, *68*, 5689–5705. [CrossRef]
32. Miki, S.; Miyajima, S.; China, F.; Yabuno, M.; Terai, H. Photon detection at 1 ns time intervals using 16-element SNSPD array with SFQ multiplexer. *Opt. Lett.* **2021**, *46*, 6015–6018. [CrossRef]
33. Webb, P.P.; Mcintyre, R.J.; Conradi, J. Properties of Avalanche Photodiodes. *RCA Rev.* **1974**, *35*, 234–278.
34. Davidson, F.M.; Sun, X. Gaussian approximation versus nearly exact performance analysis of optical communication systems with PPM signaling and APD receivers. *IEEE T. Commun.* **1988**, *36*, 1185–1192. [CrossRef]
35. Srinivasan, M.; Vilnrotter, V. Symbol-error probabilities for pulse-position modulation signaling with an avalanche photodiode receiver and Gaussian thermal noise. In *The Telecommunications and Mission Operations Progress Report*; TMO-PR: Pasadena, CA, USA, 1998; pp. 1–11.
36. Cvijetic, N.; Wilson, S.G.; Brandt Pearce, M. Performance Bounds for Free-Space Optical MIMO Systems with APD Receivers in Atmospheric Turbulence. *IEEE J. Sel. Areas Commun.* **2008**, *26*, 3–12. [CrossRef]
37. Dabiri, M.T.; Sadough, S.M.S.; Khalighi, M.A. FSO channel estimation for OOK modulation with APD receiver over atmospheric turbulence and pointing errors. *Opt. Commun.* **2017**, *402*, 577–584. [CrossRef]
38. Amico, A.A.; Morelli, M. Joint Frame Detection and Channel Parameter Estimation for OOK Free-Space Optical Communications. *IEEE Trans. Commun.* **2022**, *70*, 4731–4744. [CrossRef]
39. Ivanov, H.; Hatab, Z.; Leitgeb, E. Channel Emulation of Satellite-to-Ground APD-based FSO Link in the Presence of Atmospheric Turbulence Fading. In Proceedings of the 2021 16th International Conference on Telecommunications (ConTEL), Zagreb, Croatia, 30 June–2 July 2021; IEEE: Piscataway, NJ, USA, 2021; pp. 33–36. [CrossRef]
40. Yu, X.; Xiao, L.; Gao, S.; Song, Y.; Zhang, L.; Yao, H.; Wang, T.; Chen, X.; Jin, Z.; Zhou, C.; et al. Atmospheric Turbulence Suppression Algorithm Based on APD Adaptive Gain Control for FSO Links of 5G Floating Base Stations. *IEEE Photonics J.* **2020**, *12*, 7904011. [CrossRef]
41. Kaur, S. Analysis of inter-satellite free-space optical link performance considering different system parameters. *Opto-Electron. Rev.* **2019**, *27*, 10–13. [CrossRef]
42. Kiasaleh, K. Performance of APD-based PPM free-space optical communication systems in atmospheric turbulence. *IEEE Trans. Commun.* **2005**, *53*, 1455–1461. [CrossRef]
43. Liang, J.; Chaudhry, A.U.; Erdogan, E.; Yanikomeroglu, H.; Kurt, G.K.; Hu, P.; Ahmed, K.; Martel, S. Free-Space Optical (FSO) Satellite Networks Performance Analysis: Transmission Power, Latency, and Outage Probability. *IEEE Open J. Veh. Technol.* **2024**, *5*, 244–261. [CrossRef]
44. Ata, Y.; Alouini, M.S. Performance of Integrated Ground-Air-Space FSO Links Over Various Turbulent Environments. *IEEE Photonics J.* **2022**, *14*, 7358916. [CrossRef]
45. Born, M.; Wolf, E. *Principles of Optics: Electromagnetic Theory of Propagation, Interference and Diffraction of Light*; Elsevier: Amsterdam, The Netherlands, 2013.
46. Xu, G.; Song, Z. Amplitude fluctuations for optical waves propagation through non-Kolmogorov coronal solar wind turbulence channels. *Opt. Express* **2018**, *26*, 8566–8580. [CrossRef] [PubMed]
47. Wolfram. The Wolfram Functions Site. 2021. Available online: https://functions.wolfram.com/PDF/MeijerG.pdf (accessed on 23 May 2025).
48. McIntyre, R.J. The distribution of gains in uniformly multiplying avalanche photodiodes: Theory. *IEEE Trans. Electron Devices* **1972**, *19*, 703–713. [CrossRef]
49. Dolinar, D.S.; Divsalar, J.H.; Pollara, F. *Capacity of Pulse-Position Modulation (PPM) on Gaussian and Webb Channels*; Technical Report JPL-TMO-PR-42-142; TMO-PR: Pasadena, CA, USA, 2000.
50. Andrews, L.C.; Phillips, R.L. *Laser Beam Propagation Through Random Media*, 2nd ed.; SPIE Press: Bellingham, WA, USA, 2005. [CrossRef]

photonics

MDPI

Article

Deep Learning-Assisted High-Pass-Filter-Based Fixed-Threshold Decision for Free-Space Optical Communications

Yan Gao, Qian-Wen Jing, Min-Fang Liu, Wen-Hao Zong and Yan-Qing Hong *

School of Information Science and Engineering, Shenyang University of Technology, Shenyang 110870, China; gy230408330@smail.sut.edu.cn (Y.G.); qianwenj@smail.sut.edu.cn (Q.-W.J.); lmf0914@smail.sut.edu.cn (M.-F.L.); 13861482399@smail.sut.edu.cn (W.-H.Z.)

* Correspondence: hongy7@sut.edu.cn

Abstract: This paper proposes a deep learning (DL)-assisted high-pass-filter (HPF)-based fixed-threshold decision (FTD) for free-space optical (FSO) communication. HPF is applied to reduce the scintillation effect by filtering out the low-frequency components of the received signal. However, the performance is limited owing to the signal distortion from HPF and remnant scintillation effect due to insufficient filtering. Therefore, the DL model is adopted to improve the performance of HPF-based scintillation effect compensation. The multilayer perceptron (MLP) model is used to adaptively select the peak frequency component of the received signal as the optimized cutoff frequency of HPF. Furthermore, recurrent neural network (RNN) and long short-term memory (LSTM) models are cascaded after HPF to compensate for the remnant scintillation effect and recover the signal distortion without the optimization of HPF cutoff frequency. The simulation was conducted under different turbulence channels and data rates. Simulation results showed that MLP-assisted adaptive optimized cutoff frequency and cascaded LSTM and HPF methods were close to the adaptive-threshold decision with precise channel state information under various turbulence channel degrees.

Keywords: deep learning; high-pass filter; scintillation effect; free-space optical communications

1. Introduction

Free-space optical (FSO) communication systems are widely used in space and terrestrial communication links. These systems have the features of ease of installation, free licensed spectrum, high data rates, high security, low cost, and low power consumption compared to conventional radio frequency communication systems [1]. However, FSO systems experience intensity fluctuations when optical signals traverse atmospheric turbulence channels, that is, the scintillation effect [2]. This phenomenon significantly degrades the performance of FSO communication systems and is a challenge for FSO technology. Regarding the on–off keying (OOK)-FSO communication system, the turbulence effect leads to fluctuations in the received OOK signal intensity [3]. Several studies have been conducted to mitigate the scintillation effects of OOK-FSO links. Adaptive optics systems have been studied to reduce the scintillation effect by detecting and correcting wavefront distortions [4]. However, this method is only effective in the context of weak turbulence channels. In [5], an adaptive channel coding technique has been proposed to pre-compensate the scintillation effect under the estimation of imperfect channel state information (CSI) at the transmitter. However, the periodic CSI estimation process leads to the hardware overhead increasing and time delay. The aperture averaging method has been discussed to mitigate the scintillation effect by employing a large aperture in the receiver end [6]. Nonetheless, challenges remain in deploying a large aperture in practice.

Fixed-threshold detection (FTD) and adaptive-threshold detection (ATD) are widely utilized in FSO systems [7]. The traditional FTD method employs the mean value of the received signal intensity as the decision threshold. However, it becomes ineffective under a stronger turbulence effect [8]. ATD can mitigate the scintillation effect through symbol-by-symbol decision threshold estimation [9]. However, the optimal decision-making process requires an accurate CSI. The instantaneous CSI is filtered by a low-pass filter (LPF) with optimization of the cutoff frequency [10]. However, with respect to the low transmission speed, the performance of the ATD is degraded owing to the severe signal distortion caused by the LPF. In [11], researchers integrate a dual-transmitter system with an adaptive-threshold receiver, which can mitigate the adverse effects of turbulence and improve system performance without the need for expensive adaptive optics. However, their effectiveness may be limited in scenarios of intense turbulence. Recently, deep learning (DL) techniques have been introduced into FSO communication in order to improve the system performance [12,13]. A convolutional neural network (CNN) has been employed to identify the intensity image categories of Laguerre Gaussian beams transmitted through free-space channels under various turbulence conditions for the sake of simultaneously detecting turbulence intensity and orbital angular momentum modes [14]. Fully connected neural networks (FCNNs) have been studied to compensate for the turbulence effects by extracting turbulence channel features from the received signal through element-wise multiplication and summation [15]. In [16], a feed-forward de-noising convolutional neural network (DnCNN) scheme has been proposed to implicitly remove potential clean images from the hidden layers through a residual learning strategy. The high-pass filter (HPF) has been applied to achieve FTD by blocking atmospheric turbulence caused by low-frequency fluctuations [17]. However, the low-frequency components of the signal are truncated by HPF as well, which decreases the performance of FTD. Therefore, the DL method is studied to improve the performance of HPF-based scintillation effect compensation.

In this study, a fixed-threshold OOK detection technique based on the HPF and DL is proposed for the FSO system. The HPF was used to block the low-frequency components of the OOK signal to reduce the scintillation effect. A multilayer perceptron (MLP) model was introduced to adaptively select the peak frequency component of the received signal as the optimized cutoff frequency of the HPF. Furthermore, recurrent neural networks (RNN) and long short-term memory (LSTM) models are cascaded to the HPF to compensate for the remnant scintillation effect and recover signal distortion without the optimized cutoff frequency of the HPF. The proposed method was evaluated through a simulation. Simulation results illustrated that the performance of the proposed MLP-assisted adaptive optimized cutoff frequency, cascaded LSTM and HPF were close to that of the traditional ATD with precise CSI.

The outline of this paper is as follows: In Section 2, we describe the principle of the proposed DL-assisted HPF-based FTD technique. In Section 3, we describe the turbulence channel analysis and simulation parameter setup. The simulation results are given in Section 4. In Section 5, we describe the conclusion of this paper.

2. Operation Principle

The principle of HPF-based FTD, MLP-assisted adaptive optimized cutoff frequency, and cascaded HPF and DL-based FTD without the optimization of HPF cutoff frequency was discussed in this section.

A block diagram of the proposed HPF-based FTD OOK detection technique is shown in Figure 1. A 1550 nm laser diode (LD) was directly modulated with the OOK signal. The modulated laser beam propagates through the FSO link and experiences turbulence-induced scintillation effects. The strength of the scintillation effect is assessed by the scintillation index σ_I^2, and it is given by

$$\sigma_I^2 = \frac{\langle I^2 \rangle}{\langle I \rangle^2} - 1, \tag{1}$$

where I denotes the intensity fluctuation and $\langle \cdot \rangle$ denotes the ensemble average [1]. At the receiver end, the received optical signal is converted into an electrical signal using a photodiode (PD). The PD-received signal is converted into a digital signal $r[k]$ using an analog-to-digital converter (ADC), and it is expressed as

$$r[k] = I[k]s[k] + n[k], \tag{2}$$

where $s[k] \in \{0,1\}$ denotes the symbol information, $n[k]$ represents signal-independent additive white Gaussian noise from the PD and background [18].

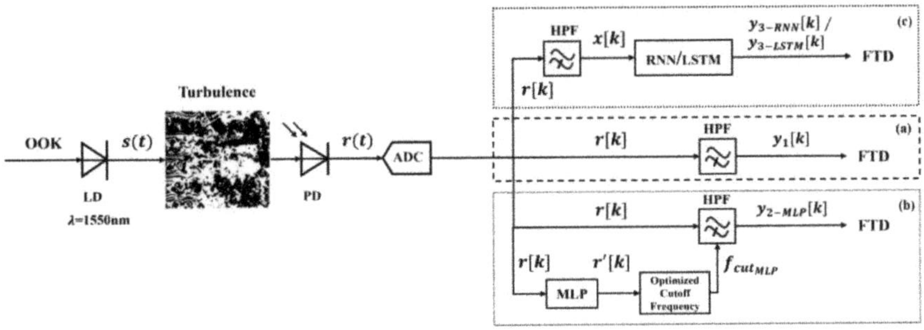

Figure 1. Block diagram of the proposed technique. (**a**) HPF-based FTD, (**b**) DL-based adaptive cutoff frequency of HPF, (**c**) cascaded HPF and DL-based turbulence effect compensation.

A. HPF-based FTD

In Figure 1a, the received signal passes through the HPF, and the scintillation-caused signal-intensity fluctuation is compensated for via low-frequency blocking. The frequency components of the received signal $r[k]$ are calculated using Discrete Fourier Transform (DFT) as follows [19]

$$R(f) = \sum_{n=0}^{N-1} r[k] \cdot e^{(-j2\pi fn/N)}, \tag{3}$$

where $R(f)$ is the DFT of the output $r[k]$, n denotes a discrete time point in the time domain, and N is the signal length. The frequency characteristics of the received signal are then calculated using the power spectral density (PSD), which is given as follows [19]

$$S(f) = \frac{1}{N}|R(f)|^2, \tag{4}$$

where $S(f)$ is the PSD of the $r[k]$. The frequency spectrum characteristics and PSD of the received signal are analyzed to determine the cutoff frequency f_{cut} of the HPF. The cutoff frequency of the HPF is calculated as [20]

$$\omega_{cut} = \frac{2\pi f_{index}}{N f_s}, \tag{5}$$

where f_{index} denotes the index of the selected cutoff frequency f_{cut} in PSD, while ω_{cut} represents the angular frequency from f_{cut}, and f_s is the sampling rate. The output signal $y_1[k]$ at the HPF is expressed as [20]

$$y_1[k] = \tan\left(\omega_{cut}\left(r[k] - 2\tan^{-1}\left(\frac{\omega_{cut} f_s}{N}\right)\right)\right). \tag{6}$$

The mean value of $y_1[k]$ is calculated as the decision threshold I_{th} of the received OOK signal to distinguish bits "1" and "0". The performance of the proposed technique is assessed by analyzing bit error rate (BER), and the BER is calculated by [21]

$$BER = \frac{1}{2}\left[\frac{1}{2}erfc\left(\frac{I_{th}}{\sqrt{2\sigma_n^2}}\right) + \int_0^{\infty}\frac{1}{2}erfc\left(\frac{I - I_{th}}{\sqrt{2\sigma_n^2}}\right)f(I)\mathrm{d}I\right], \tag{7}$$

where σ_n^2 is the variance of $n[k]$, and $erfc(\cdot)$ is the complementary error function.

B. **MLP-assisted adaptive optimized cutoff frequency**

The optimized cutoff frequency of the HPF is dependent on the data rates and turbulence degrees. Thus, a dynamic cutoff frequency is required to support the variation in data rates and turbulence channels. The DL technique is applied to adaptively select the optimized cutoff frequency of the HPF to effectively compensate for the scintillation effect. In Figure 1b, the digital signal is split into two branches, and the lower branch is used to learn the frequency characteristics of the received signal and adaptively select the appropriate cutoff frequency of the HPF using DL. Figure 2 illustrates the proposed DL model employing MLP, including two fully connected layers with 64 neurons and an output layer [22]. In the MLP model, the modified input layer only delivers the input signal $r[k]$ into the hidden layers first. The ReLU activation function $R\{\cdot\}$ learns the feature representation of $r[k]$ due to the nonlinear transformations. Then, the output layer reconstructs the spectrum of $r'[k]$ using the learned feature representation. Finally, the optimal cutoff frequency $f_{cut_{MLP}}$ is determined by identifying the main components of the received signal through the analysis of the $r'[k]$ spectrum. The processes of input and output layers are defined as follows

$$\begin{aligned} n_1[k] &= w_1 r[k] + b_1 \\ n_2[k] &= R(w_2(R(w_1 r[k] + b_1)) + b_2) \\ r'[k] &= R(w_3 n_2[k] + b_3), \end{aligned} \tag{8}$$

where w_i is the weight matrix of the layer i, b_i is the bias term of the layer i, n_1 is the input layer, n_2 represents a neural network structure composed by two hidden layers, and r' is the output layer. Figure 3 shows the PSD of the OOK signal after the turbulence channel with a σ_I^2 of 0.4286 at 10 Mbps. MLP model adaptively selects the peak frequency component in the PSD of the OOK signal, excluding the DC component. Then, the selected frequency is set to the optimized cutoff frequency of the HPF under various data rates and turbulence degrees.

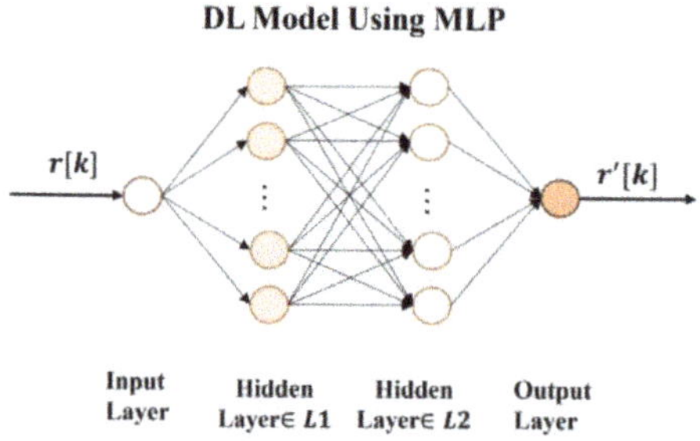

Figure 2. Proposed DL-based detector model using MLP.

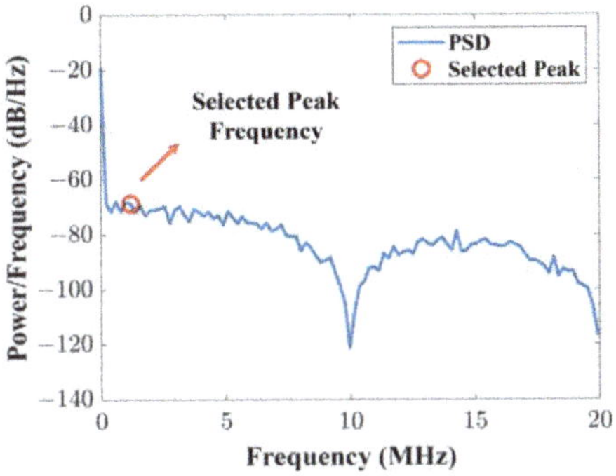

Figure 3. PSD of OOK signal after the turbulence channel with a σ_I^2 of 0.4286 at 10 Mbps.

C. Cascaded DL and HPF

RNN and LSTM are employed after HPF to further suppress the scintillation effect and recover signal distortion from the HPF without optimizing the HPF cutoff frequency, as shown in Figure 1c. Figure 4 illustrates a DL-based detector model utilizing RNN [23], characterized by a hidden layer comprising 64 neurons and an output layer. The filtered signal $x[k]$ is imported into the RNN model, and it captures the temporal dependencies using the recurrent structure by learning the features and patterns of $x[k]$. Ultimately, it utilizes the learned features and temporal information to achieve the turbulence effect compensation. The formula of the hidden state $h[k]$ is given by

$$h[k] = \tanh\{W_{ih} * x[k] + b_{ih} + W_{hh} * h[k-1] + b_{hh}\}, \quad (9)$$

where W_{ih} is the weight matrix between input and hidden layers, W_{hh} is the weight matrix between hidden and hidden layers, b_{ih} and b_{hh} are the bias terms between input and hidden layers, and tanh is the activation function. The RNN adjusts its parameters through backpropagation to improve the accuracy of turbulence prediction and compensation. Based on the hidden state $h[k]$, the output $y_{3-RNN}[k]$ is calculated as follows

$$y_{3-RNN}[k] = \text{softmax}\left\{W_{hy} * h[k] + b_{hy}\right\}, \quad (10)$$

where W_{hy} denotes the weight matrix from hidden to output layers. The RNN structure can be used to process sequential data. However, it has the issue of gradient vanishing and gradient explosion for long sequential data processing, which leads to difficulties in capturing the long-range dependencies. LSTM with complex variants can handle the longer sequence dependencies by using additional gating mechanisms [23]. Figure 5 illustrates the proposed DL-based detector model using LSTM. The input gate, forget gate, cell state, and output gate of LSTM collectively aid the network in capturing and adapting to the signal intensity variations [24]. The input gate adjusts the cell state $c[k]$ of the current time step to reflect the current input signal $x[k]$ and the previous hidden state $h[k-1]$. The signal with serious intensity fluctuations is prioritized by the input gate. The calculation formula of the input gate is as follows

$$i[k] = \sigma\{W_{xi} * x[k] + W_{hi} * h[k-1] + b_i\}, \quad (11)$$

where $i[k]$ is the output of the input gate, and σ is the sigmoid activation function. The forget gate reduces the interference and noise to ensure the cell state $c[k]$ can capture the intensity fluctuations caused by the turbulence effect. The calculation formula of the forget gate is as follows

$$f[k] = \sigma\left\{W_{xf} * x[k] + W_{hf} * h[k-1] + b_f\right\}. \quad (12)$$

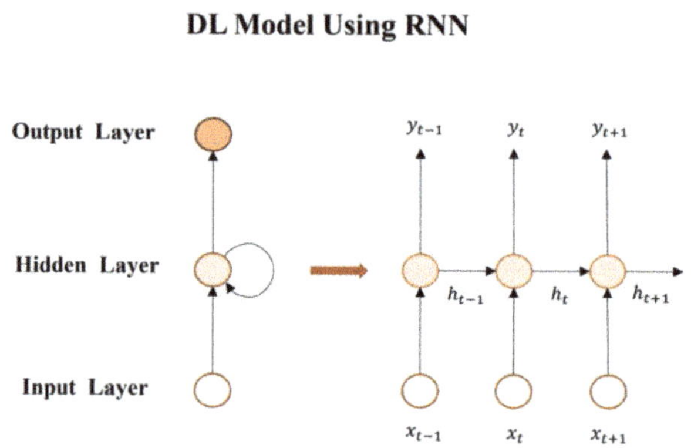

Figure 4. Proposed DL-based detector model using RNN.

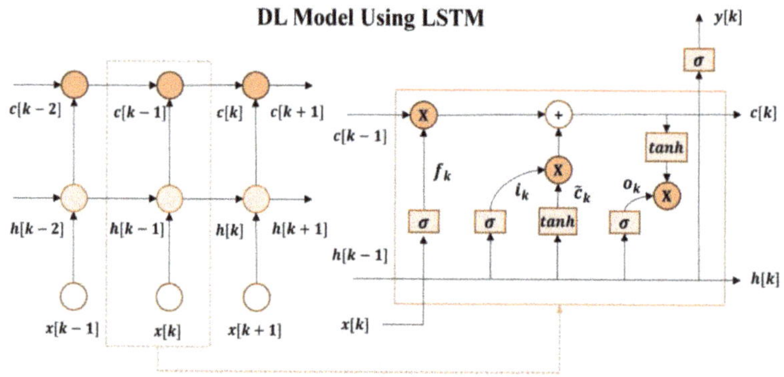

Figure 5. Proposed DL-based detector model using LSTM.

The cell state $c[k]$ is updated based on the information from the input gate and the forget gate in order to integrate the features of the $x[k]$. The specific process is as follows

$$\begin{aligned} \tilde{c}[k] &= \tanh\{W_{xc} * x[k] + W_{hc} * h[k-1] + b_c\}, \\ c[k] &= f[k] * c[k-1] + i[k] * \tilde{c}[k], \end{aligned} \tag{13}$$

where $\tilde{c}[k]$ represents the updated candidate cell state. The output gate determines the information delivered to the output layer according to the features learned at the current time step.

$$o[k] = \sigma\{W_{xo} * x[k] + W_{ho} * h[k-1] + b_o\}. \tag{14}$$

The final output layer implements a turbulence compensation process based on the features learned by the network, and the output signal $y_{3-LSTM}[k]$ is obtained as

$$y_{3-LSTM}[k] = \sigma\left\{W_{hy} * o[k] * \tanh \odot c[k]\right\}, \tag{15}$$

where $\odot$ denotes the dot product between the vectors. Finally, FTD is achieved for RNN and LSTM-based methods by using the mean values of $y_{3-RNN}[k]$ and $y_{3-LSTM}[k]$ as decision thresholds, respectively. The BERs can be calculated using Equation (7) to assess the performance of the two models. Consequently, the scintillation effect is compensated for by the proposed DL-assisted HPF-based FTD technique under different turbulence channels and data rates.

3. Channel Analysis and Simulation Parameters

A. Channel analysis

The atmospheric turbulence effect encompasses beam wandering, beam spreading, and beam scintillation effects. Among these effects, the scintillation effect causes significant temporal and spatial fluctuations in received signal intensity, which is the major concern in intensity modulation/direct detection FSO systems. Thus, this study focuses on the performance degradation induced by scintillation effects. The lognormal model, a widely recognized and simple approach, was used for modeling weak turbulence channels. The probability density function (PDF) is calculated by [25,26]

$$f(I) = 1/\sqrt{2\pi\sigma_I^2 I}\ \exp\left[-(\ln(I) - \mu)^2/\left(2\sigma_I^2\right)\right], \tag{16}$$

where μ is the mean of $\ln(I)$ [1]. The turbulence channel was modeled through phase modulation, inverse Fourier transform, and the first-order Rytov approximation. Figure 6 shows the turbulence channel modeled by employing parameters in Table 1. The modeled turbulence channel exhibits time-varying signal intensity features as illustrated in Figure 6a, and it has a low-frequency dominant spectrum as demonstrated in Figure 6b. In Figure 6c, the PDF of the modeled turbulence channel fits well with the lognormal distribution. Therefore, the turbulence channel was modeled effectively, and it was used as the dataset in this study. Turbulence channels modeled with gamma–gamma distribution, Weibull distribution, K-distribution, and negative exponential distribution will be discussed in further works.

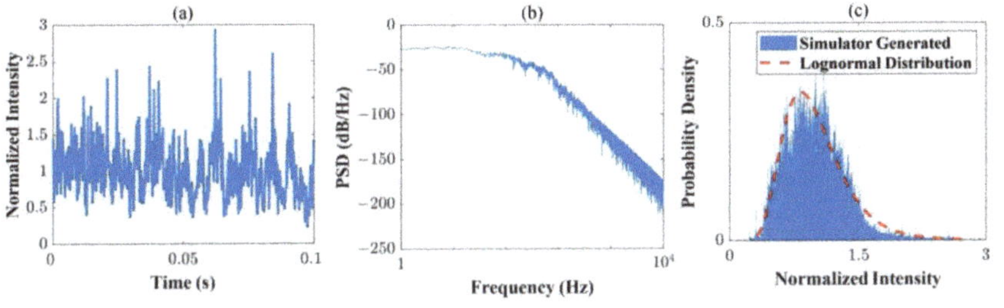

Figure 6. Turbulence channel model at σ_I^2 = 0.1256. (**a**) Intensity variation, (**b**) frequency, (**c**) probability density function.

Table 1. Parameters of the turbulence channel.

Parameters	Values
Link distance	10 km
Wavelength	1550 nm
Aperture diameter	10 cm
Wind velocity	5 m/s
C_n^2	1.5×10^{-11}
Divergence angle	10 urad
Visibility	30 km

B. Simulation parameters

The standard parameters of the DL model are outlined in Table 2. Simulations were conducted using MATLAB 2021b and PyCharm 2021.3.2. The test platform employed an NVIDIA GeForce RTX 4070 Laptop GPU. Throughout the training process, the output signals were used as the dataset. In total, 80% of the dataset was used for training, and 20% of the dataset was used for testing. The DL detector was trained using labeled training data with one-hot vector labels representing the transmitted symbols.

Table 2. The tuned hyperparameters.

Hyperparameters	Value
Size of dataset-10 Mbps	1×10^7
Size of training dataset-10 Mbps	8×10^6
Size of test dataset-10 Mbps	2×10^6
Size of dataset-100 Mbps	1×10^8
Size of training dataset-100 Mbps	8×10^7
Size of test dataset-100 Mbps	2×10^7
Loss function	MSE
Epoch	20
Batch size	2^{16}
Learning rate	0.005
Turbulence channel	σ_I^2 = 0.6850, 0.4286, 0.1256, 0.0596

4. Simulations and Results

The proposed OOK detection method using the DL-assisted HPF-based FTD was evaluated via simulation. The BER performance of the ATD with precise CSI, HPF-based FTD under different cutoff frequencies, MLP-assisted adaptive optimized cutoff frequency, and cascaded HPF and DL-based FTD without the optimization of HPF cutoff frequency techniques were compared. Additionally, the accuracy and computational complexity of the proposed DL-assisted HPF techniques were analyzed. The average signal-to-noise ratio (SNR) was used in this work due to the signal intensity variation, and it is calculated as $E[I^2]/2\sigma_n^2$ [27].

Figure 7 illustrates the BER performance of the proposed OOK detection method at different cutoff frequencies of HPF. Turbulence channels with σ_I^2 values of 0.0596, 0.1256, 0.4286, and 0.6850 were adopted in the simulation. The data rates were set to 10 Mbps and 100 Mbps, and the average signal-to-noise ratio (SNR) was set to 28 dB for PD and background noise reduction. The cutoff frequency of the HPF was set to 10 Hz-0.8 MHz due to the low-frequency features of the turbulence channel. At a data rate of 10 Mbps, the BER was initially improved by increasing the cutoff frequency of the HPF, because a much larger proportion of turbulence frequency was removed from the received signal by HPF. When the cutoff frequency of the HPF exceeded a critical value, the BER was degraded due to a severe signal distortion from HPF. A similar performance was observed in the case of 100 Mbps. Therefore, an optimization process is required for HPF under various σ_I^2 and data rates.

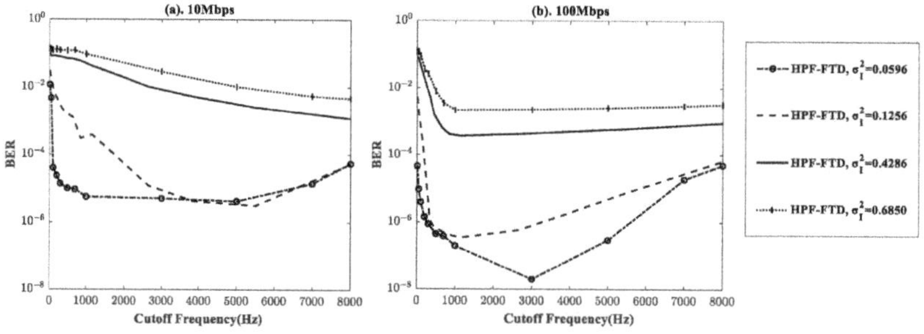

Figure 7. BER performance of HPF-based FTD under various cutoff frequencies of HPF: (**a**) 10 Mbps, (**b**) 100 Mbps.

Figure 8 illustrates the BER performance of HPF-based FTD with the optimized HPF cutoff frequency at data rates of 10 Mbps and 100 Mbps. The optimized cutoff frequency was set according to the results from Figure 7. A similar BER performance was achieved by HPF-based FTD with the optimized cutoff frequency compared to ATD with precise CSI due to an effective turbulence effect mitigation. Figure 9 depicts the BER performance of fixed-threshold OOK detection with the MLP-assisted adaptive optimized cutoff frequency of HPF at data rates of 10 Mbps and 100 Mbps. The cutoff frequency of HPF was optimized under various turbulence channels and data rates by selecting the peak frequency component in the PSD of the OOK signal excluding the DC component, as illustrated in Figure 10. The BER performance of the proposed MLP-assisted adaptive method is similar to that of the HPF-based FTD with the optimized cutoff frequencies of the HPF, as shown in Figure 8.

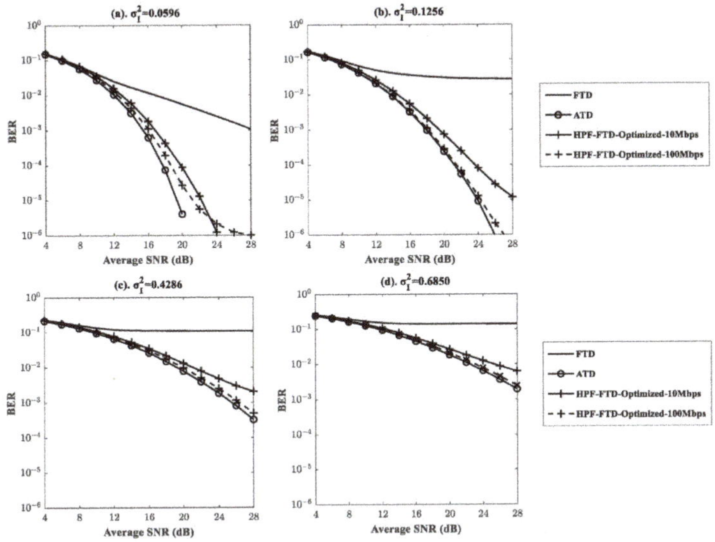

Figure 8. BER performance of HPF-based FTD with optimized cutoff frequencies under various σ_I^2 of (**a**) 0.0596, (**b**) 0.1256, (**c**) 0.4286, and (**d**) 0.6850.

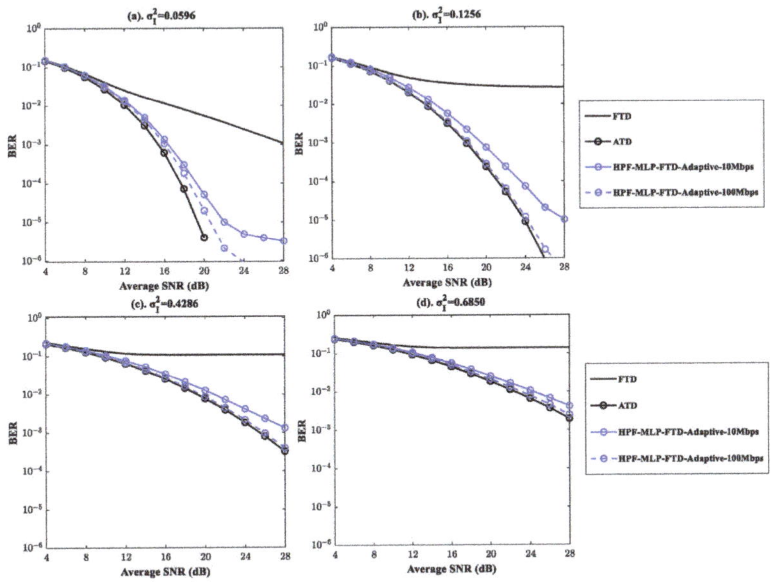

Figure 9. BER performance of proposed MLP-assisted adaptive optimized cutoff frequency of HPF under various σ_I^2 of (**a**) 0.0596, (**b**) 0.1256, (**c**) 0.4286, and (**d**) 0.6850.

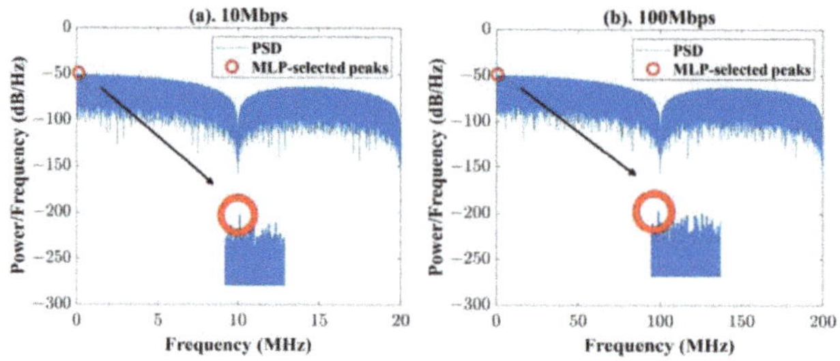

Figure 10. Peak frequency component selection in PSD with σ_I^2 of 0.4286 at various data rates. (**a**) 10 Mbps, (**b**) 100 Mbps.

Figure 11 shows the BER performance of the proposed cascaded HPF and DL-based FTD without optimization of the HPF cutoff frequency under various turbulent channels and data rates. The RNN and LSTM models were used to recover the signal distortion from the HPF and mitigate the remnant scintillation effect without the optimized cutoff frequency of HPF. The cutoff frequency of the HPF was fixed at 1 kHz. Vanishing and exploding gradient BERs were observed by the proposed cascaded HPF and RNN methods at high SNR values because of the difficulty of capturing long-term dependency. The BER performance of the proposed cascaded HPF and LSTM methods was close to that of the ATD technique owing to the introduction of gating mechanisms. These gating mechanisms control the information flow and filter unimportant information; thus, the LSTM-based method can mitigate the effect of gradient explosions. Therefore, the proposed cascaded HPF and LSTM without optimization of the cutoff frequency technique can effectively compensate for the turbulence effect and recover the signal distortion.

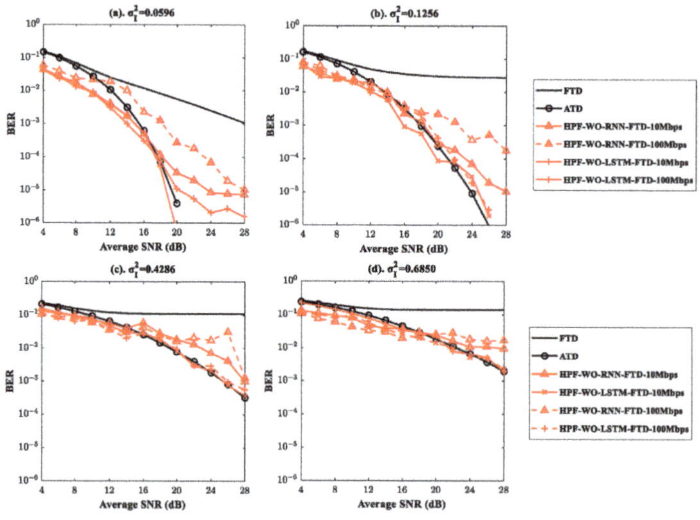

Figure 11. BER performance of proposed cascaded HPF and DL-based FTD without the optimization of HPF cutoff frequency under various σ_I^2 of (**a**) 0.0596, (**b**) 0.1256, (**c**) 0.4286, and (**d**) 0.6850. HPF-WO-RNN-FTD: RNN and HPF-based FTD without optimized cutoff frequency, HPF-WO-LSTM-FTD: LSTM and HPF-based FTD without optimized cutoff frequency.

Table 3 shows the accuracy and computational complexity of MLP-based adaptive cutoff frequency of HPF under σ_I^2 of 0.4286 at data rates of 10 Mbps and 100 Mbps. Longer training times and higher accuracy were observed by higher data rate signal transmission due to the application of larger training datasets. Table 4 illustrates the computational complexity and accuracy of cascaded HPF and DL-based FTD without optimized cutoff frequency under σ_I^2 of 0.4286. A shorter training time was obtained by using the RNN model compared to the LSTM model owing to its simpler structure and lower computational complexity. However, a higher accuracy was achieved by applying the LSTM model due to the strong memory capabilities and flexibility, which has the capability of handling long-term dependencies. The accuracy and computational complexity of MLP-based adaptive cutoff frequency of HPF and cascaded HPF and DL-based FTD without optimized cutoff frequency were analyzed under σ_I^2 of 0.0596, 0.1256 and 0.6850, as shown in Tables 5–7. A similar accuracy and computational complexity performance was obtained compared to these methods under σ_I^2 of 0.4286. Therefore, the proposed cascaded HPF and LSTM methods can effectively compensate for the turbulence effects under different data rates and turbulent channel conditions with relatively longer computational times. The application of other DL models, accuracy improvement of the RNN model, and computational complexity reduction in the LSTM model will be conducted in our future work.

Table 3. Accuracy and computational complexity of MLP-based adaptive cutoff frequency of HPF under σ_I^2 of 0.4286.

Data Rate	Training Data	Accuracy Rate	Training Time
10 Mbps	1×10^5	91.22%	57 s
100 Mbps	1×10^6	95.94%	747 s

Table 4. Accuracy and computational complexity of cascaded HPF and DL-based FTD without optimized cutoff frequency under σ_I^2 of 0.4286.

DL Model	Data Rate	Training Data	Accuracy Rate	Training Time
RNN	10 Mbps	1×10^5	92.67%	61 s
	100 Mbps	1×10^6	89.75%	793 s
LSTM	10 Mbps	1×10^5	96.63%	76 s
	100 Mbps	1×10^6	94.23%	859 s

Table 5. Accuracy and computational complexity of MLP-based adaptive cutoff frequency of HPF under various σ_I^2.

Atmospheric Turbulence	Data Rate	Training Data	Accuracy Rate	Training Time
0.0596	10 Mbps	1×10^5	91.37%	59 s
	100 Mbps	1×10^6	96.10%	740 s
0.1256	10 Mbps	1×10^5	92.48%	54 s
	100 Mbps	1×10^6	96.45%	763 s
0.6850	10 Mbps	1×10^5	92.23%	61 s
	100 Mbps	1×10^6	96.90%	803 s

Table 6. Accuracy and computational complexity of cascaded HPF and RNN-based FTD without optimized cutoff frequency under various σ_I^2.

Atmospheric Turbulence	Data Rate	Training Data	Accuracy Rate	Training Time
0.0596	10 Mbps	1×10^5	89.52%	64 s
	100 Mbps	1×10^6	87.90%	826 s
0.1256	10 Mbps	1×10^5	87.34%	62 s
	100 Mbps	1×10^6	87.12%	880 s
0.6850	10 Mbps	1×10^5	90.24%	68 s
	100 Mbps	1×10^6	89.42%	893 s

Table 7. Accuracy and computational complexity of cascaded HPF and LSTM-based FTD without optimized cutoff frequency under various σ_I^2.

Atmospheric Turbulence	Data Rate	Training Data	Accuracy Rate	Training Time
0.0596	10 Mbps	1×10^5	97.31%	78 s
	100 Mbps	1×10^6	94.90%	788 s
0.1256	10 Mbps	1×10^5	96.12%	76 s
	100 Mbps	1×10^6	95.23%	808 s
0.6850	10 Mbps	1×10^5	97.12%	81 s
	100 Mbps	1×10^6	96.81%	910 s

5. Conclusions

In conclusion, we have proposed a DL-assisted HPF-based FTD for FSO communication systems. The performance of HPF-based FTD under different cutoff frequencies, MLP-assisted adaptive optimized cutoff frequency, and cascaded HPF and DL-based FTD without the optimization of cutoff frequency techniques were analyzed and compared to the conventional FTD and ATD with precise CSI under various scintillation effects and data rates. The proposed technique was evaluated via simulation. The simulation results demonstrated that the proposed MLP-assisted adaptive optimized cutoff frequency and cascaded LSTM and HPF methods were close to the ATD with precise CSI under various turbulent channels and data rates. Therefore, it provides a promising solution to address the scintillation effect in FSO communication systems.

Author Contributions: Conceptualization, Y.-Q.H.; Data curation, Y.G., Q.-W.J., M.-F.L., W.-H.Z. and Y.-Q.H.; Formal analysis, Y.-Q.H.; Project administration, Y.-Q.H.; Writing—original draft, Y.G., Q.-W.J., M.-F.L. and Y.-Q.H.; Writing—review and editing, Y.-Q.H. All authors have read and agreed to the published version of the manuscript.

Funding: This work was supported by the Educational Department of Liaoning Province General Program (JYTMS20231214).

Institutional Review Board Statement: Not applicable.

Informed Consent Statement: Not applicable.

Data Availability Statement: Dataset available on request from the authors.

Conflicts of Interest: The authors declare no conflicts of interest.

References

1. Kaushal, H.; Kaddoum, G. Optical communication in space: Challenges and mitigation techniques. *IEEE Commun. Surv. Tutor.* **2017**, *19*, 57–96. [CrossRef]
2. Hamza, A.S.; Deogun, J.S.; Alexander, D.R. Classification framework for free space optical communication links and systems. *IEEE Commun. Surv. Tutor.* **2019**, *21*, 1346–1382. [CrossRef]
3. Chaleshtory, Z.N.; Gholami, A.; Ghassemlooy, Z.; Sedghi, M. Experimental investigation of environment effects on the FSO link with turbulence. *IEEE Photonics Technol. Lett.* **2017**, *29*, 1435–1438. [CrossRef]
4. Constantine, S.; Elgin, L.E.; Stevens, M.L.; Greco, J.A.; Aquino, K.; Alves, D.D.; Robinson, B.L. Design of a high-speed space modem for the lunar laser communications demonstration. In Proceedings of the SPIE 7923, Free-Space Laser Communication Technologies XXIII, San Francisco, CA, USA, 22–27 January 2011; Volume 7923, pp. 54–62.
5. Safi, H.; Sharifi, A.A.; Dabiri, M.T.; Ansari, I.S.; Cheng, J. Adaptive channel coding and power control for practical FSO communication systems under channel estimation error. *IEEE Trans. Veh. Technol.* **2019**, *68*, 7566–7577. [CrossRef]
6. Abou-Rjeily, C. Performance analysis of FSO communications with diversity methods: Add more relays or more apertures. *J. Sel. Areas Commun.* **2015**, *33*, 1890–1902. [CrossRef]
7. Trinh, P.V.; Pham, T.V.; Dang, N.T.; Nguyen, H.V.; Ng, S.X.; Pham, A.T. Design and security analysis of quantum key distribution protocol over free-space optics using dual-threshold direct-detection receiver. *IEEE Access* **2018**, *6*, 4159–4175. [CrossRef]
8. Chan, V.W. Free-space optical communications. *J. Light. Technol.* **2006**, *24*, 4750–4762. [CrossRef]
9. Abadi, M.M.; Ghassemlooy, Z.; Khalighi, M.A.; Zvanovec, S.; Bhatnagar, M.R. FSO detection using differential signaling in outdoor correlated-channels condition. *IEEE Photonics Technol. Lett.* **2016**, *28*, 55–58. [CrossRef]

10. Ding, S.L.; Zhang, J.K.; Dang, A.H. Adaptive threshold decision for on-off keying transmission systems in atmospheric turbulence. *Opt. Express* **2017**, *25*, 24425–24436. [CrossRef] [PubMed]
11. Louthain, J.A.; Schmidt, J.D. Synergy of adaptive thresholds and multiple transmitters in free-space optical communication. *Opt. Express* **2010**, *18*, 8948–8962. [CrossRef] [PubMed]
12. Aveta, F.; Refai, H.H.; Lopresti, P.G. Cognitive multi-point free space optical communication: Real-time users discovery using unsupervised machine learning. *IEEE Access* **2020**, *8*, 207575–207588. [CrossRef]
13. Na, Y.; Ko, D.K. Deep-learning-based high-resolution recognition of fractional-spatial-mode-encoded data for free-space optical communications. *Sci. Rep.* **2021**, *11*, 2678. [CrossRef] [PubMed]
14. Li, J.; Zhang, M.; Wang, D.; Wu, S.; Zhan, Y. Joint atmospheric turbulence detection and adaptive demodulation technique using the CNN for the OAM-FSO communication. *Opt. Express* **2018**, *26*, 10494–10508. [CrossRef] [PubMed]
15. Darwesh, L.; Kopeika, N.S. Deep learning for improving performance of OOK modulation over FSO turbulent channels. *IEEE Access* **2020**, *8*, 155275–155284. [CrossRef]
16. Zhang, K.; Zuo, W.; Chen, Y.; Meng, D.; Zhang, L. Beyond a gaussian denoiser: Residual learning of deep cnn for image denoising. *IEEE Trans. Image Process.* **2017**, *26*, 3142–3155. [CrossRef]
17. Jing, Q.W.; Yu, P.Z.; Lv, H.L.; Hong, Y.Q. Turbulence-tolerant Manchester On-off Keying Transmission for Free-space Optical Communication. *Curr. Opt. Photonics* **2023**, *7*, 345–353.
18. Dabiri, M.T.; Sadough, S.M.S. Generalized blind detection of OOK modulation for free-space optical communication. *IEEE Commun. Lett.* **2017**, *21*, 2170–2173. [CrossRef]
19. Arienzo, L. Green RF/FSO communications in cognitive relay-based space information networks for maritime surveillance. *IEEE Trans. Cogn. Commun.* **2019**, *5*, 1182–1193. [CrossRef]
20. Mondal, P.; Mandal, M.K.; Chaktabarty, A. Compact bandpass filters with wide controllable fractional bandwidth. *IEEE Microw. Wirel. Compon.* **2006**, *16*, 540–542. [CrossRef]
21. Borah, D.K.; Voelz, D.G. Pointing error effects on free-space optical communication links in the presence of atmospheric turbulence. *J. Light. Technol.* **2009**, *27*, 3965–3973. [CrossRef]
22. Pinkus, A. Approximation theory of the MLP model in neural networks. *Acta Numer.* **2008**, *8*, 143–195. [CrossRef]
23. LeCun, Y.; Bengio, Y.; Hinton, G. Deep learning. *Nature* **2015**, *521*, 436–444. [CrossRef] [PubMed]
24. Sherstinsky, A. Fundamentals of recurrent neural network (RNN) and long short-term memory (LSTM) network. *Phys. D Nonlinear Phenom.* **2020**, *8*, 132306. [CrossRef]
25. Shen, H.; Yu, L.; Fan, C. Temporal spectrum of atmospheric scintillation and the effects of aperture averaging and time averaging. *Opt. Commun.* **2014**, *330*, 160–164. [CrossRef]
26. Toyoshima, M.; Takenaka, H.; Takayama, Y. Atmospheric turbulence-induced fading channel model for space-to-ground laser communications links. *Opt. Express* **2011**, *19*, 15966–15975. [CrossRef] [PubMed]
27. Shin, W.H.; Choi, J.Y.; Han, S.K. Fixed threshold on-off keying differential detection for satellite optical communications. *Opt. Express* **2019**, *27*, 1590–1596. [CrossRef] [PubMed]

photonics

MDPI

Article

Comprehensive Optical Inter-Satellite Communication Model for Low Earth Orbit Constellations: Analyzing Transmission Power Requirements

Michail Gioulis [1,*], Thomas Kamalakis [1] and Dimitris Alexandropoulos [2]

1 Department of Informatics and Telematics, Harokopio University of Athens, 17778 Athens, Greece; thkam@hua.gr
2 Department of Materials Science, University of Patras, 26504 Patras, Greece; dalexa@upatras.gr
* Correspondence: mgioulis@hua.gr

Abstract: Free-space optical communications have emerged as a powerful solution for inter-satellite links, playing a crucial role in next-generation satellite networks. This paper introduces a comprehensive model that enables the dynamic evaluation of optical power requirements for realistic low Earth orbit satellite constellations throughout the orbital period. Our approach incorporates the constellation architecture, link budget analysis, and optical transceiver design to accurately estimate the power required for sustaining connectivity for both intra- and inter-orbit links. We apply the model considering Walker delta-type constellations of varying densities. We show that in dense constellations, even at high data rates, the required transmission power can be low enough to mitigate the need for optical amplification. Dynamically estimating the power requirements is vital when evaluating energy savings in adaptive scenarios where terminals adaptively change the emitted power depending on the link status. Our model is implemented in Python and is openly available under an open-source license. It can be easily adapted to various alternative constellation configurations.

Keywords: free space optics; link budget analysis; low earth orbit; optical inter-satellite links

1. Introduction

In recent decades, there has been a notable surge in free-space optics (FSO) driven by the escalating demand for real-time quality of service and the necessity to efficiently transfer large amounts of data both at terrestrial and satellite networks, with a pronounced focus on optical inter-satellite links (OILs) [1]. While radio-frequency (RF) technologies are still prominent in satellite applications, the emergence of OILs is quickly gaining leverage in commercial applications. OILs offer several key advantages, including a vast available spectrum, high data rates, narrow optical beams favoring security, small antenna sizes, and reduced power consumption [2–5].

Low Earth orbit (LEO) Walker-delta type constellations are widely being considered in commercial deployments, with Starlink Phase 1 V2 being a typical example [6,7]. This constellation consists of a total of 1584 mini-satellites at an altitude of 550 km. In LEO applications, satellite sizes vary from large to pico satellites, the latter commonly being referred to as cubesats [8]. OILs are a favorable communication solution in all these scenarios, leading to laser terminals that are both lightweight and compact [9].

Various OIL systems have been developed. One notable example is the Optel-μ terminal, constructed by TNO and utilized for both downlink and uplink communications [10]. Another is the LCT terminal produced by TESAT, which is employed for LEO satellite networks [11]. NASA has also developed several systems for LEO-to-LEO and LEO-to-ground communication links [12]. Similarly, the German Aerospace Center has also demonstrated mini-satellite OIL transceivers [13]. In [14], a spherical system with multiple transceivers was developed, which offers omnidirectional coverage of nearly 360 degrees and a data rate of 1 Gbps over a distance of 200 km. Another approach, as described in [15], involves equipping each satellite side with a 100-degree field-of-view OIL, which also enables omnidirectional coverage.

Motivation and Related Work

Understanding the system requirements to maintain connectivity throughout a constellation is crucial. Several studies have examined the structure of satellite constellations and communications between different satellite layers. In [16], a classification of both RF and OIL links is presented, focusing on links between satellites at the same or different altitudes and between high-altitude platforms (HAPs). In [17], a satellite network utilizing OILs is examined and compared to a terrestrial network in terms of latency. In [18], the phasing parameter F is introduced to determine the relative initial satellite positions to avoid collisions. In [19,20], OIL connection strategies are explored, focusing on efficient routing schemes with low latencies. In [20], the authors introduced a latitude parameter to examine the relationship between satellite distance and the number of connections that could be established. The authors in [6] focused on the trade-off between physical layer parameters and network latencies in intercontinental satellite links. In [21], a methodology is proposed to minimize the average network latency in intercontinental connections. In [22], the authors proposed a methodology for utilizing OILs to reduce latencies and determine the most efficient routing. In [23], the impact of atmospheric effects on optical space communications—such as signal attenuation and scintillation—is analyzed, along with their potential utilization for atmospheric measurements. The work also explores the adaptability of laser communication terminals and highlights the synergy between satellite technologies and atmospheric research. However, such factors are not considered in the present study.

The primary contribution of this work is the application of a comprehensive network model to analyze optical power requirements at OIL terminals in LEO Walker delta-type constellations [24] that are widely being considered in commercial applications. The required optical power is a crucial factor that directly affects terminal design, influencing transmitter and receiver specifications, power acquisition and tracking system precision, thermal management strategies, and overall power consumption. Previous studies have typically considered the link budget without (a) accounting for the precise constellation dynamics and (b) the actual network requirements for connectivity, which involves reliable intra- and inter-orbit communication. Our approach incorporates the full constellation dynamics and is applicable to commercial constellations such as Kuiper and Starlink [6,18]. We use the Starlink Phase 1 V2 constellation as a realistic scenario to illustrate the practical applicability of our model. We applied it in two different scenarios with varying satellite densities and estimated the power requirements for high-speed optical inter-satellite links (OILs). Our results are comparable to [19,20] in terms of minimum distances between satellites and indicate that a 100 Gbps network can be sustained across a dense constellation of approximately 6000 satellites using transmission powers in the 300 mW range. This is an important finding, since such power levels can be supported by high-power distributed feedback (DFB) lasers [25], possibly removing the need for optical amplification, simplifying the terminal design and reducing power consumption. Such levels of satellite population are well within the planning of commercial satellite internet carriers [22,26].

Our comprehensive model is implemented as a `Python` module that encompasses various constellation parameters, such as satellite count, initial positioning, inclination, and phasing, as well as terminal characteristics, including beam divergence, targeted bit error rate (BER), receiver aperture, etc., to precisely calculate physical layer performance. Also, in a straightforward way, it optimizes satellite phasing in a standard manner following [4,18]. Our results indicate that the required optical power can vary significantly over an orbital period in links between satellites in adjacent orbits. These calculations are useful in evaluating adaptive optical satellite network strategies, where link parameters such as the transmission power are dynamically adjusted. Such adaptive approaches can lead to significant savings in energy consumption, minimize thermal stress, and potentially prolong the operational lifetime [27]. The model's source code is publicly available [28] on the web to foster collaboration and facilitate the utilization of our research outcomes within the scientific community. The model is easily extendable to a wide selection of constellation types, modulation formats, and detection schemes.

The rest of the paper is organized as follows: in Section 2, we start by outlining how to account for constellation dynamics. In addition, we discuss how the initial satellite position can be optimized in order to minimize the chance of collisions. We also outline the satellite connectivity requirements assumed in our work. Section 3 describes the link budget model adopted to estimate physical layer performance. We discuss channel gain and how beam divergence can be optimized in the presence of pointing errors. We also calculate the noise power estimation, the bit error rate (BER), and the receiver sensitivity. Appendix A discusses some implementation details of the `Python` module. Section 4 presents the results obtained from our model, emphasizing the estimated power requirements for various constellation architectures. Some concluding remarks are given in Section 5.

2. Constellation Model

2.1. Constellation Dynamics

The equations describing the motion of the satellite with a circular orbit can be derived using the general approach of [29]. In three-dimensional space, the satellite's position relative to the Earth's center is described by a three-component vector $\mathbf{r} = (x, y, z)^{\mathsf{T}}$, with $(\cdot)^{\mathsf{T}}$ denoting the vector's transpose. Figure 1 shows an Earth-centered coordinate system where the xy plane contains the equatorial plane (orange dashed line), while the z-axis aligns with the Earth's rotation axis, pointing north. The figure illustrates the basic parameters involved in estimating the satellite position. In the figure, $r = \sqrt{x^2 + y^2 + z^2}$ denotes the distance of the satellite from the Earth's center. In this paper, we study circular orbits, and hence, r remains constant.

Assuming a constellation consisting of N_{orb} orbits containing N_{sat} satellites each. Every satellite (k, m) can be identified by the index k of its orbit and its index m within the orbit. Its position $\mathbf{r}_{km}(t)$ is determined by [29]

$$\mathbf{r}_{km}(t) = \mathbf{T}(\Omega_k, i)\mathbf{u}_{km}(t), \tag{1}$$

where

$$\mathbf{u}_{km}(t) = \frac{h^2}{\mu}(\cos\theta_{km}(t), \sin\theta_{km}(t), 0)^{\mathsf{T}}, \tag{2}$$

is the position of the satellite with respect to the perifocal frame of reference (i.e., the frame attached to the satellite itself), i is the inclination angle, and Ω_k is the angle between the x-axis and the node line for orbit k determined by

$$\Omega_k = (k-1)\frac{2\pi}{N_{\text{orb}}}, \tag{3}$$

while h is the magnitude of the angular momentum related to the satellite's period T through $h = 2\pi r^2/T$. The period of the circular orbit is given by

$$T = \frac{2\pi r^{3/2}}{\sqrt{\mu}}, \tag{4}$$

with $\mu = GM_{\text{e}}$ being the product of the gravitational constant, $G = 6.6743 \times 10^{-11}$ N $\times$ m^2/kg^2, and the mass of the earth, $M_{\text{e}} = 5.974 \times 10^{24}$ kg. The transformation matrix $\mathbf{T}$ in (1) is given by

$$\mathbf{T}(\Omega_k, i) = \begin{bmatrix} \cos\Omega_k & -\sin\Omega_k\cos i & \sin\Omega_k\sin i \\ \sin\Omega_k & \cos\Omega_k\cos i & -\cos\Omega_k\sin i \\ 0 & \sin i & \cos i \end{bmatrix} \tag{5}$$

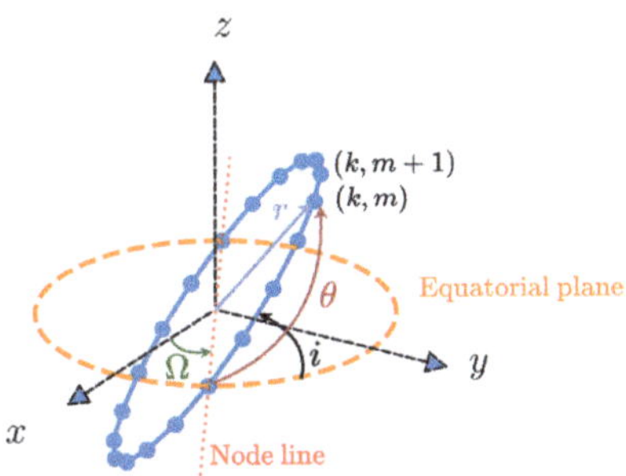

Figure 1. Satellite frame of reference and various orbital parameters used to determine the orbital dynamics.

The true anomalies $\theta_{km}(t)$ in circular orbits are determined by

$$\theta_{km}(t) = \theta_{km}(0) + \frac{2\pi t}{T}. \tag{6}$$

We can assume that on each orbit k, the satellites are evenly spaced, in which case

$$\theta_{km}(0) = \theta_k(0) + \frac{2\pi(m-1)}{N_{\text{sat}}}, \tag{7}$$

where $\theta_k(0)$ is the initial anomaly of the first satellite at the kth orbit. Given these initial anomalies, we can combine (6) and (7) to obtain the values of $\theta_{km}(t)$ for every time instance t and calculate the position of every satellite $\mathbf{r}_{km}(t)$ in the constellation using (1)–(5).

Table 1 presents the constellation parameters representative of Starlink's Phase 1 V2, which, unless otherwise specified, serves as a reference in our simulations. Figure 2 illustrates the satellite trajectories for this constellation. Due to the inclination of the constellation, the satellite trajectories are more widely spaced near the equator and progressively converge as they approach higher latitudes. Beyond a certain latitude near the poles, satellite coverage diminishes entirely.

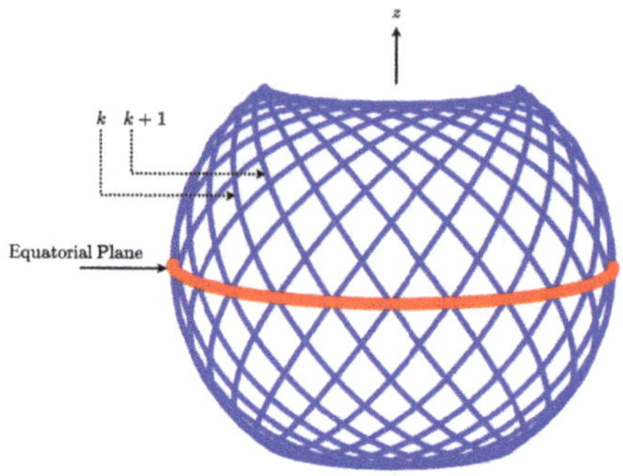

Figure 2. Satellite trajectories (colored in blue) for the Walker-delta constellation of Table 1. The equator is shown in red.

Table 1. Orbital parameters.

Parameter	Symbol	Value
Number of orbits	N_{orb}	24
Number of satellites per orbit	N_{sat}	66
Total number of satellites	N	1584
Altitude	a	550 km
Distance from Earth's center	r	6921 km
Orbital period	T	5730.11 s
Angular momentum	h	52,523.60 km^2/s
Inclination	i	53°
Optimal phasing parameter	F	13

2.2. Collision Avoidance

To avoid collisions within the constellation, we adopt the approach of [18]. For each orbit k, we assume that the satellites are placed so that

$$\theta_k(0) = (k-1)\frac{2\pi}{N}F, \tag{8}$$

where N is the total number of satellites in the constellation $N = N_{\text{sat}} \times N_{\text{orb}}$. The parameter F is an integer in the range of $0 \leq F \leq N_{\text{orb}} - 1$. This scheme is illustrated in Figure 3.

To choose the optimal value for F, we can run simulations considering various possible values of F to determine the minimum satellite-to-satellite distance r_{min} over an orbital period, $0 \leq t < T$. We can significantly speed up the computations by leveraging the uniformity of the constellation. Thus, instead of considering all possible combinations of satellites, we can focus on one particular satellite (k, m) and calculate the distances to all other satellites (p, q) within one orbital period [19,20].

$$r_{\text{min}} = \min_{pq,t} |\mathbf{r}_{km}(t) - \mathbf{r}_{pq}(t)|. \tag{9}$$

The optimal value of F, as determined in our model (Table 1), was calculated using the `Pyminisat` module (as referenced in Appendix A) considering [6] and selected for the remaining simulations of this constellation.

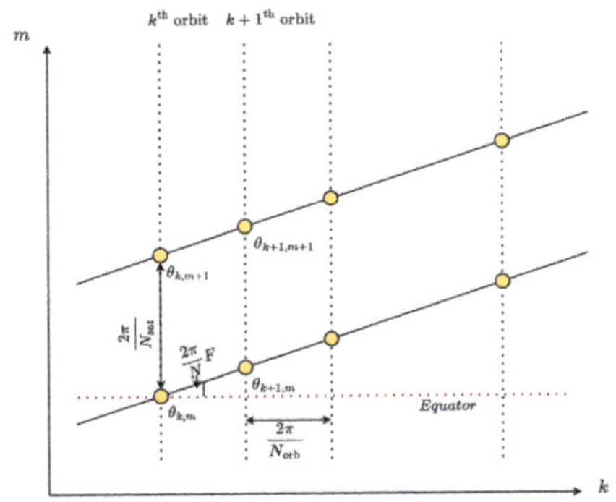

Figure 3. Illustration of the relationship between the true anomalies θ_{km} of the satellites at the same and adjacent orbits, the phasing factor F, and the number of orbits and satellites per orbit, N_{orb} and N_{sat}, respectively, on the Equatorial Plane.

2.3. Connectivity Requirements

We now discuss the requirements that need to be imposed to guarantee network connectivity throughout the constellation. This typically involves conditions imposed on both intra- and inter-orbit links. Figure 4 illustrates a typical scenario where a reference satellite C communicates with its adjacent satellites. The satellites S_1 and S_2 are the two nearest satellites in the same orbit k as C. The satellites U_1, U_2, and U_3 are the first, second, and third closest satellites to C, respectively, located at the next orbit, $k+1$. Similarly, L_1, L_2, and L_3 are the first, second, and third closest satellites to C, respectively, located in the previous orbit, $k-1$. This group of satellites forms a swarm $\mathcal{S}(\mathrm{C})$ of nine satellites around C with a constant line of sight. We assume that the connectivity of C with the rest of the satellites of $\mathcal{S}(\mathrm{C})$ is the minimum requirement to achieve reliable communication with the rest of the network, providing various alternative paths for data routing. This must be met for the swarm $\mathcal{S}$ of every satellite in the constellation, as it implies that any two satellites in the communication can communicate with each other using a single hop or multiple hops. Similar to what we discuss in Section 2.2, due to the uniformity of the constellation, it is not necessary to address the swarms formed around every satellite in the constellation. Instead, we can focus on just a single reference satellite, since the distance variation between satellites in the swarm will be the same if we change the reference satellite C, albeit with a time offset.

Since satellites in the same orbit move at a constant angular velocity, the distance between C and S_1 and S_2 will be constant with t. This is not true for the distances between C and U_i or L_i due to the difference in the angles Ω_k. As mentioned in Section 2.2, we introduce a relative initial anomaly difference between orbits. The first orbit begins from the Earth's equatorial frame with a latitude of $\Omega_k = 0$. The next orbit shifts in longitude by an amount equal to $2\pi F/N$, and so on. However, the closest neighboring satellites in the adjacent upper or lower orbits remain fixed relative to any satellite within the same orbit: the nearest upper-orbit neighbor, U_1, maintains its position relative to C throughout C's orbital period, and this pattern holds for other inter-orbit neighbors as well. At time $t = 0$, we can calculate the distances from C to all satellites in adjacent orbits and identify the three nearest satellites in each orbit to determine the swarm $\mathcal{S}(\mathrm{C})$.

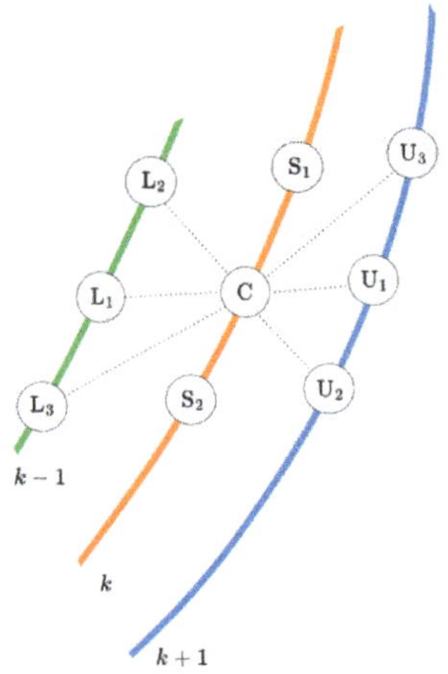

Figure 4. The satellite connectivity requirements considered in this work. Each satellite C must, at all times, communicate which its two closest neighbors in the same orbit (S_1 and S_2) as well as its three nearest satellites L_i in the lower orbit and U_i upper orbit, respectively.

3. Link Model

In this section, we elaborate on the link model adopted in order to calculate the physical layer performance and the required optical transmitter power to maintain connectivity. Figure 5 shows the basic block diagram of an OIL. The data source modulates the amplitude produced by a laser diode, and after the beam is shaped by the transmitter optics, it propagates in free space until it reaches the receiving end where a *pin* receiver converts the incoming light back to an electrical signal. The performance is impaired by the additive white Gaussian noise (AWGN) $n(t)$ added to the signal. In what follows, we assume an intensity modulation/direct detection (IM/DD) with on–off keying (OOK) and note that the model could be extended to also account for multi-level schemes [30], polarization [31] or wavelength division multiplexing [32], and coherent detection [33].

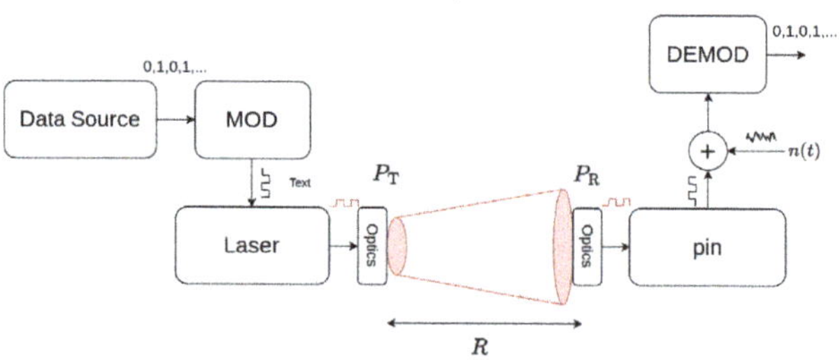

Figure 5. Block diagram of the OIL model, consisting of the data source, the modulator (MOD), the laser transmitter, the transmitter optics, the optical channel, the receiver optics and photodiode (pin), and the demodulator (DEMOD).

3.1. Link Budget

Given the satellite positions $\mathbf{r}_{km}$ at any time t, we can use a link budget model to dynamically estimate the minimum transmitted power required to maintain link connectivity over an entire orbital period T. Assuming any two satellites described by the pairs of integers (k, m) and (p, q), their distance is simply determined as

$$R = |\mathbf{r}_{km}(t) - \mathbf{r}_{pq}(t)|. \tag{10}$$

In free space, the link budget equation relating the received power P_R and the transmitted power P_T can be written as [5]

$$P_R = L_{ch} P_T, \tag{11}$$

where

$$L_{ch} = \left(\frac{\lambda}{4\pi R}\right)^2 n_T n_R G_T G_R e^{-G_T \sigma_T^2} e^{-G_R \sigma_R^2}, \tag{12}$$

is the channel gain, σ_T and σ_R are the standard deviations of the pointing errors at the transmitter and receiver sides, respectively; n_T and n_R are the optical efficiency of the transmitter and receiver, respectively; while G_T and G_R are the transmitter and receiver gains determined by

$$G_T = \frac{16}{\Theta_T^2}, \tag{13}$$

$$G_R = \left(\frac{D_R \pi}{\lambda}\right)^2, \tag{14}$$

where Θ_T is the divergence of the transmitting beam, D_R is the receiver's telescope diameter, and λ is the wavelength. The required received power is determined by the receiver sensitivity P_{sens} and the required link margin L_{LM} as follows [5]:

$$L_{LM} = \frac{P_R}{P_{sens}}. \tag{15}$$

We discuss the estimation of the receiver sensitivity in the next subsections. There is an optimal value for Θ_T lying in between that can be found by maximizing (11). Taking the first derivative of L_{ch} with respect to Θ_T, we readily see that the optimal beam divergence is

$$\Theta_T = 4\sigma_T. \tag{16}$$

We, therefore, consider this beam divergence value in our calculations.

3.2. Bit Error Rate

The bit error rate (BER), P_e, in the case of OOK is given as [34]

$$P_e = Q(\gamma), \tag{17}$$

where Q is the Q-function, and its argument γ is determined by

$$\gamma = 2\mathcal{R} \frac{r_e - 1}{r_e + 1} \frac{P_{avg}}{\sigma_1 + \sigma_0}. \tag{18}$$

In (18), P_{avg} denotes the average optical received power, $P_{avg} = (P_1 + P_0)/2$, where P_1 and P_0 are the optical received powers corresponding to bit $b = 1$ and $b = 0$, respectively. Additionally, $r_e = P_1/P_0$ denotes the extinction ratio, and $\mathcal{R}$ represents the responsivity,

$$\mathcal{R} = \frac{\eta q_e}{h_p f}, \tag{19}$$

where $q_e = 1.602 \times 10^{-19}$ Cb is the electron charge, $h_p = 6.626 \times 10^{-34}$ JHz^{-1} is Planck's constant, η is the internal quantum efficiency, and $f = c/\lambda$ is the optical frequency, with

c being the speed of light in vacuum and σ_1 and σ_0 the corresponding noise standard deviations, determined by

$$\sigma_1^2 = s_1^2 + \sigma_{\text{th}}^2 = 4q_e\mathcal{R}\frac{r_e B_e}{1+r_e}P_{\text{avg}} + \sigma_{\text{th}}^2, \tag{20}$$

$$\sigma_0^2 = s_0^2 + \sigma_{\text{th}}^2 = 4q_e\mathcal{R}\frac{B_e}{1+r_e}P_{\text{avg}} + \sigma_{\text{th}}^2, \tag{21}$$

where $s_b^2 = 2q_e\mathcal{R}P_bB_e$ represents the shot noise for $b = 1$ or $b = 0$ obtained by replacing P_b with $P_1 = 2r_eP_{\text{avg}}/(1+r_e)$ and $P_0 = 2P_{\text{avg}}/(1+r_e)$, respectively. In this equation, B_e is the electrical bandwidth, which is related to the data rate R_b by $B_e = R_b(1+\alpha)/2$, where α is the roll-off factor, assumed to be zero in this study, and σ_{th}^2 is the thermal noise power, given by [34]

$$\sigma_{\text{th}}^2 = \frac{4k_BF_nT_K}{R_L}B_e. \tag{22}$$

In (22), $k_B = 1.38 \times 10^{-23}$ JK^{-1} is Boltzmann's constant, T_K the temperature measured in Kelvin (K), F_n is the receiver amplifier noise figure, and R_L the load resistance.

3.3. Receiver Sensitivity

Given the required BER P_e, we can estimate the receiver sensitivity P_{sens} as the average power $P_{\text{avg}} = P_{\text{sens}}$ required by first inverting (17) to obtain the required value of γ. The inverse of the Q function can be written in terms of the inverse $\text{erf}^{-1}(y)$ of the error function erf, which can be estimated through standard `Python` libraries:

$$\gamma = Q^{-1}(P_e) = \sqrt{2}\text{erf}^{-1}(1-2P_e). \tag{23}$$

Once the target value of γ is obtained through (23), we can substitute Equations (21) and (20) in (18) to obtain the following solution for the required receiver sensitivity P_{sens}:

$$P_{\text{sens}} = \delta^2(a_0+a_1) + 2\delta\sqrt{\delta^2a_0a_1 + \sigma_{\text{th}}^2}, \tag{24}$$

where

$$\delta = \frac{\gamma}{2\mathcal{R}}\frac{r_e+1}{r_e-1}, a_0 = \frac{4q_e\mathcal{R}B_e}{1+r_e}, a_1 = r_ea_0. \tag{25}$$

Equations (24) and (25) can be used to determine the receiver sensitivity given the receiver parameters. One can then determine the required received power P_{sens} using (15) and then estimate the required transmitter power using (11). It is critical to note that these sensitivity values assume ideal synchronization conditions. In practice, timing recovery (e.g., Gardner loops) and adaptive equalization [35] are mandatory to mitigate BER degradation from LEO-specific impairments.

In our calculations, we deal with the uncoded BER and do not account for forward error correction (FEC). FEC may enhance the system's BER performance once the uncoded BER exceeds a given threshold value P_e^{th}. Hence, we may set the target BER in (17) equal to P_e^{th} to obtain the required optical power. In addition, we do not account for signal processing algorithms that can enhance the BER performance [35]. This is because for the constellation altitudes and inter-satellite link distances considered, the optical channel is approximately flat.

In Figure 6, we illustrate the values obtained for P_{sens}, assuming the parameters in Table 2. As expected, the sensitivity value increases with the data rate reaching just below $\cong -20$ dBm at $R_b = 10$ Gbps.

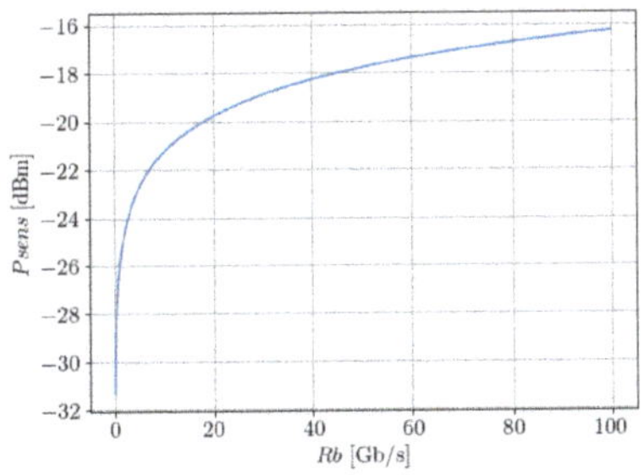

Figure 6. Receiver sensitivity with respect to the data rate obtained for the parameters of Table 2.

Table 2. System parameters.

Parameter	Symbol	Units	Value
Laser wavelength	λ	nm	1550
Transmitter optical efficiency	n_T		0.8
Receiver optical efficiency	n_R		0.8
Data rate	R_b	Gbps	1–100
Receiver telescope diameter	D_R	mm	80
Transmitter pointing error	σ_T	µrad	1
Receiver pointing error	σ_R	µrad	1
Full transmitting divergence angle	Θ_T	µrad	4
Target BER	P_e		10^{-12}
Extinction ratio	r_e	dB	20
Quantum efficiency	η		1
Amplifier noise figure	F_n	dB	3
Load resistor	R_L	Ω	100
Temperature in Kelvin	T_K	K	300
Link margin	L_{LM}	dB	3

The various model elements discussed in Sections 2 and 3 were implemented as part of the `Pyminisat Python` module [28], which is publicly available under an open-source license. Details about the inner workings and the model implementation can be found in Appendix A. The model's Github repository further describes how the initialization parameter can be adapted to the constellation at hand.

4. Results

In this section, we apply the `Pyminisat` module to study the constellation dynamics and power requirements for the system parameters summarized in Tables 1 and 2, which we refer to as Scenario A. It is interesting to point out that data rates assumed in Table 2 are in accordance with existing terminal technologies [19,36]. Pointing error levels are indicative of values reported in the literature [37,38]. A value of 20 dB is assumed for the extinction ratio, which is typical of commercially available lithium niobate electro-optic modulators. In addition, the choice of the receiver telescope diameter and wavelength is also compatible with state-of-the-art optical terminal specifications [1].

We also consider two alternative scenarios: Scenario B, where the number of satellites N_{sat} per orbit and the number of orbits N_{orb} are reduced by half, resulting in a thinner constellation with $\frac{1}{4}$ the total number of satellites N compared to Scenario A. In *Scenario C*, N_{sat} and N_{orb} are doubled compared to Scenario A, resulting in a quadrupled N, and hence, we obtain a denser constellation. The constellations remain uniform in all scenarios. Note that the total number of satellites N in Scenario C remains well within the planning of commercial satellite internet constellation deployments [22,26].

Table 3 summarizes key properties of the constellations for the three scenarios considered, derived using the `Pyminisat` module, as referenced in Appendix A. For all scenarios, we assume an inclination angle i equal to 53° and an altitude of $a = 550$ Km. The table quotes the optimal F obtained for each constellation, the inter-satellite distance r_c between two consecutive satellites in the same orbit, as well as the maximum and minimum link distances required (R_{max} and R_{min}, respectively) to achieve network connectivity, considering the links to adjacent satellites as outlined in Section 2.3.

Table 3. Scenarios considered.

Parameters	A	B	C
N_{sat}	66	33	132
N_{orb}	24	12	48
N	1584	396	6336
i	53°	53°	53°
a [Km]	550	550	550
F	13	6	19
r_c [Km]	659.3	1317.1	329.7
R_{max} [Km]	1645.6	3241.9	802.4
R_{min} [Km]	132.9	231.5	17.2

4.1. Scenario A

Scenario A is similar to Starlink's Phase 1 V2 constellation [6]. The optimal F is equal to 13 as quoted in Table 3. To ensure connectivity across the network as discussed in Section 2.3, we first need to estimate the distances between a satellite and its adjacent neighbors. Figure 7 (left, right) show the distances of the upper- and lower-orbit adjacent satellites for satellite $(k, m) = (0, 0)$. For this constellation, the distance between two consecutive satellites in the same orbit is $r_c \approx 659.3$ km. We observe that, in both cases, the distances between the satellite and its upper- and lower-orbit neighbors fluctuate significantly within one orbital period, ranging from $R_{min} \approx 132.9$ km to $R_{max} \approx 1645.6$ km. These distance fluctuations exhibit a periodic behavior as the central satellite moves through the upper and lower halves of its trajectory around the Earth. Due to the uniformity of the constellation, these fluctuations are practically the same for the upper- and lower-orbit neighbors, U_i and L_i, respectively, differing only by an initial time displacement. The figures indicate the varying conditions for inter-orbit links: while intra-orbit link distances remain constant at r_c, the inter-orbit link length varies from about one fifth of r_c to more than twice that.

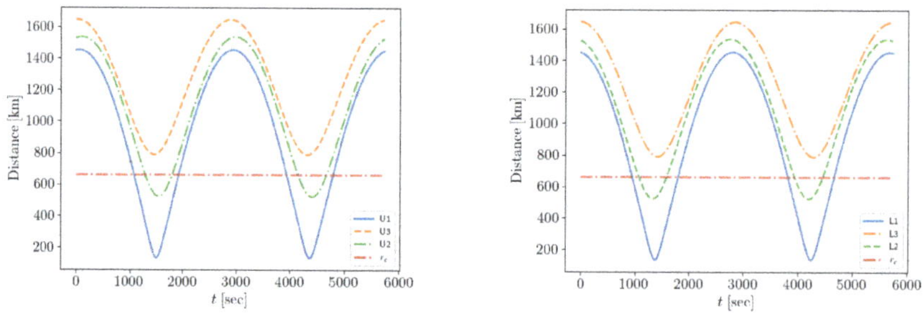

Figure 7. Distance variations between satellite $(0, 0)$ and its adjacent neighbors: (**left**) in the upper orbit and (**right**) in the lower orbit, observed over one orbital period. The intra-orbit link length between consecutive satellites remains constant throughout.

As mentioned in Section 2.2, the F value is used to determine the allocation of satellites within a constellation, ensuring a minimum safe distance between them. To achieve the safest possible configuration, we analyze how the minimum distance varies with changes in the F value, as illustrated in Figure 8. Using the Constellation class from Pyminisat, as referenced in Appendix A, we incorporate constellation dynamics to determine the optimal F value. The optimal F value is the one that maximizes the minimum distance between satellites, ensuring the most efficient and secure allocation.

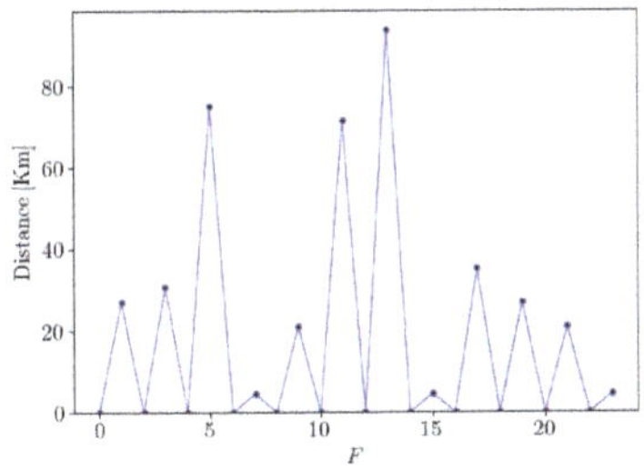

Figure 8. Minimum distance between satellites in the constellation relative to changes in the F value for Scenario A. We chose an F value of 13 to maximize the safe distance between satellites.

In Figures 9 and 10, we show the variations in path loss L_{PS} and channel gain L_{ch}, respectively, for the links between the central satellite and both the upper- and lower-orbit neighbors. Both L_{PS} and L_{ch} follow an inverse square variation ($\propto R^{-2}$) with respect to the corresponding inter-satellite distance R. For the set of parameters in Tables 1 and 2, the transmitter and receiver gains are G_T = 120 dB and G_R = 104.2 dB, respectively, while the transmitter and receiver pointing losses are L_T = −4.3 dB and L_R = −0.11 dB, respectively. High transmitter and receiver gains are necessary to compensate for the path losses shown in Figure 9. This is a clear advantage of optical inter-satellite links (OILs), which can utilize highly collimated beams to optimize the link budget. Next, we discuss the transmitter power P_T required to maintain the links between all adjacent satellites of Figure 4. Using the link budget model of Section 3 and assuming a target data rate of $R_b = 10$ Gbps, we readily obtain the results depicted in Figure 11 for the upper and lower orbital neighbors. The power turns out to be less than 500 mW in all cases, which is certainly within the reach of state-of-the-art optical transceivers [39], by combining, for example, high DFB and optical amplification [40].

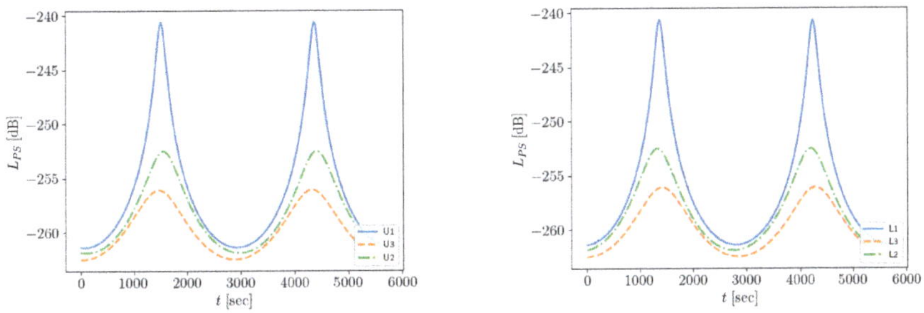

Figure 9. Path loss variations for links established between satellite $(0, 0)$ and its adjacent neighbors in the (**left**) upper orbit and (**right**) lower orbit, observed over one orbital period.

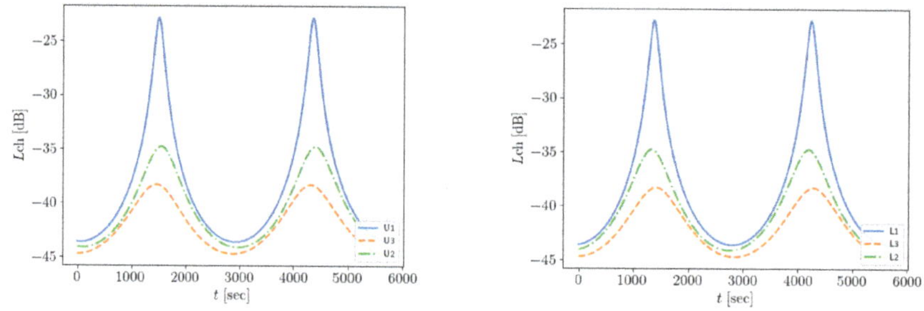

Figure 10. Channel gain for links established between satellite $(0,0)$ and its adjacent neighbors in the (**left**) upper orbit and (**right**) lower orbit, observed over one orbital period.

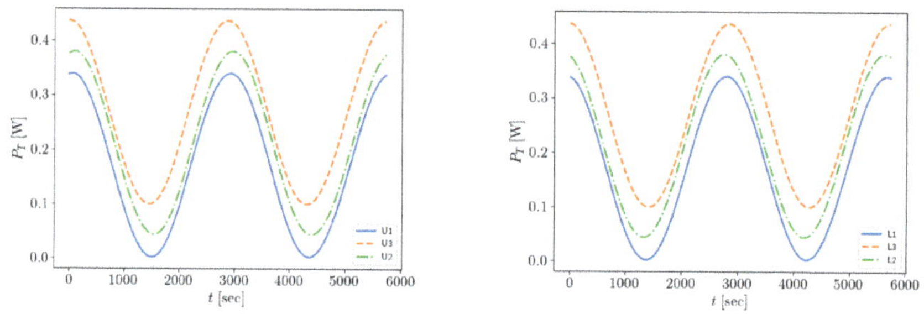

Figure 11. Required transmission power, P_T, for establishing links between satellite $(0,0)$ and its adjacent neighbors in the (**left**) upper orbit and (**right**) lower orbit, observed over one orbital period.

Another practical conclusion that can be drawn from these figures is the fact that, especially for neighbors in adjacent satellite orbits, the required power varies significantly over time, and hence, it may be advantageous to consider adaptive schemes to calibrate the transmission power to its required value rather than simply transmitting the maximum value obtained in the figures. Figure 12 shows how the required power scales with the specified data rate R_b. As expected, the required optical power increases as the data rate rises. For 100 Gbps, it surpasses 1 W [41]. Note that even at such data rates, power requirements levels remain low enough to be achieved in miniaturized terminals using optically amplified transceivers [42].

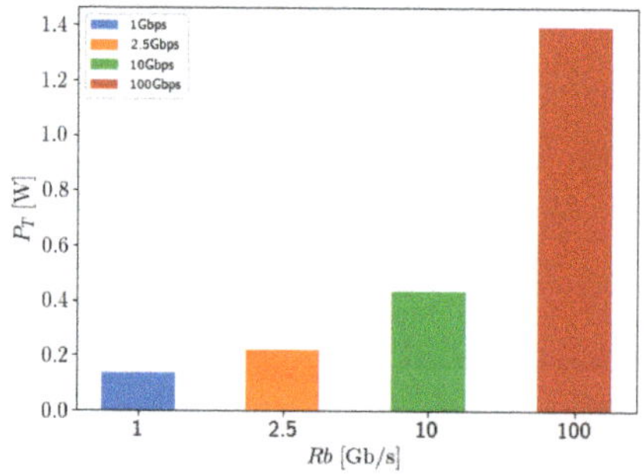

Figure 12. Maximum transmission power required to maintain connectivity between adjacent satellites as a function of the data rate, for Scenario A.

4.2. Alternative Scenarios

To further demonstrate the applicability of the `Pyminisat` module and assess the power requirements for minisat OILs, we explore the two alternative scenarios, B and

C, outlined in Table 3. For these constellations, we first need to estimate the optimal F parameters to reduce the risk of collisions, which turns out to be $F = 6$ for Scenario B and $F = 19$ for Scenario C. Note that the values of the minimum distances r_{min} are now 207.6 Km and 12.4 Km for Scenarios B and C, respectively. This is in accordance with the fact that constellation B is thinner than A while C is denser than A.

After determining the optimal F value, we applied the model to determine the link budget conditions and power requirements. Figure 13 depicts the required power to maintain the adjacent satellite links within one orbital period for both Scenarios B and C. We again assume a target data rate of $R_b = 10$ Gbps. In Scenario B, the optical power ranges from ≈400 mW to ≈1.7 W. For Scenario C, the range of the required power decreases significantly due to the higher constellation density, from ≈20 mW to ≈100 mW, respectively. The power levels of the latter scenario can, in principle, be supported using commercial high-power DFB lasers without optical amplification. This can simplify optical terminal implementation and improve power consumption.

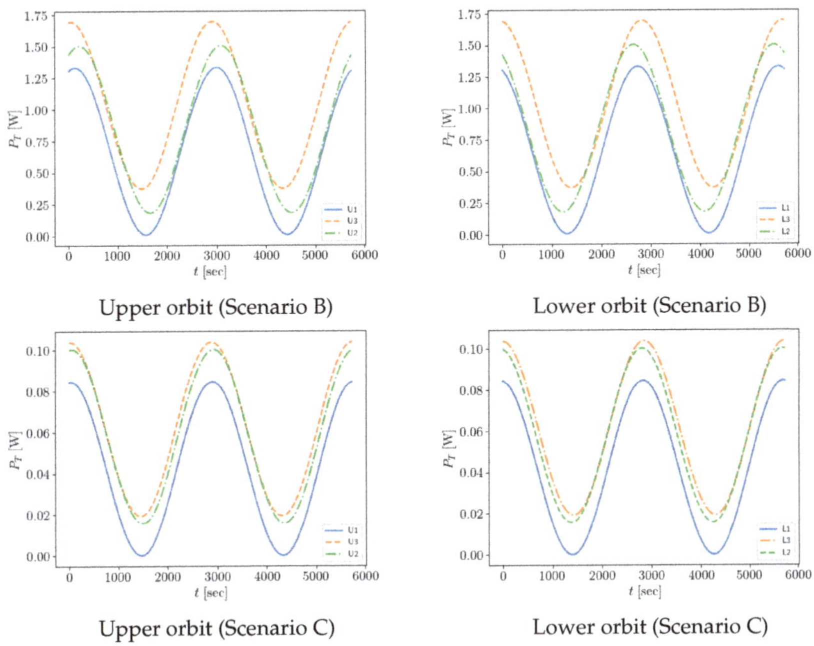

Figure 13. Transmission power required, P_T, for establishing links between satellite (0, 0) and its adjacent neighbors over one orbital period: (**left**) neighbors in the upper orbit and (**right**) neighbors in the lower orbit, for scenarios B and C.

In Figure 14, we show how the maximum required power varies with the data rate for both scenarios. The figure is indicative of the bearing of constellation density in transceiver design. For Scenario B, the optical power ranges above 1 W for R_b larger than 2.5 Gbps, reaching more than 5 W at 100 Gbps. This can pose some strict requirements at the transmitter side and especially the optical amplifier stage. Restrictions are significantly relaxed for the denser constellation of Scenario C where the transmission power reaches ≈300 mW at 100 Gbps. Such optical power levels can be supported with specially designed DFB lasers [25], thereby mitigating the need for optical amplification. The figure suggests that for this constellation, maintaining OILs at tenths of Gbps is practically feasible without excessive power requirements. One can extend the inter-satellite connection data rate using wavelength and/or polarization multiplexing to reach capacities in the Tbps realm.

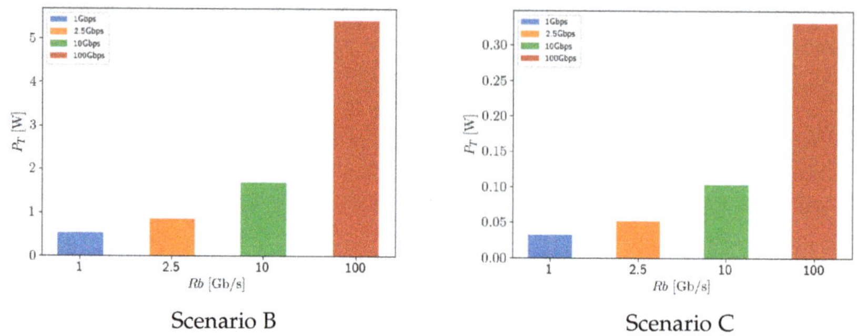

Figure 14. Maximum transmission power required to maintain connectivity between adjacent satellites as a function of the data rate, for Scenarios B and C, as described in Table 3.

5. Conclusions and Outlook

In this work, we present a complete physical layer tool for estimating the performance of OILs in LEO satellite networks. Our tool includes various aspects, including the architecture of the constellation, link budget analysis, collision avoidance optimization, transceiver optics design, etc., and it can be used to analyze practical constellation implementations. A `Python` implementation of our tool is publicly available on the world-wide web under an open-source license. Using our approach, we estimated the power requirements of high-speed OIL communication assuming LEO constellation variations. We show that, especially for dense constellations, power requirements are compatible with existing state-of-the-art optical terminals even for smaller satellite sizes, pointing out the significant potential of optical satellite communications for providing high-speed network satellite internet services. Our simulation tool can be used to ascertain the feasibility of satellite network designs accurately and efficiently.

Future research could explore several promising directions, many of which we plan to address in subsequent work. While our model assumes a standard IM/DD OOK transceiver, incorporating alternative modulation formats (e.g., multi-level amplitude modulation), as well as wavelength, spatial, or polarization multiplexing, would offer valuable insights. Extending the model to coherent systems is also a promising avenue, given the power efficiency of coherent detection [43] and its support for advanced modulation formats such as PSK and QAM. Furthermore, investigating higher-network-layer design and routing algorithms could provide key insights into performance metrics such as network latency. The evaluation of optical downlink and uplink technologies under realistic conditions is also of great importance for formulating a roadmap for the advancement of these technologies. Although our study focuses on OOK, our model, which approximates an AWGN system with a time-varying path loss, can be extended to other modulation schemes, including coherent detection. Therefore, improvements in SNR and spectral efficiency observed in conventional optical systems should also be applicable in this context. Finally, future work could investigate perturbation effects (e.g., gravitational forces, drag, and station-keeping maneuvers) on acquisition, tracking, and laser alignment, as well as the impact of satellites' structural elements such as solar panels in dense constellations, and scenarios involving uplink and downlink under turbulence effects. These studies could optimize satellite design and improve communication and tracking efficiency.

Author Contributions: M.G. was responsible for the investigation, methodology, development of models and software, original draft writing, review and editing, visualization, and data curation. T.K. contributed to the development of models and software, writing, supervision of the research activity, project administration, and participated in the review and editing process. D.A. led the conceptualization, the formulation and development of the overarching research goals and aims, and contributed to the review and editing of the manuscript. All authors have read and agreed to the published version of the manuscript.

Funding: This research received no external funding.

Institutional Review Board Statement: Not applicable.

Informed Consent Statement: Not applicable.

Data Availability Statement: The data are contained within the article.

Conflicts of Interest: The authors declare no conflicts of interest.

Appendix A. `Pyminisat` Module Implementation

The various model elements discussed in Sections 2 and 3 were implemented as part of the `Pyminisat` `Python` module [28], which is publicly available under an open-source license. In its implementation, we exclusively used standard `Python` libraries, such as `numpy`, `scipy` for numerical manipulations, and `matplotlib` for result visualization. The `Pyminisat` module implements the `satsim` class, which contains all essential attributes of the satellite network pertaining to the constellation and transceiver systems. The `satsim` class invokes two sub-modules, namely the `linkmodel` class, which can be used to study an end-to-end OIL, and the `constellation` class, which incorporates the constellation dynamics and determines the optimal F value and the minimum distances that need to be covered in order to ensure network connectivity.

To study a given constellation, e.g., the one described in Table 1, the `satsim` class carries out the simulation in the following steps: first use the `constellation` class to estimate the best value for F and, therefore, determine the initial satellite position in the constellation. Then, initialize the constellation with the optimal value of F and calculate the minimum link distances required to provide connectivity to the nearby neighboring satellites (see Figure 4) over one orbital period. Given the parameters of the receiver and the data rate, use the `linkmodel` class to calculate the receiver sensitivity. Based on this value, calculate the required transmitter power for each instance of the orbital period using the link budget in (11). The Github repository of the model [28] further describes how the initialization parameters of the `satsim` class can be changed to fit various network scenarios.

To the extent possible, we optimized the code in order to take advantage of the large speed-ups involved when `numpy` vectorization is used instead of `for` loops. By far the most computationally challenging task is to optimize the constellation in terms of the F parameter. We can take advantage of the fact that the estimation of the distances in (9) is an embarrassingly parallel problem for the different time instances $t = t_l$ assumed. We can, therefore, use `Python`'s `multiprocessing` module to distribute the load to the available central processing unit (CPU) cores. On a standard 6-core, AMD Ryzen 5 7530U CPU (with 2 threads per core), the time required to optimize the constellation in Table 1 and estimate the power requirements is ≈14.3 s. For completeness, we tested a scenario with a constellation even denser than Scenario C, tripling both the number of satellites per orbit and the total number of orbits compared to Scenario A, yielding a total of 14,256 satellites. The computation time required in the same platform was now 165.7 s.

References

1. Carrizo, C.; Knapek, M.; Horwath, J.; Gonzalez, D.D.; Cornwell, P. Optical inter-satellite link terminals for next generation satellite constellations. In Proceedings of the Free-Space Laser Communications XXXII, San Francisco, CA, USA, 1–6 February 2020; SPIE: Bellingham, WA, USA , 2020; Volume 11272, pp. 8–18.
2. Kemih, K.; Yaiche, Y.; Benslama, M. Optimization of Transmitter Aperture by Genetic Algorithm in Optical Satellite. *World Acad. Sci. Eng. Technol. Int. J. Electr. Comput. Energetic Electron. Commun. Eng.* **2007**, *1*, 1261–1267.

3. Kim, G.N.; Park, S.Y.; Seong, S.; Choi, J.Y.; Han, S.K.; Kim, Y.E.; Choi, S.; Lee, J.; Lee, S.; Ryu, H.G.; et al. Design of Novel Laser Crosslink Systems Using Nanosatellites in Formation Flying: The VISION. *Aerospace* **2022**, *9*, 423. [CrossRef]
4. Zhu, Q.; Tao, H.; Cao, Y.; Li, X. Laser inter-satellite link visibility and topology optimization for mega constellation. *Electronics* **2022**, *11*, 2232. [CrossRef]
5. Liang, J.; Chaudhry, A.U.; Erdogan, E.; Yanikomeroglu, H. Link Budget Analysis for Free-Space Optical Satellite Networks. In Proceedings of the 2022 IEEE 23rd International Symposium on a World of Wireless, Mobile and Multimedia Networks (WoWMoM), Belfast, UK, 14–17 June 2022; pp. 471–476.
6. Liang, J.; Chaudhry, A.U.; Erdogan, E.; Yanikomeroglu, H.; Kurt, G.K.; Hu, P.; Ahmed, K.; Martel, S. Free-Space Optical (FSO) Satellite Networks Performance Analysis: Transmission Power, Latency, and Outage Probability. *IEEE Open J. Veh. Technol.* **2024**, *5*, 244–261. [CrossRef]
7. Space Exploration Holdings, LLC. *SpaceX Non-Geostationary Satellite System, Attachment A, Technical Information to Supplement Schedule S*; FCC: Washington, DC, USA, 2018. Available online: https://fcc.report/IBFS/SAT-MOD-20181108-00083/1569860.pdf (accessed on 6 December 2023).
8. Carrasco-Casado, A.; Biswas, A.; Fields, R.; Grefenstette, B.; Harrison, F.; Sburlan, S.; Toyoshima, M. Optical communication on CubeSats—Enabling the next era in space science. In Proceedings of the 2017 IEEE International Conference on Space Optical Systems and Applications (ICSOS), Naha, Japan, 14–16 November 2017; pp. 46–52. [CrossRef]
9. Goorjian, P.M. Free-space optical communication for spacecraft and satellites, including cubesats in low earth orbit (LEO). In Proceedings of the Photonic Networks and Devices, Burlingame, CA, USA, 29 July–1 August 2019; Optica Publishing Group: Washington, DC, USA, 2019; p. paper NeM2D.4.
10. Saathof, R.; Crowcombe, W.; Kuiper, S.; van der Valk, N.; Pettazzi, F.; de Lange, D.; Kerkhof, P.; van Riel, M.; de Man, H.; Truyens, N.; et al. Optical satellite communication space terminal technology at TNO. In Proceedings of the International Conference on Space Optics—ICSO 2018, Chania, Greece, 9–12 October 2018; Sodnik, Z., Karafolas, N., Cugny, B., Eds.; International Society for Optics and Photonics, SPIE: Bellingham, WA, USA, 2019; Volume 11180, p. 111800K. [CrossRef]
11. Benzi, E.; Troendle, D.C.; Shurmer, I.; James, M.; Lutzer, M.; Kuhlmann, S. Optical inter-satellite communication: The Alphasat and Sentinel-1A in-orbit experience. In Proceedings of the 14th International Conference on Space Operations, Daejeon, Republic of Korea, 16–20 May 2016; p. 2389.
12. Li, L.; Zhang, X.; Zhang, J.; Xu, C.; Jin, Y. Advanced space laser communication technology on cubesats. *ZTE Commun.* **2021**, *18*, 45–54.
13. Schmidt, C.; Rödiger, B.; Rosano, J.; Papadopoulos, C.; Hahn, M.T.; Moll, F.; Fuchs, C. DLR's Optical Communication Terminals for CubeSats. In Proceedings of the 2022 IEEE International Conference on Space Optical Systems and Applications (ICSOS), Kyoto, Japan, 28–31 March 2022; pp. 175–180. [CrossRef]
14. Velazco, J.; Boyraz, O. High data rate inter-satellite omnidirectional optical communicator. In Proceedings of the 32 Annual AIAA/USU Conference on Small Satellites, Logan, UT, USA, 4–9 August 2018.
15. Zaman, I.U.; Velazco, J.E.; Boyraz, O. Omnidirectional optical crosslinks for CubeSats: Transmitter optimization. *IEEE Trans. Aerosp. Electron. Syst.* **2020**, *56*, 4556–4566. [CrossRef] [PubMed]
16. Saeed, N.; Almorad, H.; Dahrouj, H.; Al-Naffouri, T.Y.; Shamma, J.S.; Alouini, M.S. Point-to-Point Communication in Integrated Satellite-Aerial 6G Networks: State-of-the-Art and Future Challenges. *IEEE Open J. Commun. Soc.* **2021**, *2*, 1505–1525. [CrossRef]
17. Chaudhry, A.U.; Yanikomeroglu, H. Free Space Optics for Next-Generation Satellite Networks. *IEEE Consum. Electron. Mag.* **2021**, *10*, 21–31. [CrossRef]
18. Liang, J.; Chaudhry, A.U.; Yanikomeroglu, H. Phasing parameter analysis for satellite collision avoidance in starlink and kuiper constellations. In Proceedings of the 2021 IEEE 4th 5G world forum (5GWF), Montreal, QC, Canada, 13–15 October 2021; IEEE: Piscataway, NJ, USA, 2021; pp. 493–498.
19. Chaudhry, A.U.; Yanikomeroglu, H. Laser Intersatellite Links in a Starlink Constellation: A Classification and Analysis. *IEEE Veh. Technol. Mag.* **2021**, *16*, 48–56. [CrossRef]
20. Chaudhry, A.U.; Yanikomeroglu, H. Temporary laser inter-satellite links in free-space optical satellite networks. *IEEE Open J. Commun. Soc.* **2022**, *3*, 1413–1427. [CrossRef]
21. Liang, J.; Chaudhry, A.U.; Chinneck, J.W.; Yanikomeroglu, H.; Kurt, G.K.; Hu, P.; Ahmed, K.; Martel, S. Latency Versus Transmission Power Trade-Off in Free-Space Optical (FSO) Satellite Networks with Multiple Inter-Continental Connections. *IEEE Open J. Commun. Soc.* **2023**, *4*, 3014–3029. [CrossRef]
22. Handley, M. Delay is Not an Option: Low Latency Routing in Space. In Proceedings of the 17th ACM Workshop on Hot Topics in Networks, Redmond, WA, USA, 15–16 November 2018; HotNets '18, pp. 85–91. [CrossRef]
23. Perlot, N.; Sofieva, V.F. Qualifying optical inter-satellite links for atmospheric occultations. In Proceedings of the International Conference on Space Optical Systems, Tokyo, Japan, 9–12 November 2009 .
24. Walker, J.G. Satellite constellations. *J. Br. Interplanet. Soc.* **1984**, *37*, 559.

25. Wenzel, H.; Klehr, A.; Braun, M.; Bugge, F.; Erbert, G.; Fricke, J.; Knauer, A.; Ressel, P.; Sumpf, B.; Weyers, M.; et al. Design and realization of high-power DFB lasers. In Proceedings of the Physics and Applications of Optoelectronic Devices, Philadelphia, PA, USA, 25–28 October 2004; SPIE: Bellingham, WA, USA, 2004; Volume 5594, pp. 110–123.
26. McDowell, J.C. The low earth orbit satellite population and impacts of the SpaceX Starlink constellation. *Astrophys. J. Lett.* **2020**, *892*, L36. [CrossRef]
27. Kotake, H.; Abe, Y.; Fuse, T.; Kubooka, T.; Toyoshima, M. Adaptive optical satellite network architecture. In Proceedings of the International Conference on Space Optics—ICSO 2020, Online, 30 March–2 April 2021; SPIE: Bellingham, WA, USA, 2021; Volume 11852, pp. 685–693.
28. Gioulis, M.; Kamalakis, T. Pyminisat: A Python Package for Simulating Optical Intersatellite Links in a Minisat Constellation. 2024. Available online: https://github.com/thomaskamalakis/pyminisat (accessed on 22 April 2024).
29. Curtis, H.D. *Orbital Mechanics for Engineering Students*, 7th ed.; Elsevier Butterworth Heinemann: Oxford, UK, 2005.
30. Sumathi, K.; Balasaraswathi, M.; Boopathi, C.; Singh, M.; Malhotra, J.; Dhasarathan, V. Design of 3.84 Tbps hybrid WDM–PDM based inter-satellite optical wireless communication (IsOWC) system using spectral efficient orthogonal modulation scheme. *J. Ambient Intell. Humaniz. Comput.* **2020**, *11*, 4167–4175. [CrossRef]
31. Chaudhary, S.; Chaudhary, N.; Sharma, S.; Choudhary, B. High speed inter-satellite communication system by incorporating hybrid polarization-wavelength division multiplexing scheme. *J. Opt. Commun.* **2017**, *39*, 87–92. [CrossRef]
32. Betti, S.; Carrozzo, V.; Parca, G. Optical Intersatellite hybrid network links based on WDM technology. In Proceedings of the 2008 10th Anniversary International Conference on Transparent Optical Networks, Athens, Greece, 22–26 June 2008; IEEE: Piscataway, NJ, USA, 2008; Volume 4, pp. 209–212.
33. Horst, Y.; Bitachon, B.I.; Kulmer, L.; Brun, J.; Blatter, T.; Conan, J.M.; Montmerle-Bonnefois, A.; Montri, J.; Sorrente, B.; Lim, C.B.; et al. Tbit/s line-rate satellite feeder links enabled by coherent modulation and full-adaptive optics. *Light. Sci. Appl.* **2023**, *12*, 153. [CrossRef]
34. Ramaswami, R.; Sivarajan, K.N.; Sasaki, G.H. *Optical Networks, A Practical Perspective*, 3rd ed.; Elsevier: Amsterdam, The Netherlands, 2010.
35. Vieira, I.P.; Pita, T.C.; Mello, D.A.A. Modulation and Signal Processing for LEO-LEO Optical Inter-Satellite Links. *IEEE Access* **2023**, *11*, 63598–63611. [CrossRef]
36. Brashears, T.R. Achieving 99% link uptime on a fleet of 100G space laser inter-satellite links in LEO. In Proceedings of the Free-Space Laser Communications XXXVI, San Francisco, CA, USA, 27 January–1 February 2024; SPIE: Bellingham, WA, USA, 2024; Volume 12877, p. 1287702.
37. Kato, W.; Kawamoto, Y.; Kato, N.; Ariyoshi, M.; Sugyo, K.; Funada, J. Stable and Efficient Inter-Satellite Optical Wireless Communications Through Connection of Intersecting Orbits. In Proceedings of the 2024 IEEE 99th Vehicular Technology Conference (VTC2024-Spring), Singapore, 24–27 June 2024; IEEE: Piscataway, NJ, USA, 2024; pp. 1–5.
38. Zhu, Y.; Xu, G.; Gao, M.; Chu, H.; Song, Z. Average bit-error rate analysis of an inter-satellite optical communication system under the effect of perturbations. *Opt. Express* **2024**, *32*, 36796–36810. [CrossRef]
39. Stampoulidis, L.; Osman, A.; Sourikopoulos, I.; Winzer, G.; Zimmermann, L.; Dorward, W.; Rodrigo, A.S.; Chiesa, M.; Rotta, D.; Maho, A.; et al. H2020-SPACE-ORIONAS Miniaturized optical transceivers for high-speed optical inter-satellite links. *arXiv* **2022**, arXiv:2210.17304.
40. Rödiger, B.; Fuchs, C.; Nonay, J.R.; Jung, W.; Schmidt, C. Miniaturized optical Intersatellite communication terminal–CubeISL. In Proceedings of the 2021 IEEE International Conference on Communications Workshops (ICC Workshops), Montreal, QC, Canada, 14–23 June 2021; IEEE: Piscataway, NJ, USA, 2021; pp. 1–5.
41. Edmunds, J.; Thipparapu, N.; Kechagias, M.; Hall, K.; Donnot, A.; Kean, P.; Kehayas, E.; Welch, M. Designing transmit optical amplifiers for the current roll out of optical communication constellations. In Proceedings of the International Conference on Space Optics—ICSO 2022, Dubrovnik, Croatia, 3–7 October 2022; SPIE: Bellingham, WA, USA, 2023; Volume 12777, pp. 1497–1504.
42. Carrasco-Casado, A.; Shiratama, K.; Kolev, D.; Trinh, P.V.; Ishola, F.; Fuse, T.; Toyoshima, M. Development and space-qualification of a miniaturized CubeSat's 2-W EDFA for space laser communications. *Electronics* **2022**, *11*, 2468. [CrossRef]
43. Morthier, G.; Roelkens, G.; Baets, R. Optical versus RF free-space signal transmission: A comparison of optical and RF receivers based on noise equivalent power and signal-to-noise ratio. *IEEE J. Sel. Top. Quantum Electron.* **2021**, *28*, 1–8. [CrossRef]

photonics

MDPI

Article

A Hybrid Network Integrating MHSA and 1D CNN–Bi-LSTM for Interference Mitigation in Faster-than-Nyquist MIMO Optical Wireless Communications

Minghua Cao [1,*], Qing Yang [1], Genxue Zhou [1], Yue Zhang [1], Xia Zhang [2,3] and Huiqin Wang [1]

[1] School of Computer and Communication, Lanzhou University of Technology, Lanzhou 730050, China; 222085402027@lut.edu.cn (Q.Y.); 222085402011@lut.edu.cn (G.Z.); zyue940209@163.com (Y.Z.); whq1222@lut.edu.cn (H.W.)

[2] School of Physical Science and Information Engineering, Liaocheng University, Liaocheng 252000, China; wenerzhang2002@163.com

[3] Shandong Provincial Key Laboratory of Optical Communication Science and Technology, Liaocheng 252000, China

* Correspondence: caominghua@lut.edu.cn

Abstract: To mitigate inter-symbol interference (ISI) caused by Faster-than-Nyquist (FTN) technology in a multiple input multiple output (MIMO) optical wireless communication (OWC) system, we propose an ISI cancellation algorithm that combines multi-head self-attention (MHSA), a one-dimensional convolutional neural network (1D CNN), and bi-directional long short-term memory (Bi-LSTM). This hybrid network extracts data features using 1D CNN and captures sequential information with Bi-LSTM, while incorporating MHSA to comprehensively reduce ISI. We analyze the impact of antenna numbers, acceleration factors, wavelength, and turbulence intensity on the system's bit error rate (BER) performance. Additionally, we compare the waveform graphs and amplitude–frequency characteristics of FTN signals before and after processing, specifically comparing sampled values of four-pulse-amplitude modulation (4PAM) signals with those obtained after ISI cancellation. The simulation results demonstrate that within the Mazo limit for selecting acceleration factors, our proposal achieves a 7 dB improvement in BER compared to the conventional systems without deep learning (DL)-based ISI cancellation algorithms. Furthermore, compared to systems employing a point-by-point elimination adaptive pre-equalization algorithm, our proposal exhibits comparable BER performance to orthogonal transmission systems while reducing computational complexity by 31.15%.

Keywords: Faster-than-Nyquist; Bi-LSTM; deep learning; interference cancellation; optical wireless communication

1. Introduction

The rapid development of mobile communication technology has greatly improved the performance of OWC systems, which are now widely used in various applications such as intelligent transportation, the Internet of Things, and 5G and future 6G networks [1]. To achieve higher transmission rates and better signal quality, advanced technologies like MIMO have become essential. MIMO maximizes system gain by deploying multiple antennas at both ends to improve spectral efficiency, enhance interference immunity, and maintain high-quality signal transmission in complex environments [2]. Additionally, FTN technology offers new opportunities for increasing OWC system rates by enabling approximately 25% higher data transmission within the same bandwidth through shorter symbol intervals and increased signaling rates [3]. FTN is a non-orthogonal transmission technique that offers distinct advantages over Power Domain Non-Orthogonal Multiple Access (PD-NOMA) and Code Domain Non-Orthogonal Multiple Access (CD-NOMA). FTN achieves higher data transmission rates within the same time by reducing symbol interval, thereby improving spectral efficiency. In contrast to PD-NOMA, which relies on power allocation and user interference cancellation, FTN primarily focuses on optimizing information transmission in high-bandwidth-demanding scenarios. Additionally, CD-NOMA differentiates users through various coding techniques, which can enhance capacity; however, its performance may be limited by coding complexity and the number of users in the applications that require high spectral efficiency. Therefore, FTNs demonstrate superior performance in specific scenarios which necessitate high-speed data transmission and thus warrant further research and application [4,5]. The implementation of FTN technology, however, poses challenges regarding ISI. When symbol transmission rates exceed the Nyquist rate, non-orthogonal heterodyne pulses can occur, resulting in degraded demodulation performance and impaired signal identification. In high data rate transmissions, signal quality instability may significantly impact communication effectiveness [6]. Therefore, effectively mitigating the ISI introduced by FTNs in MIMO-FTN-OWC systems is crucial as it directly affects signal reliability and user experience [7].

A low-complexity equalization algorithm based on the variational method suggested in [8] exhibits linearly increasing complexity with longer block lengths due to its limited ability to eliminate the ISI caused by strong mutual interference. However, the FTN technique allows for full knowledge of the introduced ISI at the transmitter, eliminating the need for feedback of the channel state information. An alternative approach proposed in [9] is linear pre-equalization, which transforms FTN signals into orthogonal signals at higher baud rates for improved spectral efficiency and to achieve optimal BER performance. For the complete elimination of ISI, ref. [10] proposed an adaptive point-by-point elimination pre-equalization algorithm that calculates interference weights using shaping filters and transmitted symbols for various acceleration factors. It then offsets interference values after FTN shaping on a point-by-point basis based on these weights. However, this method requires increased computational complexity in exchange for higher spectral efficiency and BER performance. In recent years, there has been growing attention towards integrating DL techniques into communications. Notably, ref. [11] utilized a recurrent neural network (RNN) structure based on LSTM networks to eliminate ISI introduced by FTN, achieving similar performance to the Bahl–Cocke–Jelinek–Raviv (BCJR) algorithm. Additionally, ref. [12] introduced an equalization algorithm based on Bi-LSTM networks to address successive ISI elimination between neighboring signals, effectively enhancing system BER performance. Meanwhile, ref. [13] devised a pre-equalization network using the backpropagation (BP) algorithm to eliminate ISI caused by FTN shaping, resulting in a 5 dB improvement compared to the traditional equalization algorithms.

Researching the existing literature revealed that conventional FTN systems' ISI cancellation methods often encounter high computational complexity, necessitating intricate algorithms and substantial computational resources, which limits their practical applications. Although the recent advancements in DL-based approaches have somewhat mitigated this complexity, they still fall short of completely eliminating ISI. Therefore, effectively eradicating ISI while minimizing computational complexity remains a critical challenge. To address this issue, we propose an algorithm that combines 1D CNN, Bi-LSTM, and MHSA to eliminate ISI. This algorithm aims to uncover the hidden features in interfering data and utilize its powerful computational capabilities to directly tackle the ISI introduced by FTN [14–16]. Simulation results indicate that our proposed algorithm achieves up to a 7 dB improvement in BER within the Mazo limit compared to a conventional system without this ISI cancellation algorithm. Furthermore, our algorithm not only maintains comparable BER performance but also reduces computational complexity by 31.15% compared to the point-by-point cancellation adaptive pre-equalization algorithm, nearly matching the performance of an orthogonal system. These results demonstrate an effective enhancement of system performance through our method and provide new insights for signal processing in complex OWC links.

This manuscript is structured as follows: The first section provides a brief introduction to the study's background and significance; the second section explains the system model and the relevant theoretical foundations; the third section offers a detailed description of the experimental and simulation processes; the fourth section presents a thorough analysis of the results; and finally, the manuscript concludes with a summary.

2. System Model

The schematic diagram of the MIMO-FTN-OWC system is illustrated in Figure 1. Assuming the presence of N_t transmitting antennas and N_r receiving antennas, the binary data undergo Gray coding and subsequent mapping into a PAM signal with a modulation order of 4. The modulated 4PAM signal $g(t)$ is then passed through an FTN shaping filter to obtain the FTN signal $S_{FTN}(t)$, which is subsequently fed into a trained neural network for ISI mitigation. Following this, the signal traverses a digital-to-analog converter (DAC) and drives a laser diode (LD) to generate an optical signal $S_T(t)$ that is transmitted via multiple optical antennas through the atmospheric channel. In this system, FTN signals are directly converted to optical signals through the LD after passing through the DAC, eliminating the need for an optical amplifier (e.g., EDFA). The utilization of an optical amplifier can introduce undesirable effects such as nonlinear distortion, increased noise, and gain saturation, which may result in signal distortion, reduced signal-to-noise ratio, and compromised demodulation performance and signal reliability. Therefore, although opting out of an optical amplifier may lead to insufficient signal strength during long-distance transmission or in interference-prone environments, it enhances anti-interference capability by employing a high-precision DAC to minimize signal distortion, selecting a stable and linear laser diode and utilizing a 4PAM modulation method suitable for FTN transmission. Additionally, we apply the MHSA–1D CNN–Bi-LSTM algorithm to mitigate ISI effectively while ensuring high-quality output of optical signals. The optical signal $S_R(t)$ received by multiple optical antennas at the receiver end are converted into an electrical signal through a photodetector (PD). These electrical signals then go through an analog-to-digital converter (ADC), followed by matched filtering, sampling, and subsequent transmission to a maximum likelihood detector. Finally, the data are recovered after Gray decoding.

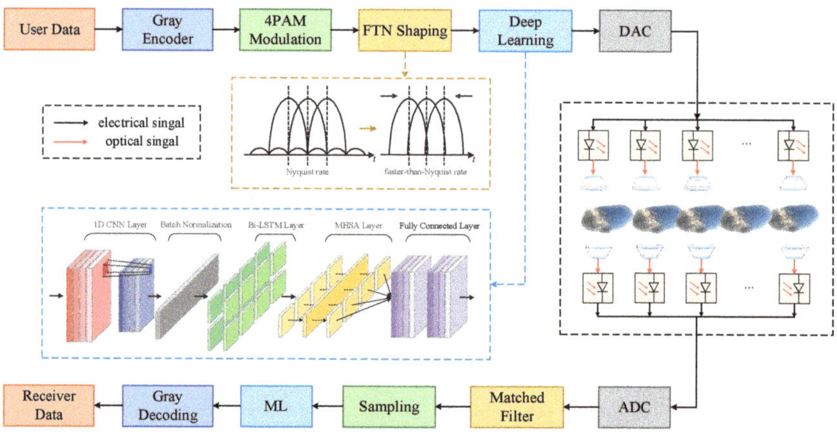

Figure 1. Schematic of MIMO-FTN-OWC system.

In the MIMO-FTN-OWC system, the PAM signal is denoted as $P = [p_1 p_2 \cdots p_n]$. The system employs a rising cosine filter $g(t)$ with a roll-off factor ψ, r denotes the number of truncation segments, p denotes the number of samples per segment, $T = p$ denotes the

Nyquist transmission interval and filter length is $L = r \times p + 1$. The vector form of the rising cosine filter $g(t)$ can be expressed as follows [17]:

$$G = [g_1, g_2, \ldots, g_{L-1}, g_L] \tag{1}$$

where G denotes the peak point of the filter coefficients, which corresponds to the sampling point of the received signal. By multiplying PAM signal values with G and superimposing them after a delay, the FTN signal $S_{FTN}(t)$ can be expressed as follows [18]:

$$S_{FTN}(t) = \sum_{i=0}^{n} p_i g(t - (i-1)\tau T), 0 < \tau < 1 \tag{2}$$

where n denotes the number of modulated symbols transmitted per frame, $p_i \in P$ denotes the i-th symbol, $P = \{a_m = 2m - 1 - U, m = 1, 2, \ldots, U\}$, U denotes the modulation order of PAM, τT is the delay interval, $\tau = p'/p$ denotes the acceleration factor, p' denotes the sampling factor when shaping FTN signal.

The FTN signal is subsequently input into a trained neural network that combines 1D CNN, the Bi-LSTM network, and the MHSA mechanism. The 1D CNN extracts crucial features from the data and arranges them into sequences for the Bi-LSTM network to learn from. Additionally, the MHSA mechanism assigns extra weight to enhance learning outcomes and ensures that the pre-processed de-interference signal $S_{FTN-DL}(t)$ closely resembles the original signal. Subsequently, $S_{FTN-DL}(t)$ is converted to the digital-to-analog format before driving the LDs to generate an optical signal $S_T(t)$. Upon propagation through the atmospheric channel, the received signal can be written as follows [19]:

$$S_R(t) = h \cdot S_T(t) + n(t) \tag{3}$$

where h denotes the channel fading coefficient, and $n(t)$ denotes the additive noise.

The channel fading coefficients exhibit a Gamma–Gamma distribution and their probability density function can be mathematically formulated as follows [20]:

$$\varphi_h = \frac{2(\alpha\beta)^{\frac{\alpha+\beta}{2}}}{\Gamma(\alpha)\Gamma(\beta)} \cdot h^{\frac{\alpha+\beta}{2}-1} \cdot N_{\alpha-\beta} \cdot (2\sqrt{\alpha\beta h}) \tag{4}$$

where $N_{\alpha-\beta}(\cdot)$ denotes the second-class modified Bessel function of order $\alpha - \beta$, $\Gamma(\cdot)$ denotes the gamma function, α denotes the large-scale scattering coefficient, and β denotes the small-scale scattering coefficient, which can be respectively expressed as follows:

$$\alpha = \left\{ \exp\left[\frac{0.49\sigma^2}{(1 + 1.11\sigma^{12/5})^{7/6}} \right] - 1 \right\}^{-1} \tag{5}$$

$$\beta = \left\{ \exp\left[\frac{0.51\sigma^2}{(1 + 0.69\sigma^{12/5})^{5/6}} \right] - 1 \right\}^{-1} \tag{6}$$

where $\sigma^2 = 0.5C_n^2 k^{7/6} l^{11/6}$ denotes the Rytov variance, C_n^2 denotes the atmospheric refractive index structure constant, $k = 2\pi/\lambda$, λ denotes the laser wavelength, and l denotes the laser transmission distance.

At the receiver end, the signal received $S_R(t)$ from multiple optical antennas is converted into an electrical signal by PDs. Subsequently, it passes through a matched filter and is sampled at intervals of τT. The vector representation of the sampled signal can be expressed as follows:

$$Y = [y_1, y_2, \ldots, y_n] = [y_{r \times p/2+1}, y_{r \times p/2+1+p'}, \ldots, y_{r \times p/2+1+(n-1) \times p'}] \tag{7}$$

where y_n denotes the sampled symbol, $Y = y_1 + y_2 + \ldots + y_n$ indicates that the signals from all the receiving antennas are combined with equal gain. Subsequently, the sampled signal

Y is subjected to maximum likelihood sequence detection, followed by the demapping process for the restoration of user data.

3. MHSA–1D CNN–Bi-LSTM Interference Cancellation Scheme

3.1. D CNN

In the MIMO-FTN-OWC system, a computational method can be employed to determine the ISI value for each signal pulse when τ is known. Utilizing this information, we initially utilize the ISI-free signal $S_{FTN}{}'(t)$ as a labeled sequence. Subsequently, a one-dimensional convolutional (Conv1D) layer is employed to train the FTN signal sequence and derive the optimal convolutional kernels that minimize loss. These kernels are then utilized to automatically extract intricate FTN-ISI features. The maximum pooling layer further consolidates and preserves these complex features extracted by the convolutional layers. Finally, the highest values from each convolutional layer's feature vectors are merged to form the ultimate feature values, as depicted in Figure 2.

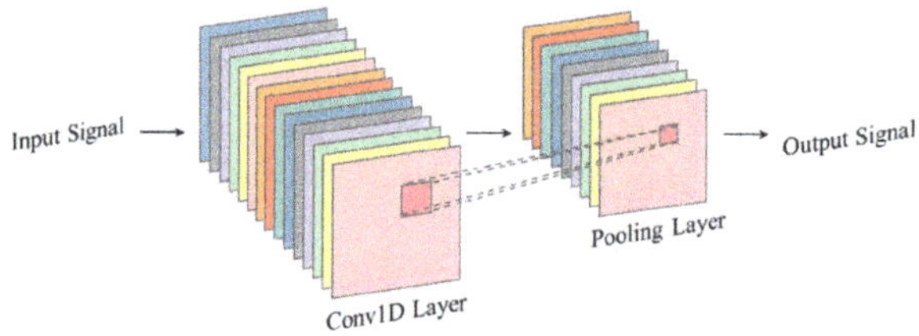

Figure 2. The structure of 1D CNN (Different colors represent different filters).

3.2. Bi-LSTM

The Bi-LSTM is an extension of the LSTM architecture designed to handle data with temporal dependencies by utilizing two independent input sequences that are fed into the same LSTM layer. The forward and reverse LSTMs of the Bi-LSTM are trained using FTN signal sequence data, resulting in a structure consisting of an input layer, a forward hidden layer, a reverse hidden layer, and an output layer as depicted in Figure 3.

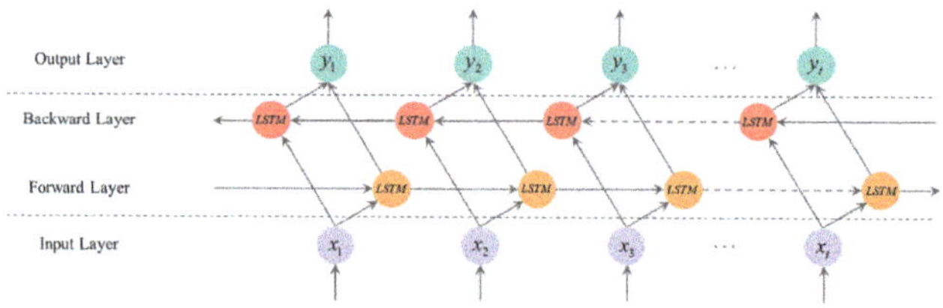

Figure 3. The structure of Bi-LSTM.

The specific implementation of the Bi-LSTM algorithm is detailed in Table 1 [21]. The Bi-LSTM algorithm processes time-series data through an input sequence $X = [x_1, x_2, \ldots, x_T]$, where each feature vector x_t is mapped to both forward and backward LSTM cells at time step t. The core of the LSTM is its gating mechanism, which includes input gates i, forget gates f, output gates o, and candidate cell states g. These gates utilize weight matrices W and U along with bias b to regulate the flow and storage of information. The forward and backward hidden states $h_{forward}[t]$ and $h_{backward}[t]$ together generate the output $h[t]$ of the bidirectional LSTM, effectively capturing both forward and backward dependencies in the sequence data while handling nonlinear relationships through activation functions such as Sigmoid and hyperbolic tangent (tanh).

Table 1. Bi-LSTM computational procedure.

Algorithm	Bi-LSTM
Start Input: Input Sequence $X = [x_1, x_2, \ldots, x_T]$, T is the length of the sequence. Parameters: LSTM cell parameters (W, U, b), W, U is weights and b is biases. Output: Output Sequence $H = [h_1, h_2, \ldots, h_T]$ **Step 1: Initialize hidden and cell states.** $h_{forward}[0] = 0,$ $c_{forward}[0] = 0,$ $h_{backward}[0] = 0,$ $c_{backward}[0] = 0.$ **Step 2: Forward propagation.** for t = 1 to T do, Loop statement that iterates from 1 to T. Calculate the output of forward LSTM: $i_{forward}[t] = \sigma(W_i * x_t + U_i * h_{forward}[t-1] + b_i),$ $f_{forward}[t] = \sigma(W_f * x_t + U_f * h_{forward}[t-1] + b_f),$ $o_{forward}[t] = \sigma(W_o * x_t + U_o * h_{forward}[t-1] + b_o),$ $g_{forward}[t] = \tanh(W_g * x_t + U_g * h_{forward}[t-1] + b_g),$ $c_{forward}[t] = f_{forward}[t] * c_{forward}[t-1] + i_{forward}[t] * g_{forward}[t],$ $h_{forward}[t] = o_{forward}[t] * \tanh(c_{forward}[t]).$ **Step 3: Backward propagation.** for $t = T$ to 1 do, Loop statement that iterates from T to 1. Calculate the output of backward LSTM: $i_{backward}[t] = \sigma(W_i * x_t + U_i * h_{backward}[t+1] + b_i),$ $f_{backward}[t] = \sigma(W_f * x_t + U_f * h_{backward}[t+1] + b_f),$ $o_{backward}[t] = \sigma(W_o * x_t + U_o * h_{backward}[t+1] + b_o),$ $g_{backward}[t] = \tanh(W_g * x_t + U_g * h_{backward}[t+1] + b_g),$ $c_{backward}[t] = f_{backward}[t] * c_{backward}[t+1] + i_{backward}[t] * g_{backward}[t],$ $h_{backward}[t] = o_{backward}[t] * \tanh(c_{backward}[t]).$ **Step 4: Combine forward and backward outputs.** for t = 1 to T do, $h[t] = concat(h_{forward}[t], h_{backward}[t])$, Splice forward and backward hidden states. return to $H = [h_1, h_2, \ldots, h_T]$. **End**	

3.3. Multi-Head Self-Attention

The MHSA mechanism is a variant of the self-attention mechanism, designed to enhance the expressive and generalization abilities of the model. Multiple attention computations are performed in parallel on the signal sequence, organized by a "head" structure. If there are h "head" structures, Q, K, and V undergo linear transformations and are allocated to each head, with $1/h$ of the original dimension during the initial modeling stage. Subsequently, each "head" independently calculates attention using its allocated data. The computed results are then concatenated along a specific dimension and transformed through a linear operation to get the final attention distribution. The process can be expressed as in the following equations [22]:

$$head_i = Attention(QW_i^Q, KW_i^K, VW_i^V) \quad (8)$$

$$MultiHead(Q, K, V) = Concat(head_1, head_2, \ldots, head_h)W^O \quad (9)$$

where W_i^Q, W_i^K, W_i^V, and W_o are learnable linear transformation parameter matrices. The two-head self-attention mechanism is depicted in Figure 4.

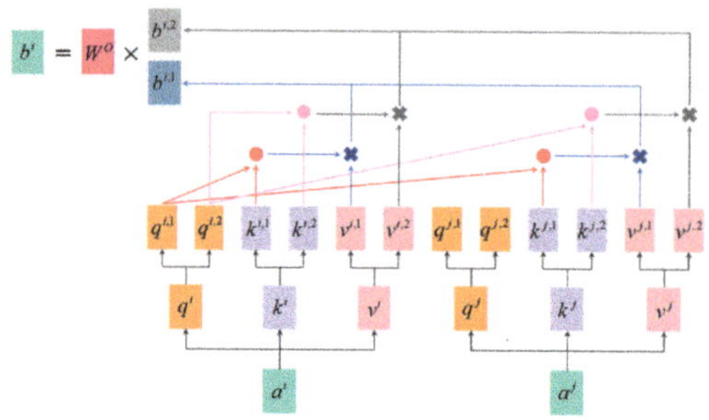

Figure 4. Calculation process of the double-head self-attention mechanism.

The distinction between the MHSA mechanism and the self-attention mechanism lies in the fact that the latter only necessitates a single W^Q, W^K, and W^V for computing Q, K, and V, whereas the former requires an additional computation of computed Q, K, and V. Taking q^i as an example, $q^{i,1} = W^{q,1}q^i$, $q^{i,2} = W^{q,2}q^i$ is initially computed, followed by the computation of $b^{i,1}$ using $q^{i,1}, k^{i,1}, v^{i,1}$ and $q^{i,1}, k^{j,1}, v^{j,1}$. Subsequently, $b^{i,2}$ is computed using $q^{i,2}, k^{i,2}, v^{i,2}$ and $q^{i,2}, k^{j,2}, v^{j,2}$. Then, $b^{i,1}$, $b^{i,2}$ are concatenated to obtain b^i through matrix multiplication with W^O. Similarly, b^j can be obtained following this procedure, thereby completing the computation of two-head self-attention mechanism [23].

3.4. MHSA–1D CNN–Bi-LSTM Interference Cancellation Algorithm

The signal sequence typically allows for extracting only a limited number of features by a single model. To effectively mitigate the ISI introduced by the FTN, we propose a hybrid model, namely MHSA–1D CNN–Bi-LSTM, which combines 1D CNN, a Bi-LSTM network, and an MHSA mechanism. The network structure is illustrated in Figure 5.

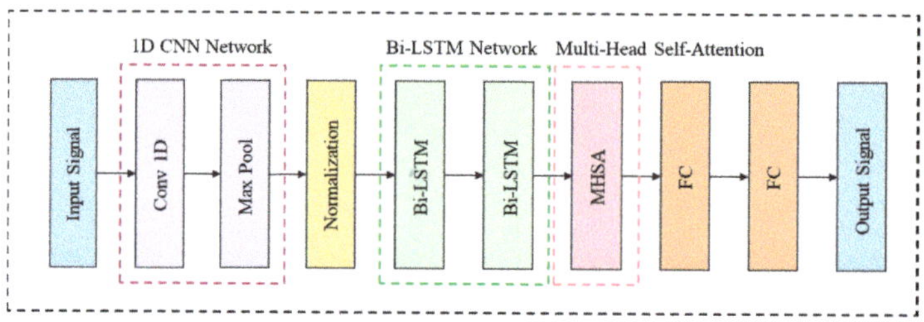

Figure 5. The structure of MHSA–1D CNN–Bi-LSTM network (Blue represents the input and output signals; purple represents the 1D-CNN; light yellow represents the normalization layer; green represents the Bi-LSTM network; red represents the multi-head self-attention mechanism layer; and orange represents the fully connected layer).

Initially, the FTN signal sequence $S_{FTN}(t)$ is utilized as input for the neural network. A 1D CNN is utilized to extract and organize feature vectors associated with information such as ISI and noise into sequences. Subsequently, the Bi-LSTM network is split into two independent LSTM networks, where these identical LSTMs generate forward and reverse hidden state sequences from the input sequences. The LSTM selectively forgets information through the forgetting gate using the previous time node's output $h_{forward}[t]$ and the current time node's input x_t. It then determines which new information will be stored in the current candidate memory cell $c_{forward}[t]$ to obtain the latest memory cell $c_{backward}[t]$. The predecessor memory cell $c_{backward}[t-1]$ is updated accordingly. By combining output value o_t of the output gate with the current moment memory cell state $c_{backward}[t]$, we get the LSTM's output value $h_{backward}[t]$. This is followed by a linear transformation with a weight matrix dimension matching that of the input data dimension, ensuring consistent dimensions for subsequent operations. An MHSA layer is introduced for a "head-splitting" operation after this step. Through the MHSA mechanism, distinct weights are assigned to each LSTM unit's output $[h_1, h_2, \ldots, h_T]$, enabling selective attention to different time steps during sequence processing. This facilitates the efficient selection of prediction information relevant to achieving current task goals. Finally, operations are performed within each self-attention mechanism and their outputs are concatenated for linear transformations before normalization.

4. Simulation and Analysis

The system is implemented in Python 3.8, utilizing Keras framework with TensorFlow serving as the backend for tensor operations. Both the training and test datasets are generated randomly using MATLAB R2022a, while the simulation parameters are presented in Table 2.

Table 2. Simulation parameters.

Parameter	Value
size of data	2×10^6
training dataset	1.6×10^6
test dataset	4×10^5
batch size	256
cycle index	100
dropout	0.3
learning rate	0.0001
loss function	MSE
optimizer	Adam
atmospheric refractive [24]	strong:1.13×10^{-13} m$^{-2/3}$ medium:1.13×10^{-14} m$^{-2/3}$ weak:1.13×10^{-17} m$^{-2/3}$

The MHSA–1D CNN–Bi-LSTM hybrid model consists of a Conv1D layer with 64 filters, kernel size of six, stride of one, and the "same" padding. ReLU activation is applied post-convolution to enhance learning performance. L2 regularization with a weight decay factor 0.01 is utilized to mitigate overfitting and enhance generalization during the training process. A subsequent max pooling layer follows the convolutional layer with a window size of two and stride of one. Power normalization is utilized to compress symbols to simulate hardware constraints accurately. Two Bi-LSTM layers ensue, where the first consists of 100 neurons and the second has only 20 neurons. In the MHSA–CNN–Bi-LSTM hybrid model, we set the dropout rate to 0.3 based on our experimental observations, striking a favorable balance between enhancing model performance and mitigating overfitting. Our cross-validation results indicate that lower dropout rates fail to adequately suppress overfitting, while higher rates may result in information loss. Therefore, by retaining adequate information and reducing the risk of overfitting, a dropout rate of 0.3 ensures the model's ability to generalize effectively to unknown data. Subsequently, an attention layer employing a two-headed self-attention mechanism employs a total of 64 neurons for enhanced representation capability. Finally, the output layer encompasses two fully connected layers, with one comprising 10 neurons and another consisting solely of 1 neuron responsible for generating the final result. During the training process, we load the labeled dataset from a CSV file, which includes both the pre-FTN shaping data and the disturbed post-FTN shaping data. To enhance the effectiveness of model training, we normalize the disturbed data to ensure a mean of 0 and a standard deviation of 1. Subsequently, we divide the dataset into training and test sets in an 80/20 ratio. Mean square error is employed as the loss function during training, optimized using the Adam optimizer. The model undergoes 100 training epochs while monitoring both training loss and validation loss to assess convergence and generalization capability. Upon completion of training, we evaluate predictive performance using metrics such as R^2, root mean square error (RMSE), and mean absolute error (MAE) to demonstrate its efficacy in handling disturbed signals and further validate its potential. Finally, for future analysis and application purposes, we save both the trained model and its normalizer.

4.1. Evaluation Index

The performance evaluation encompasses the following three selected metrics: RMSE, MAE, and R^2, as presented in Table 3.

Table 3. Model evaluation indicators.

Indicator Name	Indicator Meaning	Formula
RMSE	The square root of the ratio of the square of the deviation of the predicted value from the true value to the number of observations [25].	$RMSE = \sqrt{\frac{1}{n}\sum_{i=1}^{n}(y_i - \hat{y}_i)^2}$
MAE	The average of the absolute values of the deviations of all individual observations from the arithmetic mean [26].	$MAE = \frac{1}{n}(\sum_{i=1}^{n}\lvert y_i - \hat{y}_i\rvert)$
R^2	The degree of fit of the regression line to the observations [27].	$R^2 = 1 - \frac{\sum_{i=1}^{n}(\hat{y}_i - y_i)^2}{\sum_{i=1}^{n}(\overline{y}_i - y_i)^2}$

where y_i denotes the true value, $\hat{y}_i$ denotes the predicted value, $\overline{y}_i$ denotes the average of all true values in the test set and n denotes the number of samples in the test set within $[0, +\infty]$. A smaller RMSE and MAE indicate better prediction accuracy for the model. Similarly, a higher R^2 value indicates enhanced discriminative ability for the model.

Additionally, the comparative analysis encompasses LSTM, Bi-LSTM, CNN–Bi-LSTM models, in addition to the proposed MHSA–1D CNN–Bi-LSTM hybrid model. The specific outcomes are presented in Table 4.

Table 4. Performance comparison of different models.

Model	RMSE	MAE	R^2
LSTM	0.68733	0.56313	0.62213
Bi-LSTM	0.64597	0.55962	0.66623
CNN–Bi-LSTM	0.36303	0.31812	0.89458
MHSA–1D CNN–Bi-LSTM	0.31499	0.24220	0.92060

The superiority of the proposed MHSA–1D CNN–Bi-LSTM hybrid model over the LSTM, Bi-LSTM, CNN–Bi-LSTM models is evident, as demonstrated by the significantly improved results presented in Table 4.

4.2. System Performance Analysis

Figure 6 illustrates the amplitude–frequency characteristics of various signals. Upon observing Figure 6, it is evident that the interference cancellation algorithm significantly improves the amplitude–frequency characteristics of the signal. However, there is a slight deviation from those of the 4PAM signal. This discrepancy arises because, although the MHSA–1D CNN–Bi-LSTM interference cancellation algorithm effectively eliminates interference values at signal sampling points, residual interference values still exist at other moments outside these sampling points. Therefore, it can be concluded that the proposed algorithm exclusively eliminates interference affecting the target value of signal amplitude during transmission without altering the overall duration of FTN signals.

The sampled values of three different signals are presented in Figure 7. Figure 7a illustrates the sampled values of the 4PAM signal (prior to FTN shaping filtering), while Figure 7b displays the sampled values of the signal after FTN shaping filtering. Additionally, Figure 7c presents the sampled values of the FTN signal following the application of our proposed interference cancellation algorithm. A comparison between Figures 7a and 7b clearly reveals the presence of ISI introduced by employing FTN technique for modulation purposes. In contrast, a point-by-point comparison between Figures 7a and 7c demonstrates consistent amplitudes for almost all corresponding points, indicating that our interference cancellation algorithm effectively eliminates the ISI caused by the FTN modulation technique, thereby showcasing its potential as a robust solution for enhancing reliability in FTN communication systems.

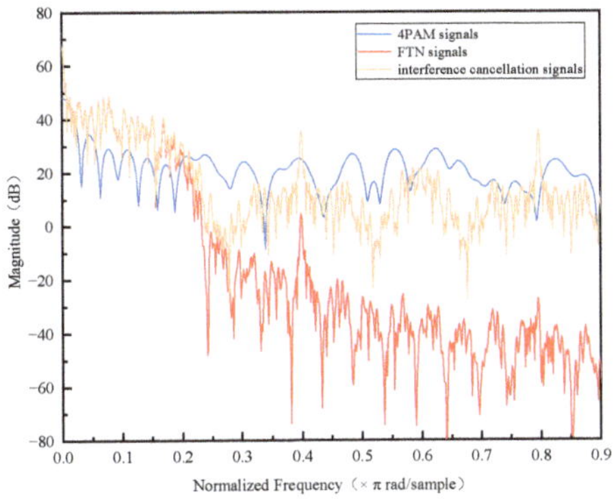

Figure 6. Magnitude–frequency characteristics.

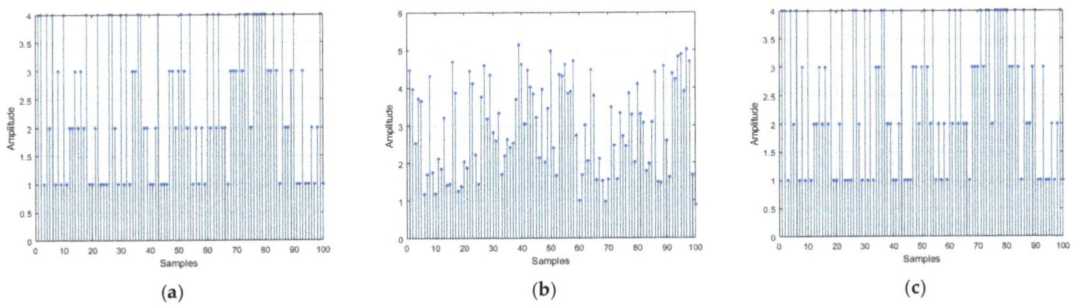

Figure 7. Comparison of signal sampling values (**a**) 4PAM signals (**b**) FTN signals (**c**) signals after ISI cancellation.

Figure 8 illustrates the BER plots for various transceiver antenna configurations, with and without the interference cancellation algorithm, under medium turbulence channel conditions with $\tau = 0.8$, $\lambda = 1550$ nm, $ipce = 0.5$. $ipce$ denotes the photoelectric conversion efficiency and is set to 0.5 to represent typical photodetector performance across diverse environmental conditions. This selection effectively accounts for optical signal attenuation and noise effects, ensuring system reliability under varying channel conditions. Moreover, opting for a value of 0.5 strikes a balance between signal quality and system complexity, thereby enhancing overall performance. Consequently, this setting guarantees stable and efficient model operation in real-world applications. Firstly, our proposed algorithm demonstrates comparable performance to the orthogonal system across all antenna configurations, indicating its effectiveness in eliminating ISI. Experimental results on BER performance with 1×1, 2×4, and 4×2 transceiver antenna configurations confirm that both the algorithms exhibit nearly identical performances, thereby showcasing their equivalent diversity gain. This equivalence is primarily attributed to their effective utilization of the spatial resources provided by multiple antennas under similar channel conditions while maintaining signal independence and employing comparable signal processing strategies. Consequently, they achieve similar BER performance with reduced signal fading effects. Notably, in the 1×1 transceiver antenna configuration, the MIMO-FTN-OWC system without the interference cancellation algorithm achieves an approximate signal-to-noise ratio (SNR) of 31 dB compared to around 27 dB when utilizing the algorithm at a $BER = 10^{-3}$. Therefore, the proposed algorithm yields a coding gain of 4 dB.

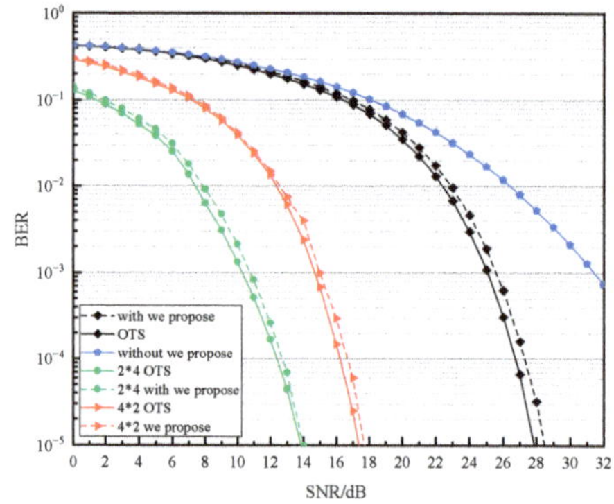

Figure 8. Relationship between BER and SNR for our proposal and orthogonal system (OTS refers to Orthogonal Transmission System).

Figure 9 demonstrates the BER performance under medium turbulence conditions within the Gamma–Gamma channel for different numbers of transceiver antennas. It compares scenarios with and without the utilization of the proposed algorithm, at $\tau = 0.8$, $ipce = 0.5$, and $\lambda = 1550$ nm. Firstly, in the Single Input Single Output (SISO) system, employing the DL ISI cancellation algorithm improves system performance by 6 dB at $BER = 10^{-3}$ compared to its absence. In the MIMO system, with a fixed number of two transmitting antennas and varying receiving antennas set to two and four, respectively, utilizing the proposed algorithm improves system performance by 4.5 dB and 3 dB at $BER = 10^{-5}$ compared to its absence. Similarly, when four transmitting antennas are fixed while the receiving antennas are set to two and four, respectively, applying the proposed ISI cancellation algorithm improves system performance by 5.5 dB and 3 dB at $BER = 10^{-5}$ compared to its non-usage scenario. All aforementioned simulation results effectively demonstrate that our proposed MHSA–1D CNN–Bi-LSTM hybrid network successfully mitigates eliminates ISI introduced by FTN across various numbers of transceiver antennas, thereby significantly enhancing overall system performance.

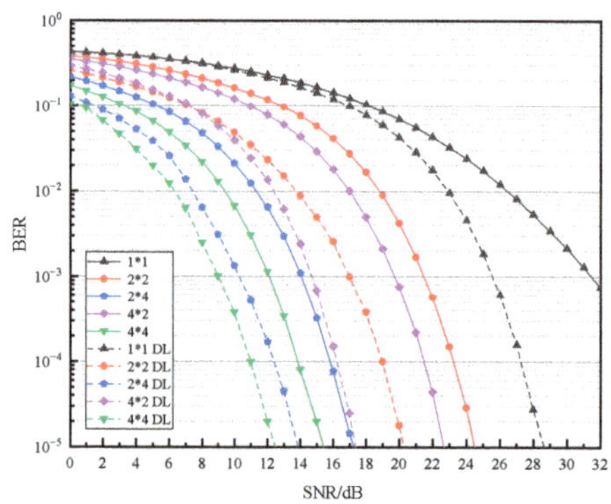

Figure 9. Relationship between BER and SNR under different numbers of antennas.

The optimal system performance is observed with a transceiver antenna configuration of 4×2, as shown in Figure 9. Therefore, this configuration is selected to investigate the algorithm's performance under varying turbulence intensities. Figure 10 presents a comparison of the BER performance with and without the proposed algorithm across weak, medium, and strong turbulence intensities, at $\tau = 0.8$, $ipce = 0.5$, and $\lambda = 1550$ nm. From the figure, it can be observed that utilizing the proposed ISI cancellation algorithm leads to an improvement in the system performance by 6.5 dB, 5.5 dB, and 7 dB, respectively, at

$BER = 10^{-5}$. This clearly demonstrates that the BER performance consistently surpasses that of the system without employing this algorithm.

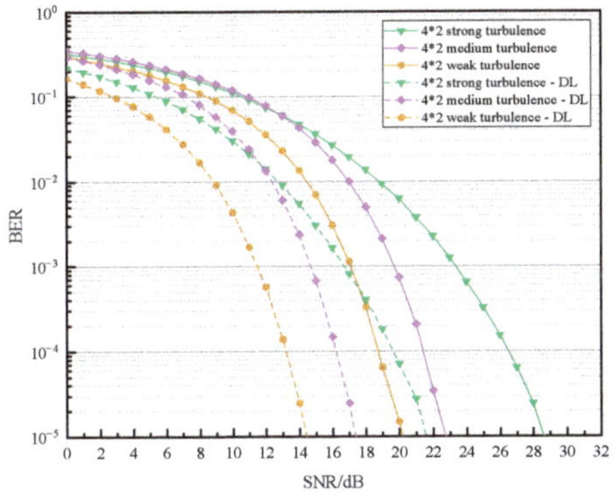

Figure 10. Relationship between BER and SNR under different turbulence intensities.

Under Gamma–Gamma medium turbulence conditions with an 4 × 2 antenna configuration, Figure 11 presents a comparison of the BER performance at different laser wavelengths with and without the proposed algorithm, when $\tau = 0.8$ and $ipce = 0.5$. At laser wavelengths of 1550 nm, 850 nm, 780 nm, and 635 nm, respectively, the system's performance is improved by 5.5 dB, 5 dB, 4.2 dB, and 3.4 dB at $BER = 10^{-5}$. The results indicate that as the laser wavelength increases, there is a decrease in BER along with an improved ability of the algorithm to eliminate interference.

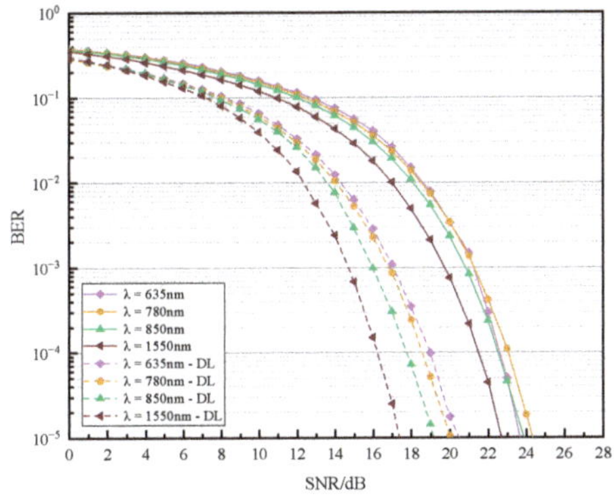

Figure 11. Relationship between BER and SNR under different laser wavelengths.

The effectiveness of the proposed ISI cancellation algorithm is verified at different acceleration factors in Figure 12, comparing the system performance with and without the proposed algorithm for four acceleration factors (0.5, 0.6, 0.7, and 0.8) under Gamma–Gamma medium turbulence conditions with a 4 × 2 antenna configuration at $ipce = 0.5$ and $\lambda = 1550$ nm. Firstly, in the absence of the ISI cancellation algorithm and for $\tau < 0.8$, a significant occurrence of ISI arises due to the close overlap of waveforms, resulting in a degradation of normal communication performance. However, upon application of the algorithm, there is a notable enhancement in the BER performance. Specifically, when $\tau = 0.8$, there is an improvement in system performance by 5.5 dB; and at $\tau = 0.7$ and $\tau = 0.6$, the ISI cancellation algorithm elevates the system SNR to 18 dB and 19.4 dB, respectively, at $BER = 10^{-5}$, thereby restoring normal communication capability. It is also noted that when $\tau = 0.5$, employing the ISI cancellation algorithm has minimal impact on system performance. The simulation results indicate that this algorithm significantly enhances system performance particularly within a range of τ values between 0.6 and 0.8, with smaller values yielding better interference cancellation.

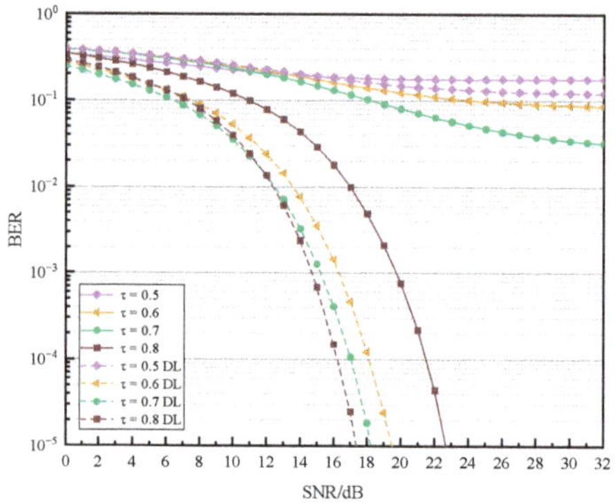

Figure 12. Relationship between BER and SNR under different acceleration factors.

To further validate the impact of the acceleration factor on the interference cancellation capability, Figure 13 shows the system BER with and without the ISI cancellation algorithm under Gamma–Gamma moderate turbulence conditions, using a 4×2 antenna configuration at $ipce = 0.5$, $\lambda = 1550$ nm, and SNR of 15 dB. The results demonstrate that for $\tau \geq 0.5$, there is an improvement in the cancellation effect of the algorithm as τ decreases. However, when $\tau < 0.5$, it is observed that the algorithm becomes less effective in mitigating ISI introduced by the FTN, resulting in significant performance degradation. The acceleration factor below 0.5 significantly increases ISI due to increased symbol overlap. A critical threshold is reached at an acceleration factor of 0.5, where the wider pulse shape reduces the symbol overlap and weakens the effect of ISI on demodulation. Conversely, when the acceleration factor drops below 0.5, the narrower pulse width leads to abnormally high ISI with weakened effectiveness of the interference cancellation algorithm. On the other hand, within the range of 0.5 to 1, our proposed algorithm effectively eliminates ISI, and its effectiveness improves as ISI increases.

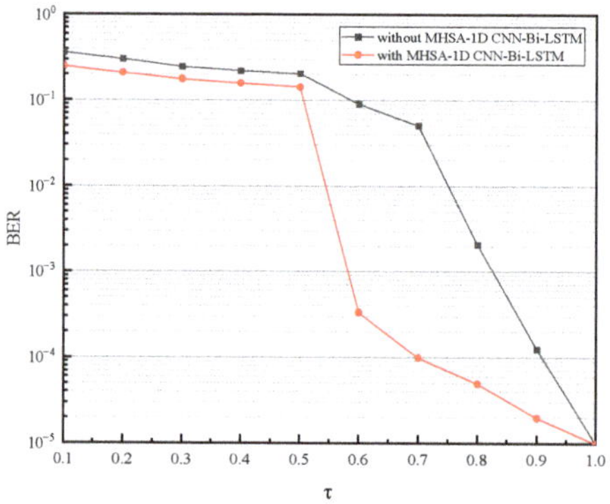

Figure 13. Relationship between acceleration factor and BER.

4.3. Computational Complexity

The number of floating-point operations serves as a complexity metric for comparing the computational complexity between the proposed algorithm and the point-by-point cancellation adaptive pre-equalization algorithm [10]. The computational complexity of the point-by-point cancellation adaptive pre-equalization algorithm is as follows:

$$\begin{aligned} Com_1 &= n \times L + (L - \tau \times p' + n \times \tau \times p') \times 2 + n \times 4 + [v + (2 \times v - 1)] \\ &\times v + (n - 2 \times v) \times 2 \times v + n \times L + (L - \tau \times p' + n \times \tau \times p') + n \\ &\times 2 + [v + (2 \times v - 2)] \times v + (n - 2 \times v) \times (2 \times v - 1) \end{aligned} \quad (10)$$

where v denotes the number of signs of the trailing influence.

The computational complexity of our proposal is as follows:

$$\begin{aligned} Com_2 &= O(n \times k) + O(3n) + O(n \times d_1) + O(n \times d_2) \\ &+ O(n \times m_1) + O(n \times m_2) + O(n \times m_3) + v \times (2 \times n - 1 - v) \\ &+ v \times (3 \times v - 2) + (n - 2 \times v) \times (2 \times v - 1) \end{aligned} \quad (11)$$

where k denotes the size of the convolution kernel of the Conv1D, d_1 and d_2 denote the number of neurons of the two Bi-LSTM, respectively. m_1 denotes the number of neurons of the multi-head attentional layer, m_2 denotes the number of neurons of the fully connected layer, and m_3 denotes the number of neurons of the output layer. When $f = 6$, $p' = 10$, $L = 61$, $k = 6$, $d_1 = 100$, $d_2 = 20$, $m_1 = 64$, $m_2 = 10$, and $m_3 = 1$, the relationship between the acceleration factor τ and computational complexity is shown in Table 5.

Table 5. Computational complexity.

Algorithm	τ			
	0.6	**0.7**	**0.8**	**0.9**
Algorithm proposed in [10]	4.21×10^8	4.17×10^8	4.12×10^8	4.09×10^8
MHSA–1D CNN–Bi-LSTM interference cancellation algorithm	2.98×10^8	2.95×10^8	2.77×10^8	2.75×10^8

The results in Table 5 demonstrate a reduction in the computational complexity of the proposed algorithm by 32.74% compared to the point-by-point elimination adaptive pre-equalization algorithm for $\tau = 0.9$, 33.33% for $\tau = 0.8$, 29.28% for $\tau = 0.7$, and 29.24% for $\tau = 0.6$. On average, the overall computational complexity decreased by 31.15%.

5. Conclusions

A novel MHSA–1D CNN–Bi-LSTM algorithm is proposed to mitigate the ISI caused by the FTN in MIMO-FTN-OWC systems. The algorithm integrates 1D CNN, Bi-LSTM, and MHSA to effectively capture interference signal features, handle long sequence data, and assign importance weights to features, respectively. Simulation results demonstrate that the proposed ISI cancellation algorithm achieves a BER improvement of up to 7 dB within the Mazo limit compared to a conventional system without the algorithm. This significant enhancement validates the effectiveness of the algorithm in addressing ISI. Moreover, the algorithm not only maintains comparable BER performance with the point-by-point cancellation adaptive pre-equalization algorithm but also reduces computational complexity by 31.15%. These findings offer new insights and practical guidance for signal processing in complex OWC channels and highlight its potential for improving system performance. We will further investigate the algorithm's applicability in more intricate environments, such as dynamic channel conditions and multi-user scenarios. Moreover, we will explore emerging approaches that integrate DL techniques to enhance system robustness and efficiency. These efforts aim to provide stronger technical support and broaden application potential for advancements in OWC systems.

Author Contributions: Conceptualization, M.C., Q.Y., G.Z. and H.W.; Methodology, M.C., H.W. and Q.Y.; Validation, M.C., Q.Y. and G.Z.; Investigation, G.Z. and Y.Z.; Resources, Q.Y. and X.Z.; Data Curation, X.Z.; Writing—Original Draft Preparation, G.Z. and Q.Y.; Writing—Review and Editing, M.C., H.W., Q.Y. and Y.Z.; Visualization, Q.Y. and H.W.; Supervision, M.C. and H.W.; Project Administration, M.C. All authors have read and agreed to the published version of the manuscript.

Funding: This research was supported by NSFC Program under Grant 62265010, 62261033 and Natural Science Foundation of Gansu Province, China (24JRRA183) and Youth Science and Technology Foundation of Gansu Province, China (24JRRA974) and Natural Science Foundation of Shandong Province, China (ZR2022MF284).

Institutional Review Board Statement: Not applicable.

Informed Consent Statement: Not applicable.

Data Availability Statement: The original contributions presented in the study are included in the article, further inquiries can be directed to the corresponding author.

Acknowledgments: We gratefully acknowledge the assistance of Genxue Zhou in preparing and debugging the program. We also thank Ruifang Yao and Yue Zhang for help with the English writing and details. All authors have consented to and are informed by this acknowledgement.

Conflicts of Interest: The authors declare no conflicts of interest.

References

1. Dash, S.P.; Kaushik, A. Ris-assisted 6g wireless communications: A novel statistical framework in the presence of direct channel. *IEEE Commun. Lett.* **2024**, *28*, 717–721. [CrossRef]
2. Jeyapal, R.; Matrouk, K.; Purushothaman, D. An efficient heuristic-aided adaptive autoencoder-based dilated DNN with attention mechanism for enhancing the performance of the MIMO system in 5G communication. *Multimed. Syst.* **2024**, *30*, 116. [CrossRef]
3. Gemeda, M.D.; Han, M.S.; Abebe, A.T.; Kang, C.G. Design of Deep Learning-based One-bit Transceiver with Oversampling and Faster-Than-Nyquist (FTN) Signaling. *IEEE Access* **2024**, *12*, 43968–43983. [CrossRef]
4. Chen, C.; Zhong, W.D.; Yang, H.; Du, P. On the performance of MIMO-NOMA-based visible light communication systems. *IEEE Photonics Technol. Lett.* **2018**, *30*, 307–310. [CrossRef]
5. Li, Y.T.; Geng, T.W.; Gao, S.J. On the signal combinations for a uniquely decodable coded MIMO-FSO communication system. *Opt. Laser Technol.* **2024**, *172*, 110533. [CrossRef]
6. Zhang, Z.; Yuksel, M.; Yanikomeroglu, H. Faster-than-Nyquist signaling for MIMO communications. *IEEE Trans. Wirel. Commun.* **2023**, *22*, 2379–2392. [CrossRef]
7. Tong, M.; Huang, X.; Zhang, J.A. Faster-Than-Nyquist Transmission with Frame-by-Frame Decision-Directed Successive Interference Cancellation. *IEEE Trans. Commun.* **2023**, *71*, 4851–4861. [CrossRef]
8. Yuan, W.; Wu, N.; Wang, H.; Kuang, J. Variational inference-based frequency-domain equalization for faster-than-Nyquist signaling in doubly selective channels. *IEEE Signal Process Lett.* **2016**, *23*, 1270–1274. [CrossRef]
9. Jana, M.; Medra, A.; Lampe, L.; Mitra, J. Pre-equalized faster-than-Nyquist transmission. *IEEE Trans. Commun.* **2017**, *65*, 4406–4418. [CrossRef]
10. Cao, M.; Zhang, W.; Wang, H.; Lv, J. Adaptive pre-equalisation algorithm for point-by-point cancellation in super-Nyquist wireless optical communications. *Acta Opt. Sin.* **2020**, *40*, 46–53.
11. Baek, M.S.; Park, W.; Lee, Y.T. Deep learning-based signal detection technique for ftn signaling-based emergency alert communication system. In Proceedings of the 2021 IEEE International Symposium on Broadband Multimedia Systems and Broadcasting (BMSB), Chengdu, China, 4–6 August 2021; pp. 1–3.
12. Lai, S.; Li, M. Recurrent neural network assisted equalization for FTN signaling. In Proceedings of the ICC 2020—2020 IEEE International Conference on Communications (ICC), Dublin, Ireland, 7–11 July 2020; pp. 1–6.
13. Cao, M.; Wu, Z.; Wang, H.; Xia, J.; Zhang, J.; Li, W. Deep Learning Assisted Pre-equalization Scheme for Faster-than-Nyquist Optical Wireless Communications. In Proceedings of the 2021 13th International Conference on Advanced Infocomm Technology (ICAIT), Yanji, China, 15–18 October 2021; pp. 118–121.
14. Oh, S.; Yu, M.; Cho, S.; Noh, S.; Chun, H. Bi-LSTM-Augmented deep neural network for multi-Gbps VCSEL-based visible light communication link. *Sensors* **2022**, *22*, 4145. [CrossRef] [PubMed]
15. Palitharathna, K.W.; Wickramasinghe, N.D.; Vegni, A.M.; Suraweera, H.A. Neural Network-Based Optimization for SLIPT-Enabled Indoor VLC Systems with Energy Constraints. *IEEE Trans. Green. Commun. Netw.* **2024**, *8*, 839–851. [CrossRef]
16. Li, L.; Zhu, Z.; Zhang, J. Neural Network-Based Transceiver Design for VLC System over ISI Channel. *Photonics* **2022**, *9*, 190. [CrossRef]
17. Zou, D.; Wang, W.; Li, F.; Sui, Q.; Yi, X.; Lu, C.; Li, Z. Colored Noise Suppressed FTN *vs.* PS in Peak Power Constrained IM-DD PAM Systems. *J. Light. Technol.* **2024**, *42*, 550–559. [CrossRef]

18. Che, D.; Chen, X. Modulation format and digital signal processing for IM-DD optics at post-200G era. *J. Light. Technol.* **2024**, *42*, 588–605. [CrossRef]
19. Ke, J.; Gao, M.; Chen, B.; Shao, W. Partial response THP for high-speed PAM-4 transmission. *Opt. Fiber Technol.* **2023**, *81*, 103487. [CrossRef]
20. Zedini, E.; Ata, Y.; Alouini, M.S. Improving Performance of Integrated Ground-HAPS FSO Communication Links with MIMO Application. *IEEE Photonics J.* **2024**, *16*, 1–14. [CrossRef]
21. Li, Q.; Wang, Y.; Dong, J.; Zhang, C.; Peng, K. Multi-node knowledge graph assisted distributed fault detection for large-scale industrial processes based on graph attention network and bidirectional LSTMs. *Neural Netw.* **2024**, *173*, 106210. [CrossRef]
22. Liang, R.; Xie, Y.; Cheng, J.; Pang, C.; Schuller, B. A non-invasive speech quality evaluation algorithm for hearing aids with multi-head self-attention and audiogram-based features. *IEEE/ACM Trans. Audio Speech Lang. Process* **2024**, *32*, 2166–2176. [CrossRef]
23. Guo, R.; Liu, H.; Xie, G.; Zhang, Y.; Liu, D. A Self-Interpretable Soft Sensor Based on Deep Learning and Multiple Attention Mechanism: From Data Selection to Sensor Modeling. *IEEE Trans. Ind. Inform.* **2023**, *19*, 6859–6871. [CrossRef]
24. Cao, M.; Wang, R.; Zhang, Y.; Zhang, X.; Wang, H. End-to-end performance of a CNN-AE based faster-than-Nyquist rate free space optical communication system. *J. Chongqing Univ. Posts Telecommun. (Nat. Sci. Ed.)* **2024**, *36*, 181–190.
25. Zhou, H.; Huang, L.; Yang, J.; Zhou, J.; Song, D.; Chen, X.; Dong, M. Low Complexity Double Vector Model Predictive Control with Minimum Root Mean Square Error for Three-phase Three-Level Inverters. *IEEE J. Emerg. Sel. Top. Power Electron.* **2023**, *11*, 5809–5819. [CrossRef]
26. Zafar, M.H.; Khan, N.M.; Abou Houran, M.; Mansoor, M.; Akhtar, N.; Sanfilippo, F. A novel hybrid deep learning model for accurate state of charge estimation of Li-Ion batteries for electric vehicles under high and low temperature. *Energy* **2024**, *292*, 130584. [CrossRef]
27. Huang, F.; Zhang, X. A new interpretable streamflow prediction approach based on SWAT-BiLSTM and SHAP. *Environ. Sci. Pollut. Res.* **2024**, *31*, 23896–23908. [CrossRef]

photonics

MDPI

Article

Performance Analysis of Soft-Switching FSO/THz-RF Dual-Hop AF-NOMA Link Based on Cognitive Radio

Rongpeng Liu [1,2], Ziyang Wang [1,2], Xuerui Wang [1,2], Jingwei Lu [1,2], Yawei Wang [1,2], Yizhou Zhuo [1,2], Ruihuan Wu [1,2], Zhongchao Wei [1,2] and Hongzhan Liu [1,2,*]

1 Guangdong Provincial Key Laboratory of Nanophotonic Functional Materials and Devices, Guangzhou 510006, China; lrp2021@m.scnu.edu.cn (R.L.)

2 School of Information and Optoelectronic Science and Engineering, South China Normal University, Guangzhou 510006, China

* Correspondence: hongzhanliu@m.scnu.edu.cn

Abstract: This paper presents a promising solution to address the scarcity of spectrum resources and enhance spectrum efficiency in the context of cognitive radio (CR)-based soft-switching free-space optical (FSO)/terahertz (THz) radio frequency (RF) dual-hop amplify-and-forward (AF)–non-orthogonal multiple access (ROMANO) links. The impact of maximum tolerable interference power in the primary network, transmit power in the secondary transmitter, and maximum relay transmission power on the link are thoroughly studied. The numerical results ultimately validate the effectiveness of this link in improving performance, and a comparative analysis is conducted with the without-CR scheme, highlighting the distinctive characteristics of the proposed link.

Keywords: free-space optics (FSO); terahertz (THz); radio frequency (RF); non-orthogonal multiple access (NOMA); cognitive radio (CR)

1. Introduction

Wireless communication has attained the utmost importance in the digital era, and the advent of the sixth-generation (6G) wireless networks has further underscored its significance [1]. However, scarce spectrum resources [2] and low spectrum efficiency (SE) [3] are two main problems that really need to be solved in radio frequency (RF) communication immediately. Despite this, the RF band is still a better choice for connecting with users because of its better bypassing capabilities [4]. Therefore, a link that addresses scarce spectrum resources and improves SE is worth investigating.

Free-space optics (FSO) and terahertz (THz) have become one of the key technologies to address the scarce spectrum resources [5]. On the one hand, FSO provides a high available bandwidth and high-speed transmission rates, and can effectively mitigate the current shortage of RF communication bands [6]. On the other hand, THz with a frequency of 0.1–10 THz have a high frequency in order to meet the requirements of the ultra-dense small cell networks, which facilitates the high data rate coverage and secure transmission in the wireless network [7]. These unique advantages have made FSO and THz indispensable in overcoming the limitations of the RF bands.

It is worth noting that FSO and THz resources have no limitations compared to RF, but both are affected by atmospheric conditions during propagation. Specifically, FSO outperforms THz in clear weather, while THz is more resistant to fog, dust, and air turbulence (scintillation only) than FSO. These effects are due to the presence of water vapor, scattering of particles of different sizes, and changes in the refractive index of the air [8]. As a result, FSO and THz are suitable as a hybrid link for high-speed transmission. In the research of hybrid link switching, two main types are currently studied: hard-switching and soft-switching. In hard-switching, both FSO and THz have only one threshold; when the transmitted signal-to-noise ratio (SNR) is higher than the threshold of one link, the link can be activated, if it is higher than the threshold of both at the same time, the corresponding link can be activated according to the preset situation, and if they are both lower than the threshold of both, the hybrid link will be outage. For soft-switching, FSO requires two thresholds to be set, while THz requires only one threshold. Therefore, THz links work the same as when hard-switching, and for two threshold FSO links, the relationship with the high and low thresholds is judged. Similarly, if both are outage, the hybrid link is outage. The proposed soft-switching hybrid link increases the complexity of the link, but it greatly improves the problem of frequent switching damage to the link by hard-switching [9].

The proposal of using non-orthogonal multiple access (NOMA) is a revolutionary solution to the issue of low SE [10]. By enabling multiple users to share frequency bands or time bands, NOMA increases the communication capacity in the new generation of wireless networks [11]. While NOMA can undoubtedly enhance SE, it still requires the allocation of a specific frequency band. This becomes particularly adverse in an age where spectrum usage remains disappointingly low. For instance, in 2005, the spectrum usage in Chicago for the 30–3000 MHz range stood at 17.4% [12]. In 2008, the spectrum usage for the 80–5850 MHz range in Singapore was 4.54% [13]. Similarly, in 2012, the spectrum usage for the 30–3000 MHz range in Beijing was 7.63% [14].

In order to solve the abovementioned problem of low spectrum usage, cognitive radio (CR)–NOMA has been proposed. There are three ways of spectrum sharing based on CR: underlay, interweave, and overlay [15]. In the case of underlay CR, the secondary transmitter (ST) and the primary transmitter (PT) may transmit simultaneously using the original frequency bands of the primary network (PN), ensuring that the interference from the secondary network (SN) to the PN is below a tolerance level. This way needs to protect the communication quality of the PN, and it has strict requirements for spectrum sharing. In interweave CR, the cognitive engine first determines the frequency bands that the PN has a license to use, and when the PN does not detect activity in these bands, the SN uses these licensed bands. The determination of these spectrum holes by the cognitive engine is called spectrum sensing. This way requires the flexibility to detect the spectrum usage of the PN at any time. In the case of overlay CR, the ST is transmitted simultaneously with the PT, but the SN needs to be used as a relay to forward the signals from the primary user (PU) to the target receiver to compensate for the interference caused to the PN. This way requires the SN to bear part of the signals' transmission in the PN. This increases the complexity of the SN. The fusion of underlay CR and NOMA has received particular attention in the last few years due to the fact that underlay CR, compared to interweave CR, can be transmitted simultaneously with the PU, increasing the flexibility of the SU, and not being responsible for the signals of the PU reduces the burden on the SN, compared to overlay CR [16].

By incorporating CR into the SN, underutilized spectrum bands in the PN can be effectively utilized, which not only improves SE and optimizes spectrum usage [17] but also contributes to a more efficient allocation of limited spectrum resources [18]. Leveraging CR, the SN can access and utilize the underutilized spectrum of the PN, an innovative approach with great potential to improve SE and optimize the use of available frequency resources [19].

Thus far, there exist scholarly researches that have explored the combination of CR with dual-hop links or NOMA. In CR-based RF-FSO dual-hop links, Neeraj et al. investigated amplify-and-forward (AF) in the presence of imperfect channel state information (CSI) [20]. Based on this, the case of multiple antennas at the relay (R), PT, and primary receiver (PR) was studied [21]. In addition, the multi-relay case introduced feedback links to determine and select the link with the best channel gain [22]. Notably, the fixed and CSI-assisted AF cases were examined to study single and two power-constrained points [23]. Shifting the focus to the fusion of CR and NOMA, CR-enabled AF-NOMA links in a two-user scenario were reviewed, considering the scaling coefficient of additive white gaussian noise (AWGN) [24]. Additionally, a CR-NOMA bidirectional cooperative relay scheme was proposed [25]. Lastly, a CR-based DF-NOMA in five modes was analyzed, considering the maximum transmit power of the ST and the maximum tolerable interference power of the

PN [26]. Through the above survey, it is found that most of the investigations only focus on a single problem, as CR with FSO/THz solves the problem of scarce spectrum resources and CR-NOMA improves SE; thus, it can be seen that the combination of both of them is essential and innovative, which arouses a great deal of research interest.

In this article, research gaps were identified, so FSO/THz-RF dual-hop links, and NOMA and CR were combined for their unique and fascinating characteristics. Consequently, a novel soft-switching FSO/THz-RF dual-hop AF-NOMA link based on CR is proposed. The link entails a soft-switching FSO/THz link serving as the first hop from the ST to the R, while the second hop exploits the potential of CR to effectively allocate the spectrum resources of the PN from the R to the secondary user (SU). The signals are transmitted via the AF-NOMA scheme, traversing an RF link to reach two users. In short, a soft-switching FSO/THz-RF dual-hop AF-NOMA link based on CR can use CR to address lower spectrum usage, NOMA can improve SE, the soft-switching FSO/THz pairing can reduce the impact of different atmospheric conditions, and RF can traverse obstacles to communicate with users.

The rest of the paper is organized as follows. Section 2 derives the probability density function (PDF) and cumulative distribution function (CDF) for FSO following the Weibull distribution, and the PDF and CDF for THz following the $\alpha-\mu$ distribution and considering pointing errors, and sets two FSO thresholds and a single THz threshold to form a soft-switching FSO/THz link. In Section 3, we first analyze the characteristics of the underlay CR and derive the relay transmission power with the maximum tolerable interference power in PN and the maximum relay transmission power. Subsequently, we analyze the properties of the NOMA system and derive the link OP by combining CR and NOMA. We deliver meaningful scenarios into the system to obtain beneficial numerical results in Section 4, and Section 5 concludes the paper.

2. System and Channel Models

The CR model comprises an ST, an R, two SUs, and a PU. Within the SN, the ST employs soft-switching FSO/THz links to propagate signals to R. As shown in Figure 1, R takes control of the spectrum band belonging to the PN using the CR and transmits signals to the two SUs via the AF-NOMA RF link. Without loss of generality, the channel gains are assumed to be $|h_{RU_1}|^2 < |h_{RU_2}|^2$ according to the inverse of the link distance because the greater the distance traveled, the greater the attenuation by atmospheric turbulence.

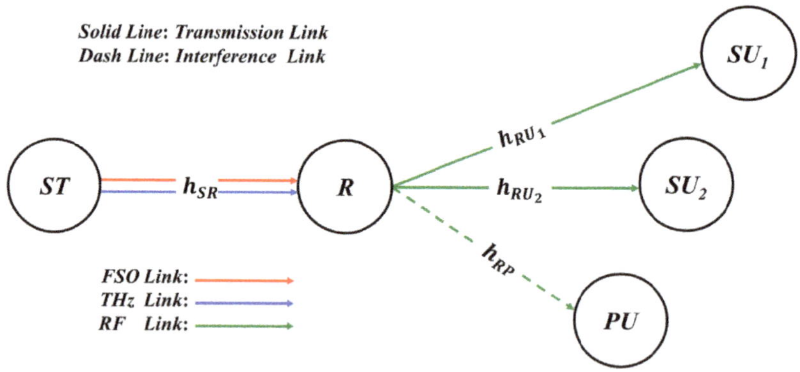

Figure 1. Soft-switching FSO/THz-RF dual-hop AF-NOMA link with CR.

2.1. Soft-Switching FSO/THz Link

The complementary characteristics of FSO and THz in coping with atmospheric conditions lead us to build hybrid FSO/THz links. Compared with hard-switching, soft-switching between FSO and THz links is used to avoid frequent switching and improve the lifetime of the links, and a more practical approach is to use two FSO thresholds and a single THz threshold for soft-switching between the links. Therefore, in this section, we utilize the complementary characteristics of FSO and THz in different atmospheric environments to build a soft-switching FSO/THz scheme in the SR link. The exponential Weibull model is a commonly used model to characterize the atmospheric turbulence of FSO links in recent years, which not only can accurately simulate the optical irradiance under weak turbulence and strong turbulence, but also more importantly, the derivation of the exponential Weibull model is relatively simple, which makes it very suitable for the soft-switching case. Therefore, the exponential Weibull model is chosen in this paper, and the channel gain x_F of the PDF is [27]:

$$f_{|h_F|^2}(x_F) = \frac{\alpha_F \beta_F}{\eta}\left(\frac{x_F}{\eta^2}\right)^{\frac{\beta_F - 1}{2}} \exp\left[-\left(\frac{x_F}{\eta^2}\right)^{\frac{\beta_F}{2}}\right]\left\{1 - \exp\left[-\left(\frac{x_F}{\eta^2}\right)^{\frac{\beta_F}{2}}\right]\right\}^{\alpha_F - 1}. \quad (1)$$

Subsequently, integrating (1) yields the CDF of the exponential Weibull model as

$$F_{|h_F|^2}(x_F) = \left\{1 - \exp\left[-\left(\frac{x_F}{\eta^2}\right)^{\frac{\beta_F}{2}}\right]\right\}^{\alpha_F}, \quad (2)$$

where α_F and β_F are the shape parameter, and η is the scale parameter.

For THz, channel fading is appropriately modeled as the $\alpha - \mu$ distribution, which is a generalized model that can be simplified to numerous important distributions such as Gamma, Nakagami-*m*, etc. [28]. THz link channel gain x_T is affected by pointing errors; it follows $\alpha - \mu$ distribution as its PDF [29]:

$$\begin{aligned} f_{|h_T|^2}(\lambda_T) = {} & \frac{g_T^2 \mu_T^{\frac{g_T^2}{\alpha_T}} \lambda_T^{\frac{g_T^2}{2} - 1}}{2\left(A_{0,T} h_{l,T} \hat{h}_{f,T}\right)^{g_T^2} \Gamma(\mu_T)} \times \\ & G_{1,2}^{2,0}\left[\frac{\mu_T}{\left(A_{0,T} h_{l,T} \hat{h}_{T,f}\right)^{\alpha_T}} \lambda_T^{\frac{\alpha_T}{2}} \middle| \begin{matrix} -;1 \\ 0, \frac{\alpha_T \mu_T - g_T^2}{\alpha_T}; - \end{matrix}\right]. \end{aligned} \quad (3)$$

The CDF of $\alpha - \mu$ distribution obtained by integrating (3) and performing some algebraic calculations is

$$F_{|h_T|^2}(x_T) = \phi_T G_{2,3}^{2,1}\left[\psi_T x_T^{\frac{\alpha_T}{2}} \middle| \begin{matrix} 1; 1 + \frac{g_T^2}{\alpha_T} \\ \frac{g_T^2}{\alpha_T}, \mu_T; 0 \end{matrix}\right], \quad (4)$$

where $\phi_T = g_T^2/[\alpha_T \Gamma(\mu_T)]$, $\psi_T = \mu_T/\left(A_{0,T} h_{l,T} \hat{h}_{f,T}\right)^{\alpha_T}$, and $G_{p,q}^{m,n}[\cdot]$ is a MeijerG function [30]. g_T is the ratio between the equivalent beam radius and the pointing error displacement standard deviation at receive (RX). $A_{0,T}$ is the fraction of collected power at the origin. $h_{l,T}$ is the path loss of the THz link. α_T and μ_T denote the fading parameters, while $\hat{h}_{f,T}$ represents the α-root mean value of the fading channel envelope.

Deriving the CDF of soft-switching FSO/THz-RF based on the CDF above for each link, the upper threshold $x_{th,U}^F$ and the lower threshold $x_{th,L}^F$ are set for the FSO link, and there is only one threshold, x_{th}^T, for the THz link, which is depicted in Figure 2 [31].

The specific operation of the soft-switching is based on the following process:

- If $x_F > x_{th,U}^F$, regardless of x_T, the FSO link is activated.
- If $x_{th,L}^F > x_F > x_{th,U}^F$, and given $x_F > x_{th,U}^F$ previously, regardless of x_T, the FSO link is activated.

- If $x^F_{th,L} > x_F > x^F_{th,U}$, and given $x_F < x^F_{th,L}$ previously, if $x_T > x^T_{th}$, the THz link is activated; if $x_T < x^T_{th}$, the hybrid link is outage.
- If $x_F < x^F_{th,U}$, if $x_T > x^T_{th}$, the THz link is activated; if $x_T < x^T_{th}$, the hybrid link is outage.

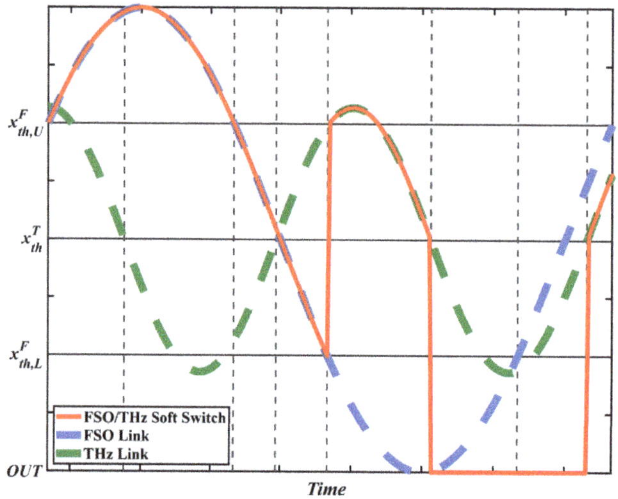

Figure 2. Description of the soft-switching FSO/THz hybrid link.

For ease of calculation, we introduce an offset Δ, which is used to describe the offset of the two thresholds for the FSO link and the single threshold for the THz link from the center threshold such that $x^F_{th,U} = (1+\Delta)x_{th}$, $x^F_{th,L} = (1-\Delta)x_{th}$, and $x^T_{th} = x_{th}$. According to the specific operation of soft-switching, the OP of the FSO link in the soft-switching system can be expressed as

$$F^{Soft}_{|h_F|^2}(x_{th}) = P^F_{low} + P^F_{med}\left(\frac{P^F_{low}}{P^F_{low} + P^F_{hig}}\right), \tag{5}$$

where, $P^F_{low} = F_{|h_F|^2}((1-\Delta)x_{th})$, $P^F_{med} = F_{|h_F|^2}((1+\Delta)x_{th}) - F_{|h_F|^2}((1-\Delta)x_{th})$, and $P^F_{hig} = 1 - F_{|h_F|^2}((1+\Delta)x_{th})$.

Because THz has only a single threshold, the OP of the THz link in the soft-switching system can be expressed as

$$F^{Soft}_{|h_T|^2}(x_{th}) = F_{|h_T|^2}(x_{th}). \tag{6}$$

Therefore, in the case of soft-switching, the OP of the hybrid FSO/THz link can be expressed as

$$F_{|h_{SR}|^2}(x_{th}) = F^{Soft}_{|h_F|^2}(x_{th}) \times F^{Soft}_{|h_T|^2}(x_{th}). \tag{7}$$

2.2. RF Link

On the other hand, the PDF and CDF of the unordered variable of the channel gain $|h_{RU_n}|^2$ in the second slot follow the Nakagami-m distribution and can be represented as [32]

$$f_{|\tilde{h}_{RU_n}|^2}(x) = \frac{m_n^{m_n} x^{m_n-1}}{\Omega_n^{m_n}\Gamma(m_n)} \exp\left(-\frac{m_n x}{\Omega_n}\right), \tag{8}$$

$$F_{|\tilde{h}_{RU_n}|^2}(x) = \frac{\Upsilon\left(m_n, \frac{m_n x}{\Omega_n}\right)}{\Gamma(m_n)}, \tag{9}$$

where m_n represents the fading coefficient and $\Omega_n = d_n^{-\alpha}$ is the average power with d_n being the distance of the nth link and α the path loss exponent, $\Gamma(a) = \int_0^\infty x^{a-1}e^{-x}dx$ denotes the Gamma function, and $\Upsilon(m, x)$ is the lower incomplete gamma function. Specifically, $\Upsilon(a, x) = \int_0^x x^{a-1}e^{-x}dx$. Additionally, it is worth noting that $\Upsilon(a, x) = (a-1)!\left[1 - e^{-x}\sum_{i=0}^{a-1}(x^i/i!)\right]$ when a is an integer greater than or equal to one. In this paper, we consider integer values for m_n; thus, we can rephrase the expression $F_{|\tilde{h}_{RU_n}|^2}(x)$ as follows:

$$F_{|\tilde{h}_{RU_n}|^2}(x) = 1 - \exp\left(-\frac{m_n x}{\Omega_n}\right)\sum_{r=0}^{m_n-1}\frac{\left(\frac{m_n x}{\Omega_n}\right)^r}{r!}. \tag{10}$$

With some algebraic calculations, the respective PDF and CDF for the ordered variable $|h_{RU_n}|^2$ can be written as [32]

$$f_{|h_{RU_n}|^2}(x) = \sum_{k=0}^{M-n}(-1)^k\binom{M-n}{k}f_{|\tilde{h}_{RU_n}|^2}(x)\left[F_{|\tilde{h}_{RU_n}|^2}(x)\right]^{n+k-1}, \tag{11}$$

$$F_{|h_{RU_n}|^2}(x) = Q\sum_{k=0}^{M-n}\frac{(-1)^k}{n+k}\binom{M-n}{k}\left[F_{|\tilde{h}_{RU_n}|^2}(x)\right]^{n+k}, \tag{12}$$

where $Q = M!/[(M-n)!(n-1)!]$. By substituting the specific parameters of the two users, we can derive the corresponding PDF and CDF.

It is worth noting that the PN spectrum band is occupied in the second slot. Assuming the channel gain $|h_{RP}|^2$ follows a Rayleigh distribution, its CDF is

$$F_{|h_{RP}|^2}(x) = 1 - \exp\left(-\frac{x}{\Omega_P}\right), \tag{13}$$

where $\Omega_P = d_p^{-\alpha}$ holds the same significance as Ω_n.

3. Performance Analysis

3.1. CR System

For the analysis in this section, the OP will be derived. The underlay CR we consider is the PowIntSCSI model with a constrained power budget at ST and computed statistical CSI at ST, which increases the practical application effect of the link [26]. Specifically, this model is restricted by both the maximum relay transmit power (i.e., P_{Peak}) and the maximum tolerable interference power of PN (i.e., I). In such a scenario, the quality of service (QoS) for the PU can be safeguarded through statistical constraints, ensuring that the probability of R causing interference to the PU surpassing the interference threshold I remains below the predetermined threshold δ. The transmitted power at R is defined as P_R, which is obtained as

$$\Pr\left(|h_{RP}|^2 P_R > I\right) \le \delta \implies 1 - F_{|h_{RP}|^2}\left(\frac{I}{P_R}\right) \le \delta \implies P_R \le \frac{I}{F^{-1}_{|h_{RP}|^2}(1-\delta)}, \tag{14}$$

where $F^{-1}_{|h_{RP}|^2}(x) = -\Omega_P \ln(1-x)$ represents the inverse function of $F_{|h_{RP}|^2}(x)$. Thus, the transmission power at R can be expressed as [26]

$$P_R = \min\left\{P_{Peak}, \frac{-I}{\Omega_P \ln\delta}\right\} = \begin{cases} P_{Peak}, & \text{if } P_{Peak} \le \frac{-I}{\Omega_P \ln\delta} \\ \frac{-I}{\Omega_P \ln\delta}, & \text{otherwise.} \end{cases} \tag{15}$$

In the subsequent derivation, P_R will take the smaller value in (15) to represent the transmit power at R. As I keeps increasing, $\frac{-I}{\Omega_P \ln \delta}$ will increase and P_R will be equal to maximum relay transmission power P_{Peak} at R.

3.2. NOMA System

In the NOMA system, it is challenging to consider a large number of users due to the occupancy of the PN spectrum band. Hence, we only focus on the superposition of signals from two users. Let us assume that the normalized signal $x_S = \sqrt{a_1 P_S} x_1 + \sqrt{a_2 P_S} x_2$. Here, x_n is the signal required by the nth user, P_S denotes the transmission power at ST, and a_n is the power allocation coefficient (PAC) for the nth user, indicating the proportion of the transmitted power allocated to x_n [32]. Following the principles of NOMA, $a_1 > a_2$ and $a_1 + a_2 = 1$. Therefore, the signal at R is

$$y_R = h_{SR}\left(\sqrt{a_1 P_S} x_1 + \sqrt{a_2 P_S} x_2\right) + n_R, \tag{16}$$

where $n_R \sim \mathcal{CN}(0,1)$ represents AWGN at R.

In the second slot, R amplifies y_R using the amplification coefficient G before propagating it to all users. The signal received by the nth user can be expressed as

$$y_{U_n} = \sqrt{P_R} G h_{RU_n} y_R + n_{U_n}, \tag{17}$$

where P_R represents the transmission power at R, and $n_{U_n} \sim \mathcal{CN}(0,1)$ denotes AWGN at the nth user. Thus, G should be determined as

$$G = \frac{1}{\sqrt{\left(P_S |h_{SR}|^2 + 1\right)}}. \tag{18}$$

For each user, the desired received signals are subject to interference from the signals of other users. Consequently, successive interference cancellation (SIC) should be performed at each user to obtain its own signal. The SIC decoding order is incrementally related to the user's channel gain $\left(|h_{RU_1}|^2 < |h_{RU_2}|^2\right)$. Therefore, at the nth user, it will detect the jth user's signal where $j < n$, and then subtract it from the nth user's received signal sequentially. The signal of the mth user with $m > n$ will be treated as noise at the nth user [32].

The generalized formula for the signal-to-interference plus noise ratio (SINR) can be computed as

$$\gamma_{j \to n} = \frac{a_j P_S P_R |h_{SR}|^2 |h_{RU_n}|^2}{P_S P_R |h_{SR}|^2 |h_{RU_n}|^2 \sum\limits_{i=j+1}^{M} a_i + P_S |h_{SR}|^2 + P_R |h_{RU_n}|^2 + 1}, \tag{19}$$

where $\gamma_{j \to n}$ represents the SINR of the nth user to decode the signal from the jth user.

3.3. Outage Probability

Hence, a user achieves successful signal reception by decoding and eliminating the signal from users with higher PAC and successfully decoding its own signal. Based on this, the OP can be derived as

$$
\begin{aligned}
P_{OUT}^{n} &= 1 - \Pr\left(\gamma_{j \to n} > \gamma_{th,j}\right) \\
&= 1 - \Pr\left(\frac{a_j P_S P_R |h_{SR}|^2 |h_{RU_n}|^2}{P_S P_R |h_{SR}|^2 |h_{RU_n}|^2 \sum\limits_{i=j+1}^{M} a_i + P_S |h_{SR}|^2 + P_R |h_{RU_n}|^2 + 1} > \gamma_{th,j}\right) \\
&= 1 - \Pr\left(\left[P_S P_R \left(a_j - \gamma_{th,j} \sum\limits_{i=j+1}^{M} a_i\right) |h_{RU_n}|^2 - P_S \gamma_{th,j}\right] |h_{SR}|^2 > \gamma_{th,j}\left(P_R |h_{RU_n}|^2 + 1\right)\right) \\
&= 1 - \Pr\left(|h_{RU_n}|^2 > \frac{\gamma_{th,j}}{P_R\left(a_j - \gamma_{th,j} \sum\limits_{i=j+1}^{M} a_i\right)} \triangleq \theta_j, |h_{SR}|^2 > \frac{\theta_j\left(1 + P_R |h_{RU_n}|^2\right)}{P_S\left(|h_{RU_n}|^2 - \theta_j\right)}\right),
\end{aligned}
\tag{20}
$$

where $\gamma_{th,j}$ denotes the threshold of the jth user and θ_j is some constant product of the jth user.

After some algebraic manipulations, the expression can be written as

$$
P_{OUT}^{n} = \int_0^{\theta_n^*} f_{|h_{RU_n}|^2}(x)dx + \int_{\theta_n^*}^{\infty} f_{|h_{RU_n}|^2}(x) F_{|h_{SR}|^2}\left(\frac{\theta_n^*\left(1 + P_R |h_{RU_n}|^2\right)}{P_S\left(|h_{RU_n}|^2 - \theta_n^*\right)}\right)dx, \tag{21}
$$

where $\theta_n^* = \max\left(\theta_j, \cdots, \theta_n\right)$. Substituting (7), (12), and (15) into (21) yields the OP for the corresponding user.

4. Numerical Results

In this section, we provide meaningful values to perform a numerical analysis of the proposed link. Unless otherwise specified, the NOMA system considered here has $M = 2$, $a_1 = 0.51$, $a_2 = 0.49$, $\gamma_{th,1} = 51/149$, $\gamma_{th,2} = 49/100$, and $\Delta = 5\%$. Regarding the frequency and wavelength, the RF frequency is taken as 60 GHz, the THz frequency is 0.1 THz, and the FSO wavelength is 1550 nm [31]. For atmospheric turbulence and pointing error scenarios, we set the moderate turbulence parameters as $\alpha_F = 1$, $\beta_F = 2$, $\eta = 2/\sqrt{\pi}$, $\alpha_T = 1.7$, $\mu_T = 1.5$, $m_1 = 1$, and $m_2 = 1$. For weak turbulence, the parameters are $\alpha_F = 1$, $\beta_F = 3$, $\eta = 2/\Gamma(4/3)$, $\alpha_T = 1.7$, $\mu_T = 1.7$, $m_1 = 2$, and $m_2 = 2$. The large pointing error is represented by $g_T = 1.6224$, whereas the small pointing error is denoted by $g_T = 2.9203$. The path loss exponent $\alpha = 2$, with distances $d_{SR} = 200$ m, $d_{RU_1} = 20$ m, $d_{RU_2} = 20$ m, and $d_{RP} = 5$ m. In the subsequent analysis, we primarily focus on the performance of OP under various conditions.

Figure 3 investigates the relationship between OP and PAC under different I in weak turbulence. We set $\gamma_{th,j} = \left(a_j \rho_m\right) / \left(1 + \rho_m \sum_{i=j+1}^{M} a_i\right)$, where $\rho_m = \text{floor}[\rho + 1]$ and ρ is a value on the ordinate of the function $\rho\left(\gamma_{th,j}\right) = \gamma_{th,j} / \left(a_j - \gamma_{th,j} \sum_{i=j+1}^{M} a_i\right)$, and floor$[\cdot]$ is rounded down. The results demonstrate that the link performance improves as the mean deviation of PAC decreases. Additionally, due to variations in channel gains, User2 outperforms User1. Comparing different I, we observe that a larger I leads to better link performance, benefiting from the ability of the PN to withstand greater interference. Therefore, we consistently adopt the optimal PAC and SINR thresholds throughout the analysis.

In Figure 4, numerous deductions may be discerned. Firstly, with the elevation of I, the link's performance exhibits enhancement. Nevertheless, owing to the constraints imposed by both I and P_{Peak} upon P_R, the link's performance reaches a state of equilibrium at a specific I, as determined by (13). Furthermore, it is noteworthy that augmenting both P_S and P_{Peak} can amplify the link's performance, albeit the enhancement brought about by increasing P_{Peak} is notably more pronounced. This discrepancy arises from the operation of the link using AF-NOMA, wherein P_S contends with more extended link distances compared to P_{Peak}. Consequently, for the ensuing analysis, we establish $P_S = 15$ dBm and $P_{Peak} = 40$ dBm.

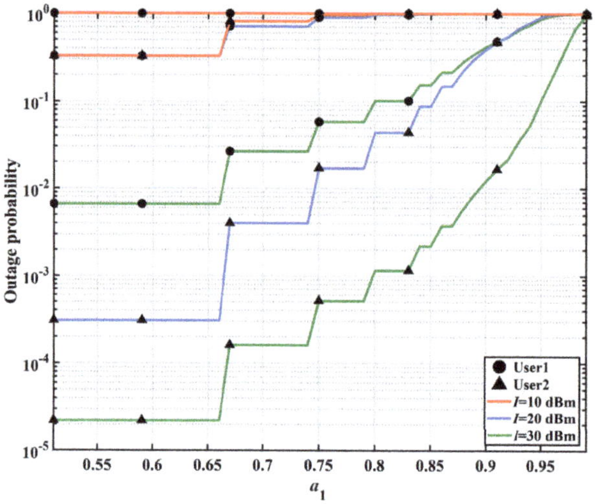

Figure 3. The relationship between OP and PAC under different I in weak turbulence.

The comparison between (a), (c), and (e) with (b), (d), and (f) in Figure 5 reveals that User2 outperforms User1 due to the difference in channel gains. Moreover, by comparing the three columns (a) (b), (c) (d), and (e) (f), it is evident that the link performance increases with the increase in P_S. Finally, we observe that a single change in either P_{Peak} or I does not significantly enhance the link performance. Only when both parameters are increased simultaneously is the link performance improved. Thus, the performance advantage is concentrated in the rectangular region in the upper right corner of the figure.

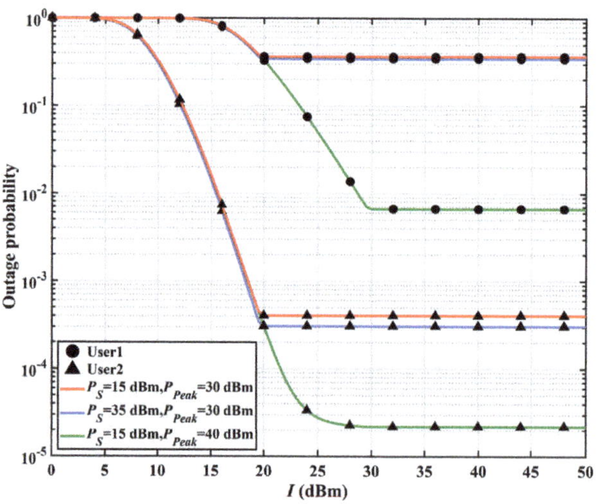

Figure 4. The impact of different values of P_S and P_{Peak} on OP under weak turbulence.

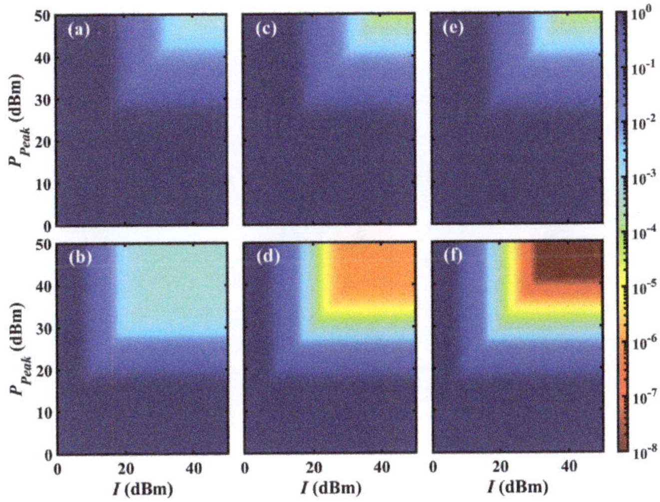

Figure 5. A comparison of P_{Peak} and I on OP under different P_S. (**a**,**b**) correspond to $P_S = 10$ dBm, (**c**,**d**) correspond to $P_S = 20$ dBm, and (**e**,**f**) correspond to $P_S = 30$ dBm. In (**a**,**c**,**e**), User1 is represented, while in (**b**,**d**,**f**), User2 is represented.

Figure 6 depicts three atmospheric scenarios: Case1 involves moderate turbulence with large pointing errors, Case2 features weak turbulence with large pointing errors, and Case3 exhibits weak turbulence with small pointing errors. Upon comparing these scenarios, it becomes evident that the link performance is superior in the cases of weak turbulence and small pointing errors. Specifically, around the smoothing point at $I = 35$ dBm, User2 achieves an OP of 4.4482×10^{-4} in Case1, 4.7948×10^{-5} in Case2, and 9.3066×10^{-6} in Case3.

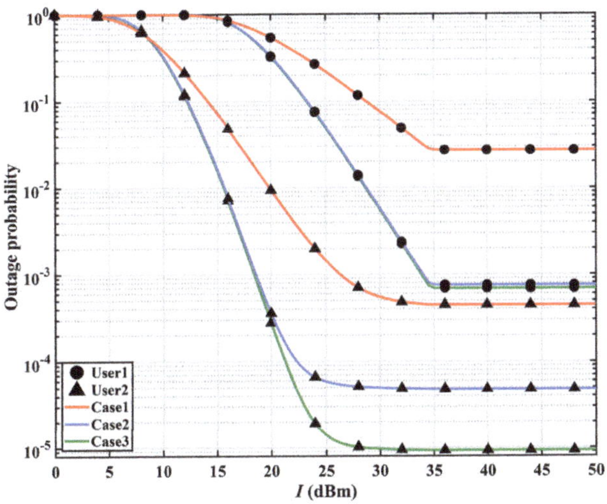

Figure 6. Impact of different turbulence and pointing errors on link performance.

In Figure 7, we maintain a constant SN power, denoted as $P_S = P_{Peak}$. We compare the OP at different I and contrast them with the scenario without CR. As the I increases, it signifies a higher tolerance to interference, leading to improved link performance. Thus, we observe a consistent performance in the low-power region in all cases. The key distinction lies in the performance that is constrained by both P_{Peak} and I with CR, whereas without CR, it is solely limited by P_{Peak}. The results reveal that when $I = 35$ dBm and the SN power ranges from 0 to 50 dBm, there is minimal disparity in performance between the with-CR and without-CR scenarios. Encouragingly, even as I continues to rise, the performance discrepancy remains insignificant.

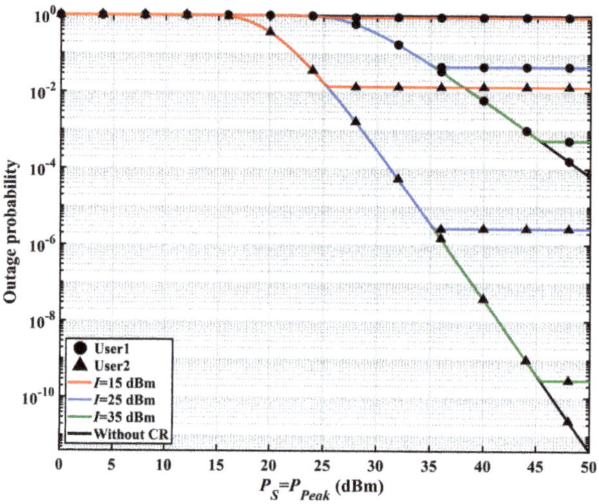

Figure 7. A comparison of the power of the SN network and OP at different I.

5. Conclusions

This paper introduces an innovative AF-NOMA-assisted soft-switching FSO/THz-RF double-hop link based on CR, addressing the scarcity of spectrum resources through FSO and THz. Firstly, a soft-switching FSO/THz system was built on the SR link, and both of them were paired with each other to largely mitigate the impact of atmospheric conditions on the SR link. In addition, NOMA allows signals of different powers to be superimposed on an identical frequency band, and CR allows that frequency band to be present in the main network, thus resolving frequency band resources and improving SE. The study primarily focuses on three key factors impacting link performance: P_S, P_{Peak}, and I. Notably, P_S has a relatively minor influence on performance improvement due to its longer propagation distance. Additionally, the SN power is maintained at a constant level compared with the case without CR. It is evident that the performance without CR surpasses that with CR, mainly due to the imposed constraints. Nevertheless, the incorporation of CR significantly enhances SE and spectrum usage. The numerical results demonstrate the superiority of this link, which offers a fresh solution for next-generation communications.

Author Contributions: Conceptualization, methodology, R.L. and H.L.; software, validation, writing—original draft preparation, R.L.; formal analysis, investigation, data curation, R.W. and Z.W. (Zhongchao Wei); writing—review and editing, visualization, Z.W. (Ziyang Wang), X.W., J.L., Y.W., and Y.Z.; supervision, project administration, funding acquisition, H.L. All authors have read and agreed to the published version of the manuscript.

Funding: This research was funded by the National Natural Science Foundation of China (No. 62175070 and 61875057), the Guangdong Basic and Applied Basic Research Foundation (No. 2021A151 5012652 and 2022A1515110752), and the Science and Technology Program of Guangzhou (No. 202201010340, and 2019050001).

Institutional Review Board Statement: Not applicable.

Informed Consent Statement: Not applicable.

Data Availability Statement: Not applicable.

Conflicts of Interest: The authors declare no conflict of interest.

References

1. Maier, M.; Ebrahimzadeh, A.; Beniiche, A. The Art of 6G (TAO 6G): How to wire Society 5.0 [Invited]. *J. Opt. Commun. Netw.* **2022**, *14*, A101–A112. [CrossRef]
2. Khalighi, M.; Uysal, M. Survey on free space optical communication: A communication theory perspective. *IEEE Commun. Surv. Tutor.* **2014**, *16*, 2231–2258. [CrossRef]
3. Ray, P.; Kumar, N.; Guizani, M. A Vision on 6G-Enabled NIB: Requirements, Technologies, Deployments, and Prospect. *IEEE Wirel. Commun.* **2021**, *28*, 120–127. [CrossRef]
4. Malik, S.; Chung, Y.; Sahu, P. On the performance of a hybrid frequency–phase-keying-based MIMO hybrid RF/FSO communication link. *J. Opt. Commun. Netw.* **2022**, *14*, 840–851. [CrossRef]
5. Li, K.; Yu, J. Photonics-Aided Terahertz-Wave Wireless Communication. *J. Light. Technol.* **2022**, *40*, 4186–4195. [CrossRef]
6. Dang, A. A closed-form solution of the bit-error rate for optical wireless communication systems over atmospheric turbulence channels. *Opt. Express* **2011**, *19*, 3494–3502.[CrossRef]
7. Boulogeorgos, A.; Alexiou, A.; Merkle, T. Terahertz technologies to deliver optical network quality of experience in wireless systems beyond 5G. *IEEE Commun. Mag.* **2018**, *56*, 144–151. [CrossRef]
8. Federici, J.; Ma, J.; Moeller, L. Review of weather impact on outdoor terahertz wireless communication links. *Nano Commun. Netw.* **2016**, *10*, 13–26. [CrossRef]
9. Yahia, O.; Erdogan, E.; Kurt, G. A Weather-Dependent Hybrid RF/FSO Satellite Communication for Improved Power Efficiency. *IEEE Wirel. Commun. Lett.* **2022**, *11*, 573 – 577. [CrossRef]
10. Li, R.; Dang, A. Multi-user access in wireless optical communication system. *Opt. Express* **2018**, *26*, 22658–22673. [CrossRef]
11. Zhuang, Y.; Zhang, J. Secrecy performance analysis for a NOMA based FSO-RF system with imperfect CSI. *J. Opt. Commun. Netw.* **2022**, *14*, 500–510. [CrossRef]
12. McHenry, M.; Tenhula, P.; McCloskey, D. Chicago spectrum occupancy measurements & analysis and a long-term studies proposal. In Proceedings of the First International Workshop on Technology and Policy for Accessing Spectrum, Boston, MA, USA, 5 August 2006 ; p. 1. [CrossRef]
13. Islam, M.; Koh, C.; Oh, S. Spectrum Survey in Singapore: Occupancy Measurements and Analysis. In Proceedings of the 2008 3rd International Conference on Cognitive Radio Oriented Wireless Networks and Communications, Singapore, 15–17 May 2008; p. 1. [CrossRef]
14. Xue, J.; Feng, Z.; Chen, K. Beijing Spectrum Survey for Cognitive Radio Applications. In Proceedings of the 2013 IEEE 78th Vehicular Technology Conference, Las Vegas, NV, USA, 2–5 September 2013; pp. 1–5. [CrossRef]
15. Goldsmith, A.; Jafar, S.; Maric, I. Breaking Spectrum Gridlock With Cognitive Radios: An Information Theoretic Perspective. *Proc. IEEE* **2009**, *97*, 894–914. [CrossRef]
16. Kader, M. A power-domain NOMA based overlay spectrum sharing scheme. *Future Gener. Comput. Syst.* **2020**, *105*, 222–229. [CrossRef]
17. Zhu, D.; Pan, S. Broadband Cognitive Radio Enabled by Photonics. *J. Light. Technol.* **2020**, *38*, 3076–3088. [CrossRef]
18. Wang, Y.; Zhan, X.; Zhan, Z. Performance of an all-links MIMO-OSTBC mixed underlay cognitive RF/FSO system. *Appl. Opt.* **2020**, *61*, 5489–5495. [CrossRef]
19. Jee, A.; Janghel, K.; Prakriya, S. Performance of Adapative Multi-User Underlay NOMA Transmission With Simple User Selection. *IEEE Trans. Cogn. Commun.* **2020**, *8*, 871–887. [CrossRef]
20. Varshney, N.; Jagannatham, A.; Varshney, P. Cognitive MIMO-RF/FSO Cooperative Relay Communication With Mobile Nodes and Imperfect Channel State Information. *IEEE Trans. Cogn. Commun.* **2018**, *4*, 544–555. [CrossRef]
21. Al-Qahtani, F.; El-Malek, A.; Ansari, I. Outage Analysis of Mixed Underlay Cognitive RF MIMO and FSO Relaying With Interference Reduction. *IEEE Photonics J.* **2017**, *9*, 7902722. [CrossRef]
22. Torabi, M.; Mohammadi, N.; Nerguizian, C. Performance analysis of an asymmetric two-hop amplify-and-forward relaying RF-FSO system in a cognitive radio with partial relay selection. *Opt. Commun.* **2020**, *505*, 127478. [CrossRef]
23. Arezumand, H.; Jafarian, H.; Nasab, E. Outage and Diversity Analysis of Underlay Cognitive Mixed RF-FSO Cooperative System. *J. Opt. Commun. Netw.* **2017**, *9*, 909–920. [CrossRef]
24. Arzykulov, S.; Nauryzbayev, G.; Tsiftsis, T. Performance Analysis of Underlay Cognitive Radio Nonorthogonal Multiple Access Networks. *IEEE Trans. Veh. Technol.* **2019**, *68*, 9318–9322. [CrossRef]
25. Bae, J.; Han, Y. Joint Power and Time Allocation for Two-Way Cooperative NOMA. *IEEE Trans. Veh. Technol.* **2019**, *68*, 12443–12447. [CrossRef]
26. Kumar, V.; Ding, Z.; Flanagan, M. On the Performance of Downlink NOMA in Underlay Spectrum Sharing. *IEEE Trans. Veh. Technol.* **2021**, *70*, 4523–4540. [CrossRef]
27. Wang, Z.; Shi, W.; Liu, W. Performance analysis of mixed RF/FSO system with spatial diversity. *Opt. Commun.* **2019**, *443*, 230–237. [CrossRef]

28. Li, S.; Yang, L. Performance analysis of dual-hop THz transmission systems over α-μ fading channels with pointing errors. *IEEE Internet Things* **2021**, *14*, 11772–1783. [CrossRef]
29. Li, S.; Yang, L.; Zhang, J. Mixed THz/FSO Relaying System: Statistical Analysis and Performance Evaluation. *IEEE Trans. Wirel. Commun.* **2022**, *21*, 10996–11010. [CrossRef]
30. Wolfram. The Wolfram Functions Site. 2001. Available online: https://functions.wolfram.com/ (accessed on 31 August 2023).
31. Singya, P.; Makki, B.; Errico, A. Hybrid FSO/THz-Based Backhaul Network for mmWave Terrestrial Communication. *IEEE Trans. Wirel. Commun.* **2023**, *22*, 4342–4359. [CrossRef]
32. Men, J.; Ge, J.; Zhang, C. Performance Analysis of Nonorthogonal Multiple Access for Relaying Networks over Nakagami-m Fading Channels. *IEEE Trans. Veh. Technol.* **2017**, *66*, 1200–1208. [CrossRef]

photonics

MDPI

Article

Tolerance-Aided Interference Degradation for Optical OFDM in Power-Constrained Systems

Yan Gao [1,2] and Jie Lian [3,*]

[1] School of Electronics and Information, Northwestern Polytechnical University, Xi'an 710072, China; yangao_5@nwpu.edu.cn
[2] School of Information Engineering, Xi'an University, Xi'an 710065, China
[3] School of Marine Science and Technology, Northwestern Polytechnical University, Xi'an 710072, China
* Correspondence: jielian@nwpu.edu.cn

Abstract: Optical wireless communications (OWCs) are power-efficient for providing short-range and high-speed data transmission. In this paper, we propose an interference degradation algorithm for power-constrained OWC systems when using orthogonal frequency division multiplexing (OFDM). In this algorithm, we introduce a decision tolerance at the receiver that can be optimally designed to reduce interference caused by peak power clipping distortion and decrease additive noise collected by the photodetectors. Although the recently proposed clipping-enhanced optical OFDM (CEO-OFDM) can transmit clipped information through extra signal frames, reconstructing the received signal introduces more noise. Using the proposed decision tolerance, we can determine whether the signals in the extra frame are required for data reconstruction. By optimally choosing the decision tolerance, the bit error rate (BER) performance is enhanced compared to those of DCO, ACO, and CEO-OFDM. Additionally, the proposed algorithm offers a wider dynamic modulation index range than DCO, ACO, and CEO-OFDM at the same BER. Using the tested parameters, the proposed algorithm with 64-QAM achieves a similar best BER performance as DCO, ACO, and CEO-OFDM using 32-QAM. Therefore, the proposed algorithm achieves a higher data rate than those of the other compared techniques.

Keywords: optical wireless communications; power-constrained systems; OFDM; clipping distortion

1. Introduction

Optical wireless communications (OWCs) have garnered significant attention from both industry and academia because of their numerous advantages over radio-frequency (RF) communication systems. With the increasing demand for high capacity to support emerging applications in networks and Internet of Things (IoT) devices, OWC is expected to be a promising candidate technique for future 6G networks [1]. By using light-emitting diodes (LEDs) or laser diodes (LDs) as transmitters, OWC is immune to RF interference, can provide higher security, potentially offers higher data rates, and consumes less power compared to RF systems [2]. However, it is important to note that LEDs are non-coherent and nonlinear optical devices with a peak transmitted power constraint.

In recent times, orthogonal frequency division multiplexing (OFDM) has gained substantial popularity in bandlimited optical systems, owing to its resistance to inter-symbol interference (ISI) and high spectral efficiency [3]. However, the adoption of conventional OFDM schemes designed for RF communication in OWC systems is not straightforward because of the non-coherent nature of light emitted from LEDs. Among the various optical OFDM schemes, DC-biased optical OFDM (DCO-OFDM) stands out as one of the most widely used options, primarily because of its simplicity [4]. In DCO-OFDM, a constant bias is added to the bipolar OFDM signal to ensure non-negativity. Another approach, asymmetrically clipped optical OFDM (ACO-OFDM), eliminates the need for adding a DC component by utilizing only the odd frequency subcarriers for modulation [5]. However, ACO-OFDM sacrifices half of its bandwidth efficiency. Alternatively, unipolar OFDM (U-OFDM), or Flip-OFDM, transmits positive and negative signals in separate adjacent frames. Nevertheless, U-OFDM utilizes an extra signal frame in each OFDM symbol, leading to a spectrum efficiency that is only half that of DCO-OFDM [6].

Given the inherent peak power constraint associated with light sources, it becomes imperative to account for the presence of clipping distortion, which can exert a pronounced adverse effect on system performance. To address this challenge, contemporary research has introduced a novel solution known as clipping-enhanced optical OFDM (CEO-OFDM), tailored specifically for power-constrained OWC systems [7]. CEO-OFDM employs an ingenious approach, whereby the clipped portion of the signal is transmitted via extra signal frames. This strategic utilization of the clipped portion effectively mitigates the nonlinear distortion stemming from peak power constraints. Derived from both analytical assessments and simulation experiments, CEO-OFDM2 exhibits superior performance, achieving a lower biterror rate (BER) or a higher data transmission rate, compared to existing state-of-the-art OFDM schemes. Nevertheless, it is important to acknowledge that the incorporation of extra signal frames necessitates a wider bandwidth, thereby introducing additional noise components at the receiver. Furthermore, the reconstruction of the intended data involves the amalgamation of multiple signal frames, a process that compounds the noise from each frame, ultimately amplifying the cumulative effective noise.

The current research focus in the field of optical OFDM revolves around addressing the issues related to clipping distortions caused by the nonlinear behavior of light sources and improving receiver design. Recent studies have explored various strategies to tackle these challenges. One approach involves optimizing power allocation for individual subcarriers in DCO- and ACO-OFDM systems, with the goal of maximizing the signal-to-noise ratio (SNR) while considering clipping distortions [8]. Furthermore, researchers have introduced variations of DCO- and ACO-OFDM, such as adaptively biased OFDM and superposition-based nonlinearity mitigation for ACO-OFDM, to mitigate clipping distortions [9,10].

To address receiver-related challenges, novel methods have been developed to combat clipping distortions. Unlike pre-processing techniques applied at the transmitter, these algorithms are designed specifically for the receiver. For example, an iterative decision method was proposed for ACO-OFDM, which improves the demodulation process by accurately estimating the sign information of the original unclipped signal, thereby enhancing overall performance [11]. Additionally, researchers introduced a semi-deterministic approach to mitigate clipping noise in DCO-OFDM systems [12]. Recently, a new decision-directed signal reconstruction (DDSR) algorithm with optimal thresholds for mitigating clipping noise was proposed in optical OFDM systems [13]. Theoretical analysis and simulation results verified that using optimal thresholds proposed in [13] significantly enhanced the DDSR algorithm's performance. Another receiver-based algorithm using a nonlinear technique to mitigate the clipping distortion in optical OFDM system was studied [14], in which the clipped signal could be extracted using a nonlinear process. These advancements represent significant progress in the ongoing efforts to optimize optical OFDM systems, especially in the presence of challenging nonlinear effects caused by the characteristics of light sources.

In this paper, we introduce a novel interference degradation algorithm, referred to as the tolerance-aided interference degradation (TAID) algorithm, designed for implementation at the receiver end. The primary objective of TAID is to mitigate both clipping distortion and partial additive noise inherent in the received signals. Specifically, when employing the CEO-OFDM technique, the proposed algorithm operates by introducing a decision tolerance at the receiver stage. This tolerance value is strategically chosen to assess the signals received in the initial two signal frames, determining whether the clipped portion transmitted in the extra frame is essential for accurate reconstruction. Through a judicious selection of an appropriate decision tolerance, our algorithm significantly reduces additive noise while concurrently compensating for clipping distortion, thereby enhancing the overall signal quality and system performance.

The remainder of the paper is organized as follows. The problem descriptions are discussed in Section 2. In Section 3, we describe the proposed TAID algorithm and analyze its performance. The numerical results of TAID and the performance comparison with DCO-, ACO-, and CEO-OFDM are discussed in Section 4. The paper is concluded in Section 5.

2. Problem Description

In this section, we provide a concise overview of the research problem at hand, encompassing the challenges posed by clipping effects arising from power-constrained light sources. Additionally, we introduce the CEO-OFDM technique, which is pertinent to addressing these challenges. Furthermore, we delineate the underlying motivation that underpins the development and proposition of our novel technique.

2.1. Clipping Effects of the Transmitted Signals

In OWC systems, the inherent characteristics of light sources dictate the utilization of only intensity modulation and direct detection (IM/DD) techniques. This necessitates the modulation of transmitted signals onto the input current of light sources, thereby confining the signals to non-negative values. Owing to stringent peak radiation power constraints, any signals surpassing these prescribed limits must undergo a process of hard clipping, inevitably engendering clipping distortions.

OFDM signals are generated through inverse fast Fourier transform (IFFT), a process known to yield a relatively high peak-to-average power ratio (PAPR). Consequently, this inherent property of OFDM signals results in the presence of both negative and high amplitude values, which invariably fall within the zones of zero clipping and peak clipping, as visually depicted in Figure 1. Therefore, the transmitted signals experiencing zero and peak power clipping are severely distorted. Figure 1 further demonstrates the nonlinearity in the light source's photocurrent-optical power response, introducing distortion during the modulation of OFDM signals. To rectify this nonlinear behavior and establish a linear relationship between output optical power and input current, pre-distortion emerges as a viable solution, as outlined in prior research [15]. Through the application of pre-distortion techniques, a linearized relationship between input current and output power is attained, as evidenced in Figure 1. In this paper, it is assumed that pre-distortion methods are employed in the system under consideration. Note that this pre-distortion is applied to linearize the current-optical power response, ensuring the accurate transmission of the desired waveform.

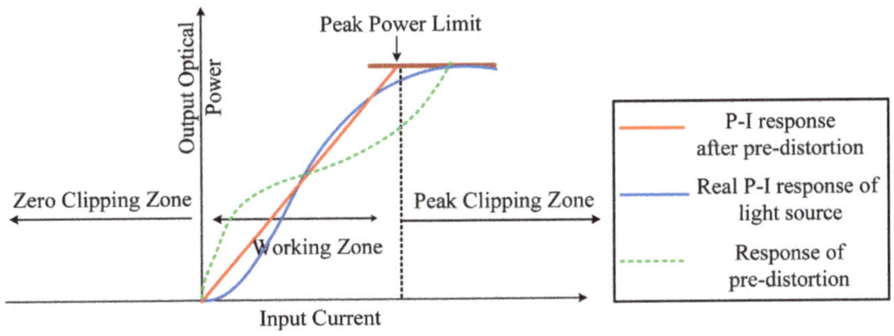

Figure 1. An illustration of the nonlinearity of the light source for OWC systems.

2.2. Clipping-Enhanced Optical OFDM Technique

CEO-OFDM stands out as a recently introduced technique specifically devised to mitigate the challenges associated with signal clipping beyond peak power limits as well as zero clipping. Empirical evidence supports its superior performance in terms of bit error rate (BER) and power efficiency when compared to conventional optical OFDM methods, such as DCO-OFDM, ACO-OFDM, and U-OFDM [7]. This section offers a brief introduction to the fundamental principles underpinning CEO-OFDM, delves into its inherent limitations, and provides insight into the motivation of the proposed TAID algorithm.

To mitigate the adverse effects of zero and peak power clipping, CEO-OFDM employs a distinctive structure, where each CEO-OFDM symbol comprises three distinct signal frames, as illustrated in Figure 2. These three frames are dedicated to transmitting the magnitudes of the positive, negative, and clipped portions of the original OFDM symbol. It is assumed that the peak radiation power is denoted as $P_{\max}$, and any positive and negative signals exceeding this magnitude threshold are subject to clipping, being constrained to $P_{\max}$. Importantly, the clipped portions are efficiently transmitted through frame 3, as shown in Figure 2. Notably, within the CEO-OFDM scheme, the occurrence of clipping distortion is confined exclusively to the signal within frame 3.

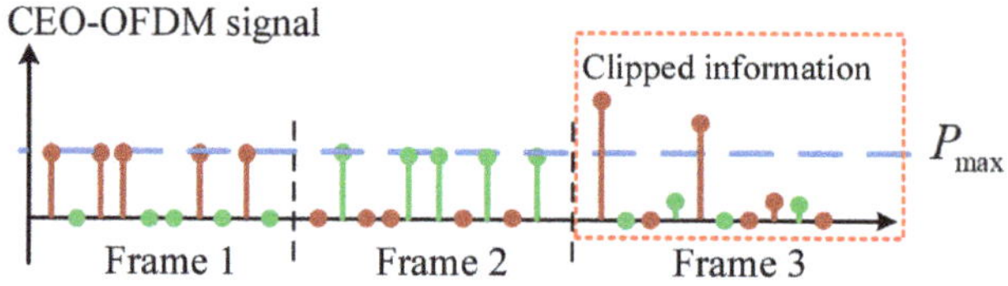

Figure 2. An example of CEO-OFDM signal. Red and green symbols represent the amplitude of the positive and negative signal, respectively

After subjecting the discrete OFDM signals within frames 1 and 2 to hard clipping at $P_{\max}$, it is noteworthy that, owing to the inherent effects of bandlimited Nyquist–Shannon interpolation [16], certain analog transmitted signals within frames 1 and 2 may still exceed the stipulated peak power constraint. In the context of this study, we adopt the assumption that the discrete signals across all three frames are uniformly clipped at a threshold of $\beta P_{\max}$, where a suitable value of $\beta < 1$ can be determined to ensure that the corresponding analog transmitted signals remain within the prescribed range of $[0, P_{\max}]$.

2.3. Motivation of TAID Algorithm

At the receiver, the task of reconstructing the original OFDM symbol involves a linear combination of the three signal frames. However, it is essential to note that the additive noise becomes intertwined across all three frames during this process, leading to an amplification of the noise power. Furthermore, the utilization of three signal frames to represent a single original OFDM symbol necessitates additional channel bandwidth, exacerbating the presence of noise with increased power levels.

Hence, this article introduces a novel tolerance-aided interference degradation algorithm to diminish the effective additive noise and facilitate the compensation of clipped signals during the demodulation process. The algorithm is designed to operate by establishing a decision tolerance, thereby enabling the examination of signals received within frames 1 and 2, as illustrated in Figure 2, to determine the necessity of incorporating signals from frame 3. Additionally, it is noteworthy that the signals within frames 1 and 2 can also undergo compensation while contending with the influence of additive noise.

3. Principles of Tolerance-Aided Interference Degradation Algorithm

This section provides an introduction to the principles underpinning the proposed TAID algorithm. It is important to note that TAID is exclusively designed for received signal processing, and thus, we adhere to the CEO-OFDM signal structure at the transmitter. To facilitate a comprehensive understanding of the principles governing TAID and to enable a thorough performance analysis, we sequentially delve into discussions related to the transmitted signal, the channel model, and the received signal model.

3.1. Transmitted Signal Model

For the sake of brevity and clarity in notation, we conduct our analysis within the confines of a single OFDM symbol duration. In our pursuit of high spectral efficiency, we employ M-ary quadrature amplitude modulation (M-QAM). For the ith subcarrier, we denote the modulated data as X_i, where M signifies the size of the modulation constellation. It is essential to emphasize that, in order to maintain a real-valued OFDM signal, X_i and X_{N-1-i} must constitute a conjugate pair, with N denoting the total number of subcarriers, as articulated in [17]. Following the N-point inverse fast Fourier transform (IFFT) operation and the parallel-to-serial conversion, the mth sample of the OFDM symbol can be expressed as

$$x_s[m] = \frac{c}{N}\sum_{k=0}^{N-1} X_i \exp\left(\frac{i2\pi km}{N}\right),\ m = 0, 1, \cdots, N-1, \tag{1}$$

where c is the modulation index controlling the scale of the transmitted signal magnitude. It is worth noting that c can be subject to optimization, effectively striking a balance between signal power and the mitigation of clipping distortion arising from the peak radiation power constraint imposed by the light sources.

At the transmitter, the bipolar signal $x_s[m]$ is reformed to be unipolar by the CEO-OFDM rules. Then, the nth sample of the transmitted signal is denoted [7]:

$$\begin{aligned} x_t[n] = &\underbrace{\varphi(x_s[n])}_{\text{frame 1}} + \underbrace{\varphi(-x_s[n-N])}_{\text{frame 2}} \\ &+ \underbrace{\varphi(|x_s[n-2N]| - P_{\max})}_{\text{frame 3}} \end{aligned} \tag{2}$$

where

$$\varphi(x) = \begin{cases} P_{\max}, & x \geq P_{\max} \\ x, & 0 < x < P_{\max} \\ 0, & x \leq 0 \end{cases}, \tag{3}$$

which represents the nonlinear response of the light sources with the peak power constraint. The response of the working zone is modeled as a linear function of the input current. $P_{\max}$ is the radiation power limit.

3.2. Channel Model

OWC systems usually use LDs, LEDs, and infrared (IR) diodes as light sources for signal transmission. In addition to the peak transmitted power constraint, it is important to recognize that these light sources inherently exhibit low-pass characteristics. Typically, the behavior of these light sources can be effectively approximated through the utilization of first-order low-pass filter models, with the parameters often estimated using pilot signals within OFDM systems, as elucidated in [18].

In this paper, we emphasize that channel estimation lies outside the scope of our research focus. We make the fundamental assumption that the channel has been accurately estimated, and for the purpose of compensating channel losses on each subcarrier, a single-tap equalizer is deployed. In the context of OFDM systems, the mitigation of intersymbol interference necessitates the inclusion of a cyclic prefix (CP), a standard component that can be meticulously designed based on prevailing channel conditions, as elaborated in [17,19,20]. For the purposes of this article, our analysis is centered on the system following the extraction of the cyclic prefix at the receiver stage. To streamline both the calculations and the ensuing analysis, we simplify our modeling by assuming an ideal channel with unity loss.

3.3. Received Signal Model and Effective Noise Models

At the receiver, the incoming optical signals undergo a conversion process into electrical signals, followed by discrete-time sampling. The representation of the nth sample of the received discrete-time electrical signal can be expressed as

$$y[n] = \rho \cdot h[n] * x_t[n] + n_y[n],\ n = 0, 1, \cdots, 3N - 1, \tag{4}$$

where ρ represents the responsivity of the photodetector (PD), which denotes the optical-to–electrical power conversion ratio. $n_y[n]$ is the additive noise, including thermal and shot noises. In this paper, we assume $n_y[n]$ is a Gaussian white noise with zero mean. Its variance can be calculated as $\sigma_n^2 = \frac{3R_s}{2N}\mathcal{N}_o$, where R_s is the transmitted M-QAM symbol rate and $\mathcal{N}_o$ is the single-sided noise power spectral density. $h[n]$ is the nth sample of the discrete-time end-to-end system impulse response, which can be pre-estimated. The operation "$*$" represents the discrete-time convolution.

Drawing upon the CEO-OFDM signal model, the received signal in one OFDM symbol contains three signal frames. Each frame is mixed with the additive noise, and the peak power clipping distortion happens in frame 3 only. Since the additive noise is mixed in all three signal frames, simply combining the three frames dramatically amplifies the effective noise. As shown in Figure 3, with the interference of the noise mixed in frames 1 and 2, the signals that should have been clipped at $P_{\max}$ are greater or less than the peak power.

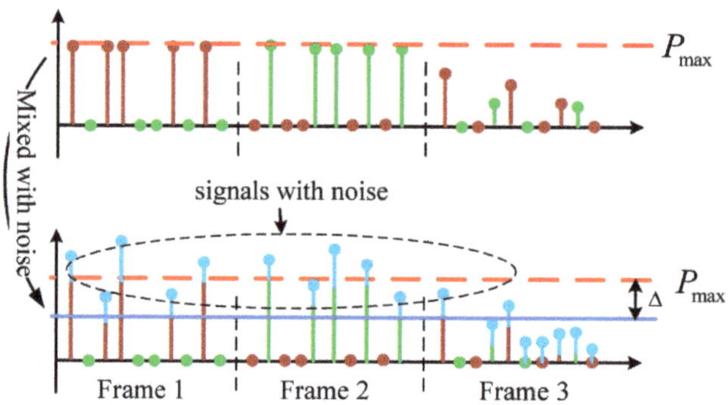

Figure 3. An illustration of the received signals with noise. Red and green symbols represent the amplitude of the positive and negative signals, respectively. Blue symbols represent the additive noise.

To address this challenge, we introduce a decision tolerance parameter denoted as Δ. By applying decision criteria to each received sample within frames 1 and 2, we can effectively reduce the impact of additive noise. Specifically, if the signals in frames 1 and 2 surpass the threshold of $P_{\max} - \Delta$, we enforce their value to be $P_{\max}$. Consequently, the corresponding signals in frame 3 are regarded as the clipped portion that necessitates combination during the demodulation process. Conversely, when the signals in frames 1 and 2 fall below the threshold of $P_{\max} - \Delta$, it is assumed that they have not been distorted by the peak power limit, and their associated clipped portion within frame 3 is deemed unnecessary.

A comprehensive block diagram illustrating this decision process is provided in Figure 4 as a detailed representation, where sign$(\cdot)$ denotes the unit sign function. It is imperative to recognize that the choice of the decision tolerance parameter Δ can significantly impact the trade-off between the noise introduced in frame 3 and the interference stemming from clipping distortion within frames 1 and 2.

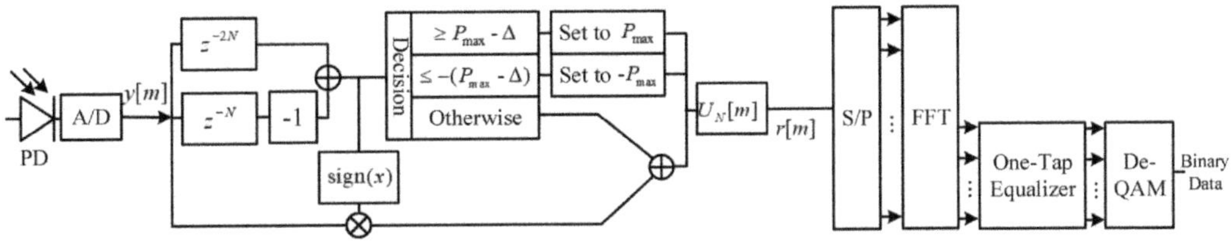

Figure 4. The block diagram of the receiver for TAID algorithm.

After applying the proposed TAID algorithm, the nth sample of the reconstructed signal is represented as

$$r[n] = r_s[n] + r_c[n], \;\; n = 0, 1, \cdots, N-1, \tag{5}$$

where

$$r_s[n] = \Big(\phi_\Delta(y[n]) - \phi_\Delta(y[n+N])\Big)U_N[n], \tag{6}$$

$$r_c[n] = \begin{cases} 0, & y[n] < (P_{\max} - \Delta) || y[n+N] < (P_{\max} - \Delta) \\ -y[n+2N]U_N[n], & y[n] < y[n+N] \\ y[n+2N]U_N[n], & y[n] > y[n+N] \end{cases}, \tag{7}$$

and $U_N[n] = u[n] - u[n-N]$. $u[n]$ is the discrete-time unit step function. $\phi_\Delta(\cdot)$ is a decision function controlled by the defined tolerance Δ, which is represented as

$$\phi_\Delta(x) = \begin{cases} P_{\max}, & x \geq P_{\max} - \Delta \\ x, & 0 < x < P_{\max} - \Delta \end{cases}. \tag{8}$$

Then, after the N-point fast Fourier transform (FFT) of the reconstructed signal $r[n]$, the demodulated M-QAM symbol on the ith subcarrier can be modeled as

$$\hat{X}_i = \frac{\alpha(c)\cdot\rho}{N}|H_i|X_i + n_{\text{eff}}, \; i = 1, 2, \cdots, N \tag{9}$$

where $\alpha(c)$ and $|H_i|$ represent the clipping coefficient and the channel frequency magnitude of the ith subcarrier, respectively. H_i is the Fourier transform of $h[m]$, which is estimated using the pilot signals. Based on Bussgang's theorem, the clipping coefficient is a function of the modulation index, which can be calculated from [17]:

$$\begin{aligned} \alpha(c) &= \frac{1}{\sigma_x^3\sqrt{2\pi}}\int_{-\infty}^{\infty} x\psi(x)\exp\left(-\frac{x^2}{2\sigma_x^2}\right)dx \\ &= 1 - \text{erfc}\left(\frac{\sqrt{2}NP_{\max}}{c\sigma_d}\right), \end{aligned} \tag{10}$$

where $\text{erfc}(\cdot)$ is the complementary error function, which is defined as $\text{erfc}(x) = 2/\sqrt{\pi}\int_x^\infty \exp(-y^2)dy$. $\psi(x, L)$ is a nonlinear function that represents the peak clipping distortion, represented as

$$\psi(x) = \begin{cases} -2P_{\max}, & x < -2P_{\max} \\ x, & -2P_{\max} < x < 2P_{\max} \\ 2P_{\max}, & x > 2P_{\max} \end{cases}. \tag{11}$$

3.4. Effective Noise Model and the Performance Analysis

In (9), n_{eff} represents the effective additive noise after reconstructing the received signal using TAID, which is modeled as a combination of the peak power clipping noise, additive noise at the receiver, and the noise of using the decision tolerance. In this paper, we derive n_{eff} as

$$n_{\text{eff}} = n_{\text{clip}} + n_{\text{add}} + n_{\text{tol}}, \tag{12}$$

where n_{clip}, n_{add}, and n_{tol} represent the clipping noise, effective additive noise, and noise introduced by the decision tolerance, respectively.

Despite CEO-OFDM being applied, it is noteworthy that the transmitted signals remain susceptible to peak clipping distortion. The severity of this peak power constraint-induced distortion escalates with the modulation index, denoted as c, which governs the scale of the transmitted signals. To evaluate the system performances using TAID, we adopt an approach wherein we represent the clipping distortion as an additional source of noise. This distortion is characterized as a Gaussian-distributed random variable with a zero mean. Its variance can be calculated as in [15], which is a function of the modulation index, as

$$\begin{aligned}\sigma_{\text{clip}}^2(c) = {} & \frac{2cP_{\max}\sigma_d}{N\sqrt{2\pi}} \exp\left(\frac{-2N^2P_{\max}^2}{c^2\sigma_d^2}\right) \\ & + \frac{1}{2}\text{erfc}\left(\frac{2NP_{\max}}{c\sqrt{2\sigma_d^2}}\right)\left(4P_{\max}^2 + \frac{c^2\sigma_d^2}{N}\right),\end{aligned} \tag{13}$$

where σ_d is the variance of X_i, $\forall\, i$.

By implementing the proposed TAID algorithm, the reduction of additive noise across the three signal frames is remarkably achieved. Leveraging the central limit theorem (CLT), the effective noise subsequent to the application of the TAID algorithm can be suitably approximated as a Gaussian random variable characterized by a zero mean. The variance of this effective noise is intricately linked to the decision tolerance, which is a composite of two constituent components, represented as

$$\sigma_{\text{tol}}^2(\Delta) = \sigma_{t1}^2(\Delta) + \sigma_{t2}^2(\Delta). \tag{14}$$

In Equation (14), $\sigma_{t1}^2(\Delta)$ pertains to the scenario where the transmitted signals within frames 1 and 2 have not undergone peak clipping. However, in practice, the influence of noise can lead to instances where these signals, upon reception, exceed the defined decision threshold of $P_{\max} - \Delta$. In such situations, it becomes necessary to artificially consider these signals as having been clipped, requiring us to enforce a value of $P_{\max}$ for the received signals within frames 1 and 2. Consequently, the introduction of interference stemming from the decision tolerance must be taken into account, and $\sigma_{t1}^2(\Delta)$ can be calculated as

$$\begin{aligned}\sigma_{t1}^2(\Delta) = {} & \int_0^{P_{\max}}\int_{P_{\max}-\Delta-r}^{\infty}(P_{\max}-r)^2\frac{\exp\left(\frac{-r^2}{2\sigma_d^2}+\frac{-n^2}{2\sigma_n^2}\right)}{2\pi\sigma_d\sigma_n}\,dn\,dr \\ = {} & \sum_{k=0}^{\infty}\sum_{v=0}^{\infty}\left(\sqrt{\sigma_n^2}P_{\max}^2 2^{-v}\left(-\frac{P_{\max}^2}{\sigma_d^2}\right)^v\cdot\left(\frac{\sqrt{2\pi}P_{\max}}{(2v^2+5v+3)k!(\Delta-P_{\max})^2(2v!v+v!)}\right.\right. \\ & + \frac{2^{-k}\sqrt{\sigma_n^2}\Gamma(2v+2)\left(-\frac{\Delta^2}{\sigma_n^2}\right)^k\left(\frac{\Delta}{\Delta-P_{\max}}\right)^{-2k}\left((P_{\max}-\Delta)((2v+1)\Delta+(2k+1)P_{\max})\cdot{}_2\tilde{F}_1\left(-2k,2(v+1);2v+3;-\frac{P_{\max}}{\Delta-P_{\max}}\right)\right)}{k!(\Delta-P_{\max})^2(2vv!+v!)} \\ & \left.\left. + \frac{2^{-k}\sqrt{\sigma_n^2}\Gamma(2v+2)\left(-\frac{\Delta^2}{\sigma_n^2}\right)^k\left(\frac{\Delta}{\Delta-P_{\max}}\right)^{-2k}\left(\Delta(2\Delta v+\Delta+2kP_{\max})\cdot{}_2\tilde{F}_1\left(1-2k,2v+2;2v+3;-\frac{P_{\max}}{\Delta-P_{\max}}\right)\right)}{k!(\Delta-P_{\max})^2(2vv!+v!)}\right)\right),\end{aligned} \tag{15}$$

where $\Gamma(\cdot)$ and ${}_2\tilde{F}_1$(a,b;c;z) represent the Gamma function and the regularized hypergeometric function, respectively [21].

In Equation (14), $\sigma_{t2}^2(\Delta)$ corresponds to the scenario wherein the transmitted signals within frames 1 and 2 are set to $P_{\max}$, indicating that the corresponding signals within these frames have been subject to hard clipping. In practice, because of the presence of noise, the signals declared as $P_{\max}$ within frames 1 and 2, when received, may actually fall below the threshold of $P_{\max} - \Delta$. Consequently, as per the TAID algorithm, the associated signals within frame 3 must be excluded during the data reconstruction process. This introduces peak clipping distortion, the variance of which, denoted as $\sigma_{t2}^2(\Delta)$, is calculated as

$$
\begin{aligned}
\sigma_{t2}^2(\Delta) =& \int_{P_{\max}}^{\infty}\int_{-\infty}^{P_{\max}-\Delta-r} (P_{\max}-r)^2 \frac{\exp\left(\frac{-r^2}{2\sigma_d^2}+\frac{-n^2}{2\sigma_n^2}\right)}{2\pi\sigma_d\sigma_n} dndr \\
=& \sum_{k=0}^{\infty}\sum_{v=0}^{\infty}\left(\frac{\sigma_n^2\sqrt{\sigma_d^2}2^{-k-v-\frac{1}{2}}\left(\Gamma(2v+1)\left(\Delta^2(v+1)(2v+1)+2\Delta kP_{\max}(2v+1)+k(2k+1)P_{\max}^2\right)\right)}{k!v!(\Delta-P_{\max})^2(k+v+1)}\right. \\
&\left(\frac{2\left(-\frac{1}{\sigma_n^2}\right)^k\left(-\frac{1}{\sigma_d^2}\right)^v\Gamma(-2k-2v-1)\left(\frac{1}{\Delta-P_{\max}}\right)^{-2(k+v+1)\left(\left\lfloor\frac{\left(\frac{P_{\max}}{\Delta-P_{\max}}\right)}{2\pi}\right\rfloor+1\right)}(\Delta-P_{\max})^{-2(k+v+1)\left\lfloor\frac{\left(\frac{P_{\max}}{\Delta-P_{\max}}\right)}{2\pi}\right\rfloor}}{\Gamma(1-2k)k!v!(\Delta-P_{\max})^2(k+v+1)}\right. \\
&+\frac{\Delta P_{\max}\left(-\frac{\Delta^2}{\sigma_n^2}\right)^k\left(-\frac{P_{\max}^2}{\sigma_d^2}\right)^v {}_2\tilde{F}_1\left(1,2(k+v+1);2(v+1);-\frac{P_{\max}}{\Delta-P_{\max}}\right)}{kk!v!(\Delta-P_{\max})^2(k+v+1)} \\
&\left.\left.+\frac{\Delta P_{\max}(P_{\max}-\Delta)\left(-\frac{\Delta^2}{\sigma_n^2}\right)^k\left(-\frac{P_{\max}^2}{\sigma_d^2}\right)^v(\Delta v+\Delta+kP_{\max})}{kk!v!(\Delta-P_{\max})^2(k+v+1)}\right)\right),
\end{aligned}
\tag{16}
$$

where $\lfloor x \rfloor$ represents the greatest integer less than or equal to x.

For a certain modulation index and a pre-designed decision tolerance, the received signal-to-interference plus noise ratio (SINR) for the ith subcarrier can be represented as a function of c and Δ, which is

$$
\begin{aligned}
\gamma_i(c,\Delta) &= \frac{c^2\alpha(c)^2|H[i]|^2\rho^2\sigma_d^2}{N\left(\sigma_{\text{tol}}^2(\Delta)+\sigma_{\text{add}}^2+\sigma_{\text{clip}}^2(c)\right)}, \\
i &= 0,1,\cdots,N-1
\end{aligned}
\tag{17}
$$

where σ_{add}^2 denotes the power of the additive noise, and it can be computed as $\sigma_{\text{add}}^2 = N_0B$. Here, N_0 signifies the noise spectral power density, and B corresponds to the receiver's bandwidth. Given the SINR, we can approximate the bit error rate (BER) for the ith subcarrier by using the expression from [22], that is

$$
\text{BER}_i(c,\Delta) \approx \frac{\sqrt{M_i}-1}{\sqrt{M_i}\log_2\left(\sqrt{M_i}\right)}\text{erfc}\left(\sqrt{\frac{3\gamma_i(c,\Delta)}{2(M_i-1)}}\right),
\tag{18}
$$

which is a function of the modulation index and the detection tolerance. Based on (18), the optimal modulation index and the decision tolerance can be found to minimize the system's BER by solving the following objective function:

$$
[c^*,\Delta^*] = \arg\min_{c,\Delta}\frac{1}{N}\sum_{i=1}^{N}\text{BER}_i(c,\Delta).
\tag{19}
$$

In (19), the optimal modulation index c^* and the optimal decision tolerance Δ^* can be obtained by using the exhaustive search method for the sake of convenience. Since the objective function is non-convex and contains hyper-geometric terms, numerical results are derived. For this problem, the exhaustive search method is a feasible way to find the optimal solution for the two parameters, c and Δ.

3.5. Computational Complexity Analysis

In this section, we conduct an analysis of the computational complexity of the proposed algorithm, which is directly related to processing time. To ensure a fair comparison of computational complexity, we evaluate it based on the transmission of an equivalent number of data. In the case of DCO-OFDM, a transmitter employs an N-point IFFT module, resulting in a computational complexity of $\mathcal{O}(N \log_2 N)$. In contrast, ACO-OFDM uses only half of the available subcarriers for data modulation, leading to a computational complexity of $\mathcal{O}(2N \log_2 N)$ when modulating the same data volume as DCO-OFDM.

Considering the proposed TAID algorithm transmits the same signal structure as CEO-OFDM, which involves sending two additional signal frames compared to DCO-OFDM, its computational complexity matches that of CEO-OFDM, which can be expressed as $\mathcal{O}(N(\log_2 N + 2))$.

It is worth noting that, for the proposed TAID algorithm, the receiver is responsible for setting the decision tolerance. Therefore, only a decision process is necessary, eliminating the need for the additional complexity associated with CEO-OFDM.

4. Numerical Results and Comparison

In this section, we present the numerical results pertaining to the system performance when employing the proposed TAID algorithm. Unless explicitly stated otherwise, we utilize the parameters as outlined in Table 1, which is typical for indoor spaces and used as a benchmark [7,15]. To facilitate an equitable comparison of system performance, we assume a channel loss factor of one, as it does not significantly impact the analysis and ensures consistency across the test scenarios. Given the inherent characteristics of light sources, we employ a first-order low-pass filter to emulate the transmitter. As per preliminary investigations, the 3 dB bandwidth of such filters typically spans a range of several tens of megahertz for commonly used lighting LEDs, as documented in [23]. Conversely, LDs and IR LEDs often exhibit larger bandwidths, reaching approximately a few hundred megahertz, as confirmed by studies such as [24–26]. In this study, we evaluate scenarios encompassing both large and small bandwidths. Furthermore, it is imperative to acknowledge that the constraints imposed by the light source's driving current lead to limitations of the peak transmitted optical power. This constraint results in the clipping of signals with magnitudes exceeding the power constraint, thereby introducing nonlinear distortion into the system.

In this study, we employ the direct current–biased optical OFDM (DCO-OFDM), asymmetrically clipped optical OFDM (ACO-OFDM), and clipping-enhanced optical OFDM (CEO-OFDM) techniques as the benchmark for evaluating system performance. To ensure an equitable and meaningful comparison, identical parameters are adopted across all tested techniques. For DCO-OFDM, a bias is introduced with a magnitude of $\frac{P_{\max}}{2}$, striking a balance between mitigating peak power and zero clipping. It is important to note that the transmitted data symbol rate provided in Table 1 reflects the effective data symbol rate, accounting for spectral utilization efficiency considerations. Comparing the performances of DCO-, ACO-, and CEO-OFDM and the proposed algorithm at the same transmitted data rate, we can calculate the required bandwidths for the compared techniques as

$$B_{\text{DCO}} = \frac{1}{2} B_{\text{ACO}} = \frac{1}{3} B_{\text{CEO}} = \frac{1}{3} B_{\text{proposed}}. \tag{20}$$

From this relationship, the ACO- and CEO-OFDM require twice and three times larger bandwidth than DCO-OFDM, respectively.

Table 1. Parameters Used for Numerical Results.

Responsivity, ρ	0.5 A/W
Peak optical power limit, $P_{\max}$	5 mW
Number of subcarriers, N	64
Noise spectral density, N_0	5×10^{-10} mW/Hz
Transmitted data symbol rate, R_s	18 Msps
Channel loss	1

Figure 5 presents a comparative analysis of BER performances among the proposed TAID algorithm, DCO-OFDM, ACO-OFDM, and CEO-OFDM techniques operating through an ideal channel, as the modulation index increases. For the proposed TAID algorithm, the figure showcases the minimum BER attained for various modulation indexes, with this minimum achieved through an exhaustive search for the optimal decision tolerance parameter Δ. In general, as the modulation index increases, the BER performance improves across all the discussed techniques. However, a notable trend emerges whereby a further increase in signal scale results in more pronounced clipping distortion. When the clipping distortion becomes dominant over the additive noise, the BER performance deteriorates. Because of ACO-OFDM's utilization of only odd-numbered subcarriers for signal modulation, its bandwidth efficiency is reduced by half compared to DCO-OFDM. Consequently, ACO-OFDM necessitates twice the bandwidth of DCO-OFDM for equivalent data throughput. Similarly, CEO-OFDM requires three times the bandwidth of DCO-OFDM for an equitable comparison. The numerical results depicted in this figure demonstrate that the proposed algorithm consistently achieves the lowest BER among the discussed techniques while employing the same modulation constellation size. Moreover, the proposed algorithm exhibits a broader dynamic modulation index range than those of DCO-OFDM, ACO-OFDM, and CEO-OFDM, maintaining a BER below 10^{-3}. Our findings reveal that, with 64-QAM, the proposed algorithm yields a comparable best BER performance to DCO-OFDM and ACO-OFDM using 32-QAM. Given that the effective transmitted data symbol rate remains consistent for fair comparison, the proposed algorithm attains a 20% higher data rate than the other compared techniques while achieving a similar BER.

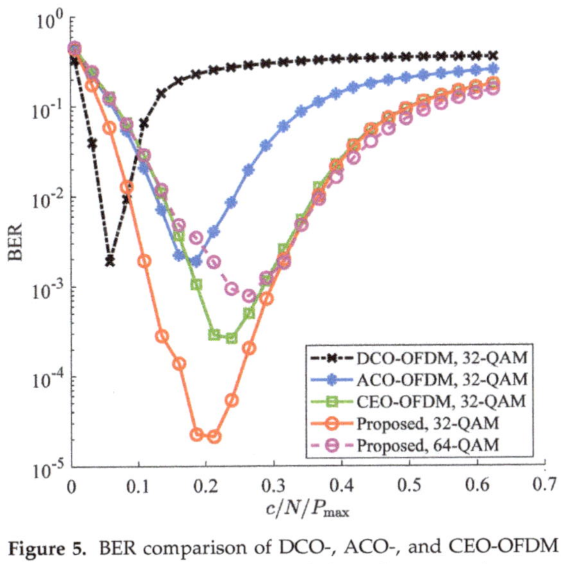

Figure 5. BER comparison of DCO-, ACO-, and CEO-OFDM and the proposed algorithm with different modulation indexes. Ideal channel is assumed.

As discussed in this paper, CEO-OFDM and the proposed technique require the widest bandwidth among the compared techniques. In the numerical results illustrated in Figure 5; we specifically compare the outcomes obtained with 64-QAM for the proposed algorithm against those with 32-QAM for DCO-OFDM and ACO-OFDM. Given the capability to achieve a comparable best BER performance through modulation index adjustments, we can deduce that the bandwidth efficiency of DCO-OFDM, denoted as η_{DCO}, is twice that of ACO-OFDM, denoted as η_{ACO}. Furthermore, we observe that η_{DCO} is approximately 2.5 times that of the proposed algorithm, denoted as η_{proposed}, based on the equation $\eta_{\text{DCO}} \approx 3 \cdot \frac{\log_2 32}{\log_2 64} \eta_{\text{proposed}}$.

In the context of a bandlimited channel, we extend our analysis to compare the BER performances of the proposed algorithm with DCO-OFDM, ACO-OFDM, and CEO-OFDM. The corresponding numerical results are presented in Figure 6. Given that ACO-OFDM and CEO-OFDM exhibit lower bandwidth efficiency compared to that of DCO-OFDM, their BER performances deteriorate when targeting the same transmitted data rate. Leveraging the benefits of the proposed TAID algorithm, the effective added noise is substantially reduced, leading to an enhanced SINR with TAID. Consequently, the BER performance surpasses that of DCO-OFDM, ACO-OFDM, and CEO-OFDM when utilizing the given parameters.

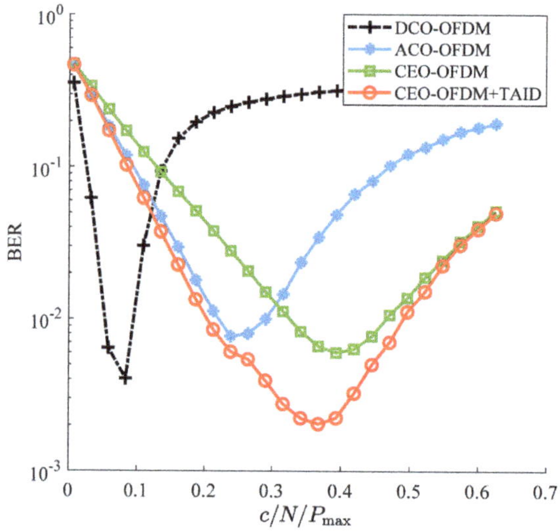

Figure 6. BER comparison of DCO-, ACO-, and CEO-OFDM and the proposed algorithm with different modulation indexes. The 3 dB bandwidth of the channel is 15 MHz. 32-QAM is applied.

The decision tolerance stands as a crucial design parameter that demands careful adjustment to attain optimal BER. Figure 7 presents the BER performances for various modulation indexes with increasing values of Δ. In this depiction, we introduce the ratio $\frac{P_{\max}-\Delta}{P_{\max}}$ to denote the separation between the peak power and the decision threshold. When $\frac{P_{\max}-\Delta}{P_{\max}} \leq 1$, the decision threshold is lower than $P_{\max}$. Across all scenarios, we observe a trend wherein as the ratio escalates, the BER performance improves. This behavior is attributed to the fact that the decision tolerance, as it increases, effectively mitigates a portion of the additive noise. However, when the ratio is further increased, bringing Δ closer to $P_{\max}$, it leads to the introduction of more added noise, resulting in a deterioration of the BER.

In this paper, we delve into the relationship between the scale of the transmitted signal and the optimal decision tolerance. The numerical results, depicted in Figure 8, elucidate the evolving pattern of the ratio $\frac{P_{\max}-\Delta}{P_{\max}}$ as the transmitted signal scale increases.

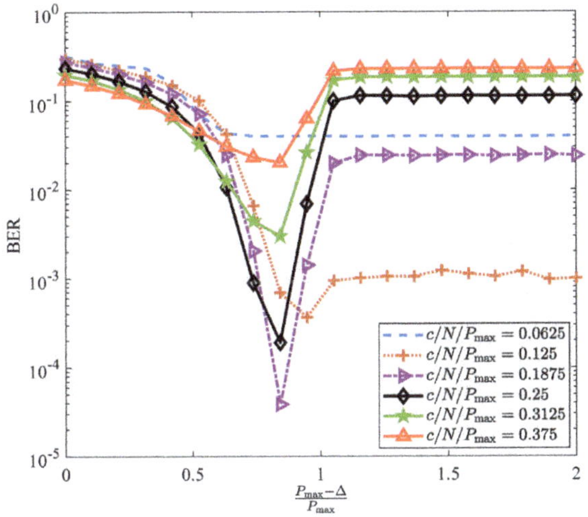

Figure 7. BER performance of the proposed algorithm with a varing decision tolerance Δ. 32-QAM is applied, and an ideal channel is assumed.

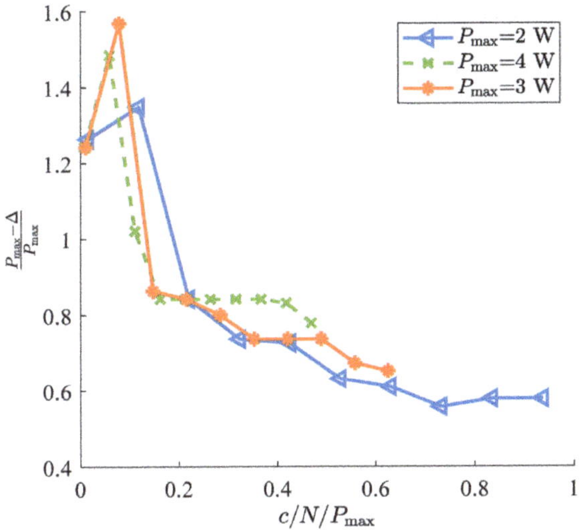

Figure 8. BER performance of the proposed algorithm with a varing decision tolerance Δ. 32-QAM is applied.

From this figure, it becomes apparent that the optimal ratio $\frac{P_{max}-\Delta}{P_{max}}$ for achieving the best BER exceeds one when the signal scale is small. This observation aligns with the fact that, for small-scale signals, the impact of peak power clipping can be marginalized, allowing us to position the decision threshold beyond the peak power value. This strategic placement effectively aids in noise reduction. Conversely, when the transmitted signals exhibit a relatively larger scale, the presence of peak power clipping distortion becomes non-negligible. Consequently, it necessitates a smaller decision threshold, falling below the peak power level, in order to effectively mitigate a portion of the added noise. This empirical result underscores the fact that, with a smaller peak power constraint imposed on the light sources, a lower value of the ratio $\frac{P_{max}-\Delta}{P_{max}}$ is needed to achieve optimal BER.

5. Conclusions

In this article, we present a novel tolerance-aided interference degradation algorithm tailored for power-constrained optical OFDM systems. At the receiver end, we introduce a decision tolerance and establish criteria to effectively mitigate both clipping distortion and additive noise. In the context of CEO-OFDM, this algorithm aids in determining whether the signals received in the first two signal frames, which are subsequently re-transmitted in the extra frame, are essential for data reconstruction. Furthermore, it enables compensation for clipping distortion. Our extensive analysis and numerical results showcased that, by judiciously selecting the decision tolerance, the bit error rate (BER) performance can be significantly enhanced by two orders of magnitude compared to conventional DCO-, ACO-, and CEO-OFDM techniques. Additionally, our proposed algorithm boasts a broader dynamic modulation index range, ensuring a BER below 10^{-3}.

Utilizing the specified parameters, our algorithm, employing 64-QAM modulation, achieves a comparable best BER performance as DCO-, ACO-, and CEO-OFDM with 32-QAM. Consequently, it delivers a remarkable 20% higher data rate than the other aforementioned techniques. This substantial performance enhancement positions TAID as a promising solution for power-constrained optical OFDM systems, offering superior BER performance and data rate efficiency.

Author Contributions: Conceptualization, Y.G.; methodology, Y.G.; software, Y.G.; validation, Y.G.; formal analysis, Y.G. and J.L.; investigation, Y.G.; resources, Y.G.; data Y.G.; writing—original draft preparation, Y.G. and J.L.; writing—review and editing, Y.G. and J.L.; visualization, J.L.; supervision, J.L.; project administration, J.L.; funding acquisition, Y.G. and J.L. All authors have read and agreed to the published version of the manuscript.

Funding: This work was supported by the Young Scientists Fund of the National Science Foundation of China (NSFC) (Program No. 62001392), the Key Research and Development Program of Shaanxi (Program No. 2022GY-097), the Scientific Research Plan of Education Department of Shaanxi Province (Grant No. 20JK0873), Xi'an Science and Technology Association Youth Talent Lifting Plan (Grant No. 959202313081), Xi'an Science and Technology Plan (Grant No. 23GXFW0079), and Xi'an University Three-Year Action Plan (Grant No. 2021XDJH30).

Institutional Review Board Statement: Not applicable.

Informed Consent Statement: The study did not involve humans.

Data Availability Statement: Data may be obtained from the authors upon request.

Conflicts of Interest: The authors declare no conflict of interest.

References

1. Chi, N.; Zhou, Y.; Wei, Y.; Hu, F. Visible Light Communication in 6G: Advances, Challenges, and Prospects. *IEEE Veh. Technol. Mag.* **2020**, *15*, 93–102. [CrossRef]
2. Komine, T.; Nakagawa, M. Fundamental analysis for visible-light communication system using LED lights. *IEEE Trans. Consum. Electron.* **2004**, *50*, 100–107. [CrossRef]
3. Azhar, A.; Tran, T.; O'Brien, D. A Gigabit/s Indoor Wireless Transmission Using MIMO-OFDM Visible-Light Communications. *IEEE Photon. Technol. Lett.* **2013**, *25*, 171–174. [CrossRef]
4. Carruthers, J.B.; Kahn, J.M. Multiple-subcarrier modulation for nondirected wireless infrared communication. *IEEE J. Sel. Areas Commun.* **1996**, *14*, 538–546. [CrossRef]
5. Armstrong, J.; Lowery, A.J. Power efficient optical OFDM. *Elect. Lett.* **2006**, *42*, 370–372. [CrossRef]
6. Tsonev, D.; Sinanovic, S.; Haas, H. Novel Unipolar Orthogonal Frequency Division Multiplexing (U-OFDM) for Optical Wireless. In Proceedings of the 2012 IEEE 75th Vehicular Technology Conference (VTC Spring), Yokohama, Japan , 6–9 May 2012; pp. 1–5.
7. Lian, J.; Brandt-Pearce, M. Clipping-Enhanced Optical OFDM for Visible Light Communication Systems. *J. Lightw. Technol.* **2019**, *37*, 3324–3332. [CrossRef]
8. Jiang, Z.; Gong, C.; Xu, Z. Clipping Noise and Power Allocation for OFDM-Based Optical Wireless Communication Using Photon Detection. *IEEE Wirel. Commun. Lett.* **2019**, *8*, 237–240. [CrossRef]
9. Li, B.; Xu, W.; Li, Z.; Zhou, Y. Adaptively Biased OFDM for IM/DD-Aided Optical Wireless Communication Systems. *IEEE Wirel. Commun. Lett.* **2020**, *9*, 698–701. [CrossRef]

10. Jiang, Y.; Wang, M.; Zhu, X.; Liang, C.; Wang, T.; Sun, S. Superposition Based Nonlinearity Mitigation for ACO-OFDM Optical Wireless Communications. *IEEE Wirel. Commun. Lett.* **2021**, *10*, 469–473. [CrossRef]
11. He, C.; Lim, Y. Iterative Decision-Directed Diversity Combining in Asymmetrically Clipped Optical OFDM. In Proceedings of the 2022 31st Wireless and Optical Communications Conference (WOCC), Shenzhen, China, 1–12 August 2022; pp. 121–125. [CrossRef]
12. Bera, K.; Bakaul, M.; Karmakar, N. A Semi-deterministic Approach for Clipping Noise Mitigation in DCO-OFDM Systems. In Proceedings of the 2019 25th Asia-Pacific Conference on Communications (APCC), Ho Chi Minh City, Vietnam, 6–8 November 2019, pp. 122–125. [CrossRef]
13. He, C.; Lim, Y. Clipping Noise Mitigation in Optical OFDM Using Decision-Directed Signal Reconstruction. *IEEE Access* **2021**, *9*, 115388–115402. [CrossRef]
14. He, C.; Armstrong, J. Clipping Noise Mitigation in Optical OFDM Systems. *IEEE Commun. Lett.* **2017**, *21*, 548–551. [CrossRef]
15. Noshad, M.; Brandt-Pearce, M. Hadamard-Coded Modulation for Visible Light Communications. *IEEE Trans. Commun.* **2016**, *64*, 1167–1175. [CrossRef]
16. Bai, R.; Hranilovic, S. Kramers-Kronig Optical OFDM for Bandlimited Intensity Modulated Visible Light Communications. *J. Light. Technol.* **2021**, *39*, 7135–7145. [CrossRef]
17. Armstrong, J. OFDM for Optical Communications. *J. Lightw. Technol.* **2009**, *27*, 189–204. [CrossRef]
18. Liao, J.C.; Chen, W.C.; Chung, C.D. Performance Analysis of SIMO–OFDM Systems With Pilot-Aided Channel Estimation. *IEEE Syst. J.* **2022**, 1–12. [CrossRef]
19. Yang, F.; Zhang, K.; Zhai, Y.; Quan, J.; Dong, Y. Artificial Noise Design in Time Domain for Indoor SISO DCO-OFDM VLC Wiretap Systems. *J. Light. Technol.* **2021**, *39*, 6450–6458. [CrossRef]
20. Niu, S.; Wang, P.; Chi, S.; Liu, Z.; Pang, W.; Guo, L. Enhanced Optical OFDM/OQAM for Visible Light Communication Systems. *IEEE Wirel. Commun. Lett.* **2021**, *10*, 614–618. [CrossRef]
21. Dwork, B. *Generalized Hypergeometric Functions*; Clarendon Press: Oxford, UK, 2019.
22. Ghassemlooy, Z.; Popoola, W.; Rajbhandari, S. *Optical Wireless Communications: System and Channel Modeling with Matlab*; CRC Press: Boca Raton, FL, USA, 2013.
23. Minh, H.L.; O'Brien, D.; Faulkner, G.; Zeng, L.; Lee, K.; Jung, D.; Oh, Y. 80 Mbit/s Visible Light Communications using pre-equalized white LED. In Proceedings of the 2008 34th European Conference on Optical Communication, Brussels, Belgium, 21–25 September 2008; pp. 1–2.
24. Kazemi, H.; Sarbazi, E.; Soltani, M.D.; El-Gorashi, T.E.H.; Elmirghani, J.M.H.; Penty, R.V.; White, I.H.; Safari, M.; Haas, H. A Tb/s Indoor MIMO Optical Wireless Backhaul System Using VCSEL Arrays. *IEEE Trans. Commun.* **2022**, *70*, 3995–4012. [CrossRef]
25. Gunawan, W.H.; Liu, Y.; Chow, C.W.; Chang, Y.H.; Peng, C.W.; Yeh, C.H. Two-Level Laser Diode Color-Shift-Keying Orthogonal-Frequency-Division-Multiplexing (LD-CSK-OFDM) for Optical Wireless Communications (OWC). *J. Light. Technol.* **2021**, *39*, 3088–3094. [CrossRef]
26. Wang, K. Remote-Powered Infrared Indoor Optical Wireless Communication Systems. *IEEE Photonics Technol. Lett.* **2022**, *34*, 455–458. [CrossRef]

photonics

MDPI

Article

Design and Analysis of Enhanced IM/DD System with Nonorthogonal Code Shift Keying and Parallel Transmission

Nobuyoshi Komuro [1,*] and Hiromasa Habuchi [2]

[1] Digital Transformation Enhancement Council, Chiba University, Chiba 263-8522, Japan
[2] Department of Information and Sciences, Ibaraki University, Hitachi 316-8511, Japan; hiromasa.habuchi.hiro@vc.ibaraki.ac.jp
* Correspondence: kmr@faculty.chiba-u.jp

Abstract: Providing Optical Wireless Communications (OWCs) is desirable for high data transmission efficiency. Intensity Modulation and Direct Detection (IM/DD) is widely adopted for its simplicity and practicality. Among various modulation schemes, Code Shift Keying (CSK) has demonstrated superior transmission efficiency compared to On-Off Keying (OOK) and Pulse Position Modulation (PPM). Prior research has shown that CSK performance can be further enhanced through parallel transmission and code concatenation techniques. However, the direct application of concatenated CSK to parallel transmission reduces the number of available code combinations as the concatenation level increases, potentially lowering modulation efficiency. This study proposes an advanced transmission scheme that integrates parallel transmission with a multi-level intensity adjustment mechanism. The proposed method preserves a high number of distinguishable transmission symbols, thereby achieving higher data transmission rates. Analytical derivations for transmission efficiency are provided for single-user scenarios, and numerical simulations validate the effectiveness of the proposed system. The key contributions of this work include mitigating symbol reduction in nonorthogonal CSK with parallel transmission and adjusting the multi-level intensity to enhance overall system performance. The results confirm that the proposed scheme significantly improves the efficiency and scalability of nonorthogonal CSK in OWC applications.

Keywords: optical wireless communication (OWC); intensity modulation and direct detection (IM/DD); code shift keying; parallel transmission

1. Introduction

The rapid evolution of wireless communication technologies over the past few decades has catalyzed their widespread integration into diverse applications [1–21]. Within this technological surge, Optical Wireless Communication (OWC) systems have emerged as a promising solution for high-speed and energy-efficient data transmission, gaining notable traction in recent years. Among the various implementation techniques, the Intensity Modulation with Direct Detection (IM/DD) approach has become a dominant methodology because of its simplicity and practicality. Unlike radio communication, where the modulating signal is a complex value, IM/DD modulates the optical intensity to encode information. In this scheme, the modulating signal (current) remains real and positive, as the received light intensity is directly detected at the receiver to recover the transmitted data. The IM/DD method, particularly in free-space optics links, is anticipated to support diverse applications due to its minimal deployment and infrastructure requirements. One notable example is LiFi (Light-Fidelity) [22,23], which leverages LED-based communication. By integrating illumination and data transmission, LiFi enhances energy efficiency. Furthermore, IM/DD has potential applications in underwater optical communication for marine [24], and vehicle-to-vehicle communication [25]. Research has been conducted on the relationship between code length and error rate in IM/DD systems [26], as well as on channel capacity modeling in the IM/DD [27] and modulation techniques [5,11,12,16–18,20,21,28,29].

Modulation schemes for IM/DD-based OWCs can be broadly categorized into On-Off Keying (OOK) [11,12], Pulse Position Modulation (PPM) [5,28], Sequence Inversion Keying (SIK) [16–18], and Code Shift Keying (CSK) [20,21]. Although OOK and SIK achieve a data transmission efficiency of 1 bit per code, PPM provides enhanced bit error rate performance by leveraging orthogonal signal structures. However, PPM's efficiency, $\frac{\log_2 N}{N}$ (bits/code), decreases as the number of slots N increases. In contrast, CSK achieves superior transmission efficiency, $\log_2 M$ (bits/code), where M is the number of distinct codes. Given these advantages, CSK represents a compelling candidate for further exploration. The optical codes are desired to be suitable for the communication quality.

The design criteria for optical codes can be classified into two main categories. The first criterion focuses on avoiding pulse collisions between codes. This approach is commonly applied in optical communications, with examples including optical orthogonal codes and extended prime codes [14,15]. The second criterion allows pulse collisions between codes but ensures that, over the code period, the correlation characteristic remains zero (orthogonal or pseudo-orthogonal). Pseudo-orthogonal M-sequence pairs [19] are a notable example of optical codes designed under this criterion. In this study, we adopt codes based on the latter design criterion.

Previous research on CSK [20,21,30] has established that CSK systems improve transmission efficiency by mapping data to one of M codes, with efficiency scaling as M increases. Additional performance gains have been achieved by employing parallel transmission techniques that leverage the orthogonality or pseudo-orthogonality of pseudo-orthogonal M-sequences [31], as well as concatenated CSK code structures. We have focused on the latter—the concatenated CSK code structures—and demonstrated the effectiveness of this design approach [32–34]. Combining these two performance-enhancing technologies is expected to further improve performance; however, the extent of the improvement remains unclear. Additionally, while conventional CSK with parallel code combination transmission significantly enhances information transfer by expanding the set of available transmission candidates, the proposed nonorthogonal CSK method, which is constructed by concatenating orthogonal or pseudo-orthogonal codes, may experience performance degradation due to a reduced number of code combinations under equal code length conditions. Therefore, further improvements are necessary to achieve the highest possible transmission rate in a parallel transmission environment.

This study proposes an advanced transmission scheme that extends the previous nonorthogonal code construction strategy [34] by introducing a novel parallel transmission mechanism. By intelligently selecting and combining multiple codes from the previously designed code set, our method increases the number of available transmission code combinations, thereby enhancing the overall data transmission rate. Furthermore, to address the limitations of directly applying the previous approach to parallel transmission, we introduce a multi-level intensity adjustment strategy for parallel transmission procedures comparable to those of conventional parallel transmission, thereby achieving higher data transmission rates. Since the nonorthogonal CSK method proposed by the authors is influenced by multiple factors—such as improved noise immunity with an increased optical scale level, increased inter-code interference with a higher number of concatenations, the expansion of the available code set through the use of nonorthogonal CSK codes, and a decrease in the number of viable transmission combinations—we determine the optimal number of concatenations. Analytical derivations for data transmission efficiency are presented for single-user scenarios, with numerical simulations validating the efficacy of the proposed system under various configurations. This study builds upon the previously proposed nonorthogonal CSK scheme. The key advancements are as follows:

Preventing Symbol Reduction in Parallel Transmission:Applying the previous method to parallel transmission reduces available code combinations as concatenation increases, lowering modulation levels. The proposed method incorporates multi-level intensity adjustment to maintain a high number of available symbols for transmission.

Achieving Higher Data Transmission Rates: While the previous method improves efficiency in single-code transmission, its direct extension to parallel transmission leads to diminishing returns. The proposed method addresses this by combining parallel code combination transmission with multi-level intensity adjustment, enhancing transmission rates.

2. Proposed System Framework

2.1. Design of Nonorthogonal Codes

The proposed method generates nonorthogonal codes by concatenating M_{con} pseudo-orthogonal M-sequences (POMs). The POM set, denoted as $\boldsymbol{OS}$, consists of M-sequences augmented by an additional chip, as defined below:

$$\boldsymbol{OS} = \begin{pmatrix} \boldsymbol{OS}_1 \\ \boldsymbol{OS}_2 \\ \vdots \\ \boldsymbol{OS}_{M_{pom}} \end{pmatrix} = \begin{pmatrix} a_1 & a_2 & a_3 & \cdots & a_{L_M} & -1 \\ a_2 & a_3 & \cdots & a_{L_M} & a_1 & -1 \\ & & \vdots & & & \\ a_{L_M} & a_1 & a_2 & \cdots & a_{L_M-1} & -1 \end{pmatrix}, \tag{1}$$

where a_i $(i = 1, 2, \cdots, L_M)$ is a binary value of $+1$ or -1, and L_M represents the M-sequence length. At the transmitter, all -1 values are replaced with 0 for transmission purposes. In contrast, the receiver uses POMs in their original $\{+1, -1\}$ form.

The cross-correlation between the i-th POM used by the transmitter, $\boldsymbol{OS}_{T_i}$, and the j-th POM at the receiver, $\boldsymbol{OS}_{R_j}$, can be expressed as

$$\frac{1}{T}\int_0^T \boldsymbol{OS}_{T_i}(t)\boldsymbol{OS}_{R_j}(t)dt = \begin{cases} \frac{L_M}{2} & \text{if } T_i = R_j, \\ 0 & \text{if } T_i \neq R_j, \end{cases} \tag{2}$$

where T denotes the POM cycle period.

For M_{pom} POMs and M_{con} concatenations, the resulting nonorthogonal code set NS is structured as follows:

$$NS = \begin{pmatrix} NS\,1 \\ NS\,2 \\ \vdots \\ NS\;M_{pom} \cdot 2^{M_{con}} \end{pmatrix} = \begin{pmatrix} \boldsymbol{OS}^{(1)} & \boldsymbol{OS}^{(2)} & \cdots & \boldsymbol{OS}^{(M_{con}-1)} & \boldsymbol{OS}^{(M_{con})} \\ \boldsymbol{OS}^{(1)} & \boldsymbol{OS}^{(2)} & \cdots & \boldsymbol{OS}^{(M_{con}-1)} & \overline{\boldsymbol{OS}^{(M_{con})}} \\ \vdots & \vdots & \ddots & \vdots & \vdots \\ \overline{\boldsymbol{OS}^{(1)}} & \overline{\boldsymbol{OS}^{(2)}} & \cdots & \overline{\boldsymbol{OS}^{(M_{con}-1)}} & \boldsymbol{OS}^{(M_{con})} \\ \overline{\boldsymbol{OS}^{(1)}} & \overline{\boldsymbol{OS}^{(2)}} & \cdots & \overline{\boldsymbol{OS}^{(M_{con}-1)}} & \overline{\boldsymbol{OS}^{(M_{con})}} \end{pmatrix}, \tag{3}$$

where $\boldsymbol{OS}^{(i)}$ represents the i-th sequence, and $\overline{\boldsymbol{OS}^{(i)}}$ is its negation. Since the proposed method constructs nonorthogonal codes in an organized manner, it can be constructed with the same level of complexity as conventional CSK.

In the proposed system, M_{pom} POMs are selected based on $\lfloor \log_2 \binom{M_{pom}}{m} \rfloor$ bits of data. The selected i-th POM is concatenated according to $M_{con}^{(i)}$ bits and the intensity of concatenated code is scaled by $M_a^{(i)}$, representing its modulation level, where the intensity scaling can be described as a parallel transmission of concatenated part. This structure allows a frame to encode $\left(\lfloor \log_2 \binom{M_{pom}}{m} \rfloor + m(M_{con} + \log_2 M_a^{(i)})\right)$ bits of information.

2.2. System Architecture

The overall system structure is shown in Figure 1, while Figure 2 provides an example of the transmitted signal format. The transmitter begins by mapping the input data into $\left(\lfloor \log_2 \binom{M_{pom}}{m} \rfloor + m(M_{con} + \log_2 M_a)\right)$ bits. In this paper, we refer to $\lfloor \log_2 \binom{M_{pom}}{m} \rfloor$ (bits), M_{con} (bits), and $\log_2 M_a^{(i)}$ (bits) as Data 1, Data 2, and Data 3, respectively. Next, m POMs are selected based on the data, and each sequence is concatenated according to its corresponding $M_{con}^{(i)}$ pattern. Light intensity levels of the concatenation pattern are then adjusted using $M_a^{(i)}$, and all -1 chips are converted into 0 before transmission.

At the receiver, shown in Figure 3, $\{+1, -1\}$-valued POMs are utilized. The received optical signal is converted into an electrical signal by an Avalanche Photo-diode (APD). The receiver calculates correlation values for each POM and identifies the m POMs with the highest absolute correlations. From these, it extracts data encoded in POM selection, concatenation patterns, and light intensity levels.

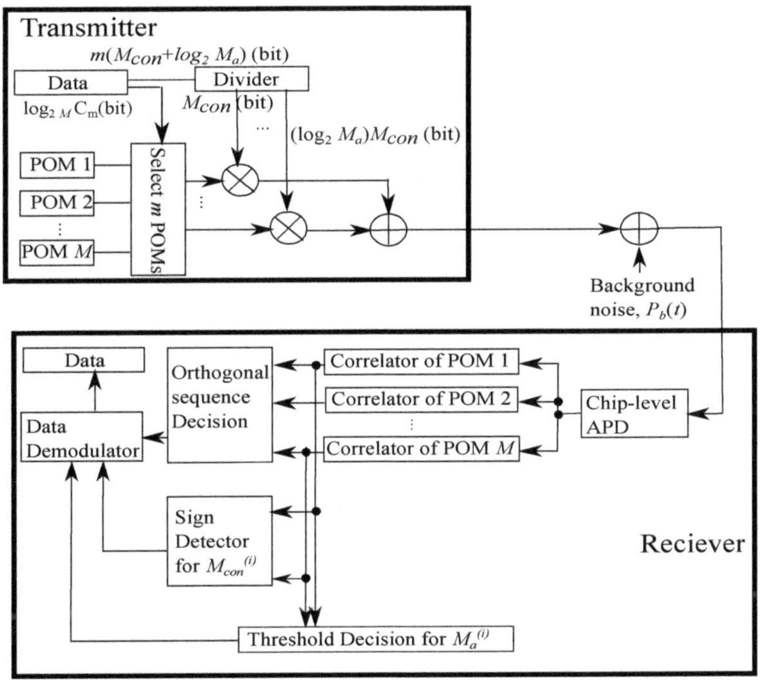

Figure 1. Proposed system architecture.

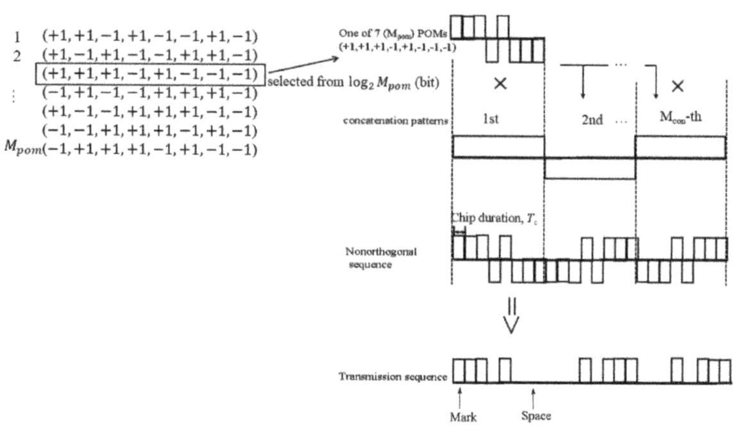

Figure 2. Example of transmitted signal structure.

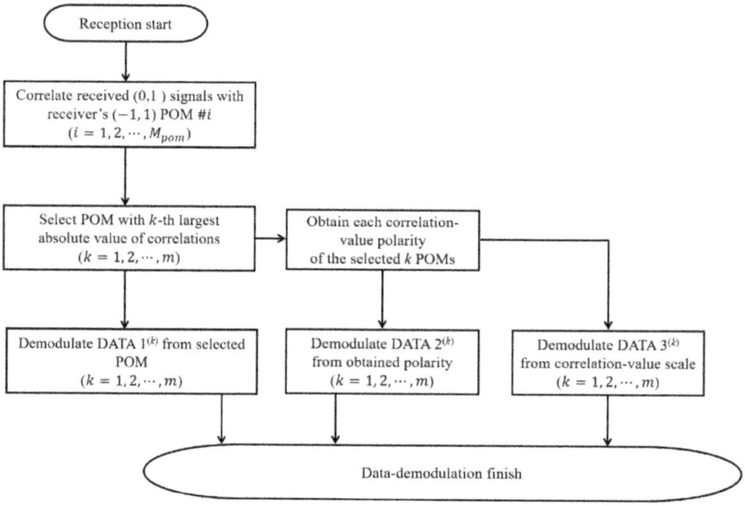

Figure 3. Receiver workflow.

3. Theoretical Analysis

This study evaluates the data transmission efficiency of the proposed system. Following the approach of previous work on IM/DD channel modeling [1,28], this analysis incorporates the effects of background noise and APD noise, where the noise variance depends on signal intensity in the optical communication channel. To extract the fundamental characteristics of the proposed method, the analysis largely follows the framework in [1,28].

In this paper, data transmission efficiency is defined as the total number of correctly transmitted bits from all users per chip duration.

The mean number of photons absorbed during T_c is expressed as

$$\lambda_s = \frac{\eta_A P_w}{hf}, \tag{4}$$

where λ_s denotes the photon absorption rate, h is Planck's constant, η_A is the efficiency of the avalanche photo-diode (APD), and P_w represents the received laser power without accounting for scintillation and background light. The total photon absorption rate λ—accounting for signal, background light, and APD bulk leakage current—is given by

$$\lambda = \begin{cases} \lambda_s + \lambda_b + \frac{I_b}{e}, & \text{for a mark,} \\ \frac{\lambda_s}{Me} + \lambda_b + \frac{I_b}{e}, & \text{for a space,} \end{cases} \tag{5}$$

where I_b is the bulk leakage current of the APD, e is the elementary charge, Me is the modulation extinction ratio of the laser diode between mark and space states, and λ_b represents the photon absorption rate due to background light, which is calculated as $\lambda_b = \frac{\eta_A P_b}{hf}$ when P_b is the background noise per chip duration.

When m represents the number of parallel transmitted POMs, the probability of a frame being correctly demodulated, $P_c(m)$, can be described as

$$P_c(m) = m\left[1 - \frac{1}{M_a}\sum_{i=1}^{M_a}\frac{1}{2}\mathrm{erfc}\left(\frac{Th_i - \mu_a(i)}{\sqrt{2\sigma_m^2(i)}}\right)\right]^{mM_{con}} \cdot \left(1 - P_{pom}(m)\right), \tag{6}$$

where $\mathrm{erfc}(x) = \frac{2}{\sqrt{\pi}}\int_x^\infty \exp(-t^2)dt$ is the complementary error function, $\mu_a(i)$ and $\sigma_s(i)$ represent the mean and variance of the largest correlation output, and Th_i is threshold for determining the light intensity levels of the concatenation pattern. $P_{pom}(m)$ is the symbol error rate of POM when the number of parallel transmitted POMs is m. If a concatenated POM code c_1 is sent, $P_{pom}(m)$ is given by

$$P_{pom}(m) = 1 - \int_{-\infty}^{\infty} f(c_1) - \int_{-\infty}^{c_1} f(c_{ii>m})\, dc_i M_{pom} - \int_{-\infty}^{c_1} f(c_m)\, dc_m m - 1\, dc_1, \tag{7}$$

where $f(c_j)$ is the PDF of correlation output for the j-th code c_j, which is

$$f(c_j) = g\left(\left|q_j^{(1)}\right|\right) * \cdots * g\left(\left|q_j^{(M_{con})}\right|\right), \tag{8}$$

$$g(q_j) = \frac{1}{\sqrt{2\pi\sigma_j^2}}\exp\left[-\frac{(q_j - \mu_j(i))^2}{2\sigma_j^2 M_{con}}\right]. \tag{9}$$

Here, $g(|q_j^{(j)}|)$ is the PDF of the absolute value of correlation output for the j-th POM q_j, $*$ represents the convolution integral, and μ_j and σ_j^2 are the mean and variance for the j-th correlation output, given by

$$\mu_1(i) = \cdots = \mu_m(i) = GT_c\left(\frac{L_f + 1}{2} i\lambda_s + \frac{L_f - 1}{2}\frac{i\lambda_s}{M_e} + \lambda_b\right), \tag{10}$$

$$\mu_{m+1}(i) = \cdots = \mu_{M_{pom}}(i) = GT_c\left(\frac{L_f i\lambda_s}{M_e} + \lambda_b\right), \tag{11}$$

and

$$\sigma_1^2(i) = \cdots = \sigma_{M_{pom}}^2(i) = G^2FT_c\left[m\frac{L_f + 1}{2} i\lambda_s + m\frac{L_f^2 - 1}{2}\frac{i\lambda_s}{M_e} + L_f\lambda_b + \frac{2I_b}{e}\right] + \frac{2I_sT_c}{e} + 2\sigma_{th}^2, \tag{12}$$

respectively. In Equations (10)–(12), F and σ_{th}^2 are

$$F = k_{eff}G + \left(1 - k_{eff}\right)\frac{2G - 1}{G}, \tag{13}$$

and

$$\sigma_{th}^2 = \frac{2k_B T_r T_c}{e^2 R_L}, \tag{14}$$

respectively.

When the number of parallel transmitted codes is m, the data transmission efficiency of the proposed system is then given by

$$S_{sys}(m) = \frac{K N_{bit} P_c(m)}{L_f}. \tag{15}$$

4. Numerical Results

This study evaluates the proposed system's performance through analysis and simulation. Figure 4 shows the simulation diagram. This study conducted the simulation by Monte Carlo simulation. In the simulation, first, each transmitter generates data bits and transmits them. Then, the simulation multiplies each light intensity of transmission signals by a random value based on a log-normal distribution (i.e., scintillation). Also, the simulation adds a random value based on a normal distribution to those signals (i.e., noise). In other words, the simulation artificially gives scintillation and noise, similar to the references [1,28]. The simulation calculates the data transmission efficiency from the demodulated signal at the receiver. The simulation trials were conducted 10,000,000 times in order to calculate data transmission efficiency. Table 1 shows the parameters which are used in this paper. This paper uses typical APD parameters [1,28]. Parameters in Table 1 are determined according to the values used in [1,28,35].

Table 1. Parameter settings.

Parameter	Value
Chip duration, T_c	4.0×10^{-4}
APD effective ionization ratio, k_{eff}	0.02
Average APD gain, G	100
Modulation extinction ratio of the laser diode output power in the mark and space states, Me	100
Boltzmann's constant, k_B	1.38^{-23}
Receiver noise temperature, T_r	300 (K)
Receive load resistor, R_L	1030 (Ω)
Elementary charge, e	1.6×10^{-16} (C)
Planck's constant, h	6.626×10^{-34}
APD bulk leakage current, I_b	0.1 (nA)
APD surface leakage current, I_s	10 (nA)
Quantum efficiency, η	0.6
Laser wavelength, λ_l	830 (nm)

Figure 5 illustrates the variation in data transmission efficiency as a function of the parallel level for both the conventional CSK system and the proposed scheme. The parallel level is defined as the product of the number of selected POMs, m, and the light intensity levels of the concatenation pattern, M_a, representing the total intensity levels of the transmission signals. The solid lines represent the analytical results, while the plots correspond to the simulation results obtained using Monte Carlo simulations. This evaluation assumes an ideal scenario, excluding both background and thermal noise. Under these conditions, the proposed system achieves data transmission efficiencies exceeding 1.0, a performance level unattainable by the conventional CSK system. Notably, the configuration $(M_{oc} = 7, M_{con} = 3)$ demonstrates the highest efficiency in this ideal scenario.

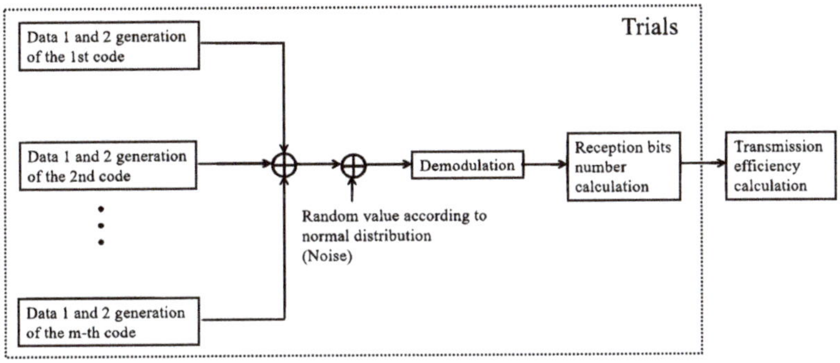

Figure 4. Simulation diagram.

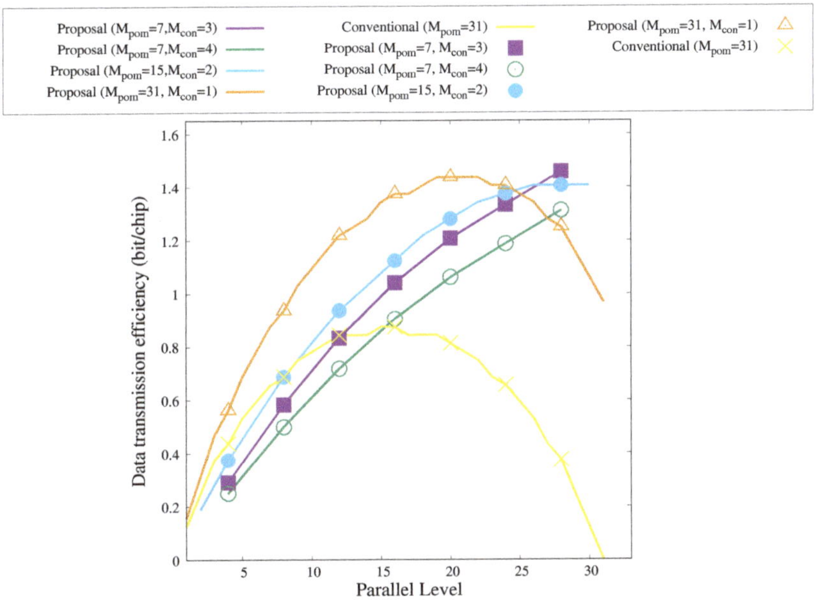

Figure 5. Data transmission efficiency as a function of the parallel level.

Figure 6 illustrates the relationship between data transmission efficiency and the transmit laser power per bit (P_w) for the conventional CSK system and the proposed scheme with light intensity scaling, evaluating the effect of background light noise. Here, the proposed method sets the threshold value Th_i to $\frac{2i+1}{2}P_w$. The evaluation assumes $L_{bit} = 4000$ bits. In the proposed scheme, the combinations of (M_{oc}, M_{con}) are $(7,3)$, $(7,4)$, $(15,2)$, and $(31,1)$. Each method sets the number of selected POMs (m) to the number at which the combined transmission reaches its maximum. The solid lines represent the analytical results, while the plots correspond to the simulation results obtained using Monte Carlo simulations. The analytical results are in qualitative agreement with the simulation results, confirming the validity of the analysis. For the conventional CSK system, the frame length is 31 chips, with the number of POMs also set to 31, and the number of bits per frame is given by $\left\lfloor \log_2 \binom{M_{oc}}{m} \right\rfloor$. In the proposed system, the frame lengths are 21, 28, 30, and 31 chips for $(M_{oc}, M_{con}) = (7,3), (7,4), (15,2)$, and $(31,1)$, respectively. As shown in Figure 6, the configuration $(M_{oc} = 7, M_{con} = 3)$ achieves the highest data transmission efficiency.

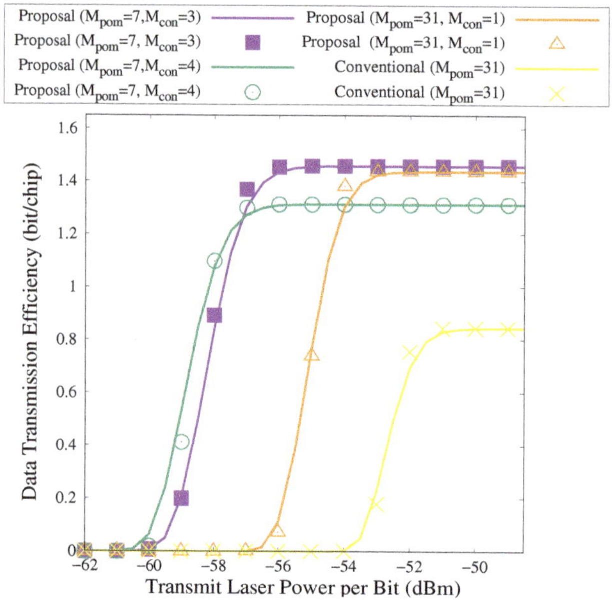

Figure 6. Data transmission efficiency as a function of the transmit laser power per bit.

Figure 7 illustrates the relationship between data transmission efficiency and the APD effective ionization ratio (k_{eff}) for the proposed scheme with light intensity scaling, evaluating the influence of APD shot noise. In the proposed method, the threshold value Th_i is defined as $\frac{2i+1}{2}P_w$, with an evaluation assuming $L_{bit} = 4000$ bits. The proposed scheme considers the parameter combinations (M_{oc}, M_{con}) as $(7,3)$, $(7,4)$, $(15,2)$, and $(31,1)$. Each method sets m to the value where the combined transmission achieves its peak performance. The solid lines represent analytical results, while the plotted points correspond to Monte Carlo simulation results, showing good agreement and confirming the validity of the analysis. In the proposed system, the frame lengths are 21, 28, 30, and 31 chips for $(M_{oc}, M_{con}) = (7,3), (7,4), (15,2)$, and $(31,1)$, respectively. The performance of the proposed method is influenced by multiple factors: noise immunity improves with an increased optical scale level, inter-code interference increases with a higher number of concatenations, the use of nonorthogonal CSK codes expands the available code set, and the number of viable transmission combinations decreases. Figure 7 indicates that configurations with fewer concatenations generally achieve higher performance due to a greater number of code combination patterns but are more vulnerable to noise. As shown in Figure 7, the configuration $(M_{oc} = 7, M_{con} = 3)$ achieves the highest data transmission efficiency.

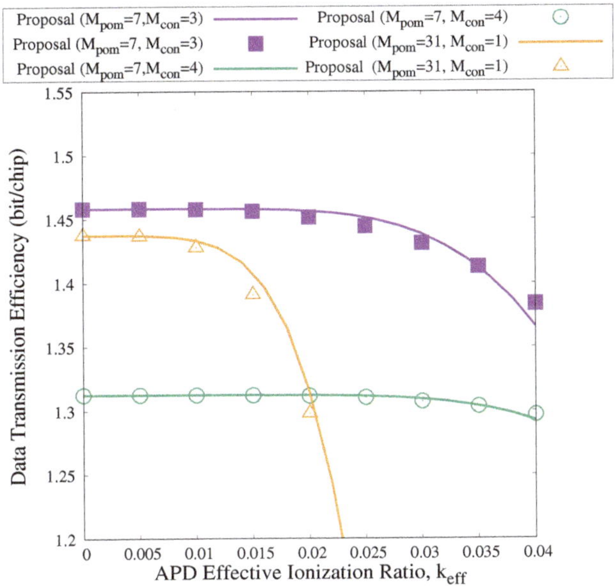

Figure 7. Data transmission efficiency as a function of the APD effective ionization ratio, k_{eff}.

Figure 8 depicts the data transmission efficiency of the proposed and conventional systems. In this figure, the combinations of (M_{pom}, M_{con}) in the proposed system are (127, 1), (63, 2), (31, 3), and (31, 4). The frame lengths for the cases of (127, 1), (63, 2), and (31, 4) are 128, while the frame length for the case of (31, 3) is 96. Additionally, Figure 9 illustrates the relationship between data transmission efficiency and frame length for both the conventional CSK system and the proposed scheme, where the number of bits per frame increases proportionally with the frame length.

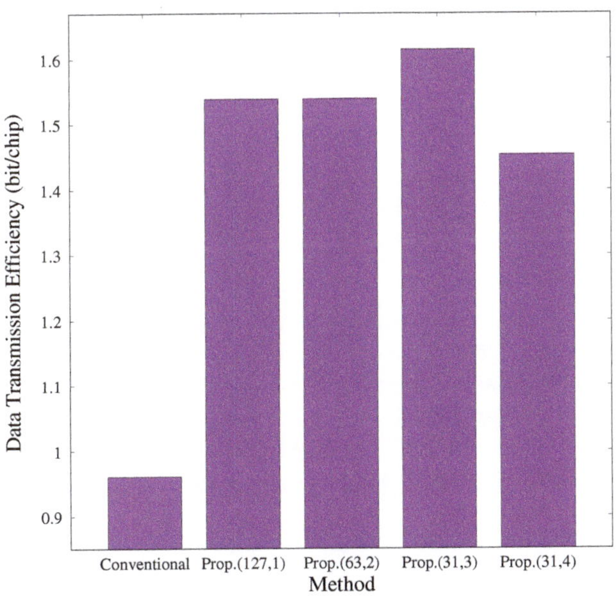

Figure 8. Comparison of data transmission efficiency as the proposed and conventional systems.

As in the evaluations shown in Figure 5, an ideal noise-free scenario is assumed. The results in Figures 8 and 9 reveal that the proposed scheme achieves data transmission efficiencies exceeding 1.0, a performance that is unattainable by the conventional CSK system. Among the evaluated configurations, the case of $M_{con} = 3$ achieves the highest data transmission efficiency, highlighting the effectiveness of combining nonorthogonal CSK with parallel transmission strategies. Furthermore, the findings indicate that the optimal concatenation number, M_{con}, for the proposed system, is 3.

A POM is constructed by extending an M-sequence with one additional chip. In the proposed method, the number of generated POMs is one less than the code length of the POM (i.e., the number of chips comprising the POM). This discrepancy leads to a reduction in data transmission efficiency, particularly when the code length is short, as shown in Figure 9. Furthermore, while the number of additional chips increases as the number of concatenations grows, this effect becomes negligible when the code length of the POM is sufficiently large relative to the number of concatenations.

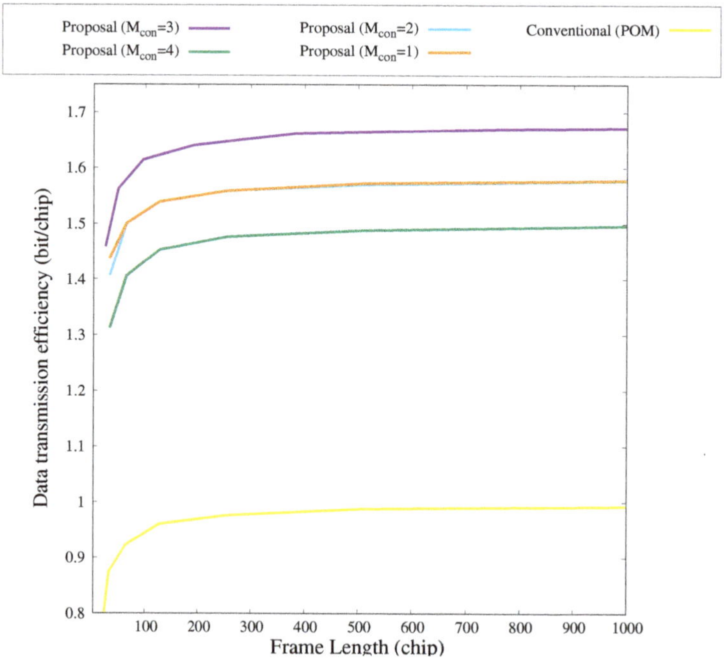

Figure 9. Data transmission efficiency as a function of the frame length.

5. Conclusions

This paper presented a novel strategy to enhance data transmission efficiency in IM/DD-based OWCs and evaluated its effectiveness. The proposed system integrated a parallel OWC structure with a nonorthogonal CSK approach, where CSK codes were generated from concatenated POMs. To assess the fundamental performance of the proposed system, its behavior was analyzed under the influence of background noise. Analytical expressions for data transmission efficiency in a single-user scenario were derived, and numerical results confirmed the improved performance of the proposed framework.

The proposed scheme enhances data transmission efficiency with increasing frame length and the number of parallel codes, demonstrating its adaptability. The optimal concatenation number was identified as $M_{con} = 3$, which provides the best balance between efficiency and performance. The results obtained confirm that the systematically constructed nonorthogonal code effectively enhances the performance of IM/DD OWCs, offering valuable insights for future research and development in this field.

Since the nonorthogonal CSK method proposed by the authors is influenced by multiple factors—such as improved noise immunity with an increased optical scale level, increased inter-code interference with a higher number of concatenations, an expanded available code set through the use of nonorthogonal CSK codes, and a reduced number of viable transmission combinations—we investigated the effects of the proposed scheme and determined the optimal number of concatenations. The optimal concatenation number was identified as $M_con = 3$, providing the best balance between efficiency and performance. The results confirm that the systematically constructed nonorthogonal code effectively enhances the performance of IM/DD OWCs, offering valuable insights for future research and development in this field.

While this study investigated the proposed system in a single-user environment, multiple primitive polynomials can generate an M-sequence, allowing multiple connections to be established by assigning the resulting sequences to different users. This is expected to enable multiple access in the proposed system. Future work includes evaluating the multi-connection performance of the proposed method while considering the effects of atmospheric turbulence, directional errors, turbulence-induced fading, and other factors.

Author Contributions: Conceptualization, N.K. and H.H.; methodology, N.K. and H.H.; formal analysis, N.K.; investigation, N.K.; data curation, N.K.; writing—original draft preparation, N.K. and H.H.; writing—review and editing, project administration, N.K. and H.H. All authors have read and agreed to the published version of the manuscript.

Funding: This work was supported by JSPS KAKENHI Grant Number 22K19770.

Institutional Review Board Statement: Not applicable.

Informed Consent Statement: Not applicable.

Data Availability Statement: Data underlying the results presented in this paper are not publicly available at this time but can be obtained from the authors on reasonable request.

Conflicts of Interest: The authors declare no conflicts of interest.

References

1. Kwon, H. Optical Orthogonal Code-division Multiple-access system – Part I: APD Noise and Thermal Noise. *IEEE Trans. Commun.* **1994**, *42*, 2470–2479. [CrossRef]
2. Qiu, Y.; Chen, S.; Chen, H.; Meng, W. Visible Light Communication based on CDMA Technology. *IEEE Wirel. Commun.* **2018**, *25*, 178–185. [CrossRef]
3. Liu, M.; Wang, T.; Tseng, S. Throughput Performance Analysis of Asynchronous Optical CDMA Networks with Channel Load Sensing Protocol. *IEEE Photonics J.* **2017**, *9*, 7202313. [CrossRef]
4. Chen, S.; Chow, C. Color-shift Keying Code-division Multiple-access Transmission for RGB-LED Visible Light Communications using Mobile Phone Camera. *IEEE Photonics J.* **2014**, *7*, 7904106. [CrossRef]
5. Kiasaleh, K. Performance of APD-based, PPM Free-space Optical Communication Systems in Atmospheric Turbulence. *IEEE Trans. Commun.* **2014**, *53*, 1455–1461. [CrossRef]
6. Zhu, X.; Kahn, J. Free-space Optical Communication through Atmospheric Turbulence Channels. *IEEE Trans. Commun.* **2002**, *50*, 1293–1300.
7. Kumari, M.; Sharma, A.; Chaudhary, S. High-Speed Spiral-Phase Donut-Modes-Based Hybrid FSO-MMF Communication System by Incorporating OCDMA Scheme. *Photonics* **2023**, *10*, 94. [CrossRef]
8. Hadi, M.; Pakravan, M. Analysis and Design of Adaptive OCDMA Passive Optical Networks. *J. Light. Technol.* **2017**, *35*, 2853–2863. [CrossRef]
9. Hien, T.; Pham, T.; Dang, N.T. Analog network coding aided multiuser visible light communication networks using optical CDMA. *OSA Contin.* **2019**, *2*, 2569–2580.

10. Yeh, B.C. Utilizing States of Polarization in One-Dimensional Corite Codes with Two-Code Keying for Optical Code-Division Multiple Access. *Photonics* **2024**, *11*, 819. [CrossRef]
11. Li, J.; Uysal, M. Optical Wireless Communications: System Model, Capacity and Coding. *Proc. IEEE VTC* **2003**, *1*, 168–172.
12. Babar, Z.; Izhar, M.; Nguyen, H.; Botsinis, P.; Alanis, D.; Chandra, D.; Ng, S. Unary-Coded Dimming Control Improves On-Off Keying Visible Light Communication. *IEEE Trans. Commun.* **2018**, *66*, 255–264. [CrossRef]
13. Fuji-Hara, R.; Miao, Y. Optical orthogonal codes: Their bounds and new optimal constructions. *IEEE Trans. Inf. Theory* **2000**, *46*, 2396–2406.
14. Kwong, W.; Perrier, P.; Pruncnal, P. Performance comparison of asynchronous and synchronous code-division multiple-access techniques for fiber-optic local area networks. *IEEE Trans. Commun.* **1991**, *39*, 1625–1634. [CrossRef]
15. Matsushima, T.; Nagano, T.; Ochiai, N.; Teramachi, Y. Generalization of modified prime sequence codes and its properties. *IEICE Trans. Fundam.* **2008**, *E91-A*, 559–573.
16. Yamashita, T.; Hanawa, M.; Tanaka, Y.; Takahara, M. An Optical Code Division Multiplexing System using Hadamard Codes and SIK. *IEICE Tech. Rep.* **1998**, *OCS98-1*, 1–6.
17. Kozawa, Y.; Habuchi, H. Theoretical Analysis of Atmospheric Optical DS/SS with On-off Orthogonal M-sequence Pairs. In Proceedings of the 2007 6th International Conference on Information, Communications & Signal Processing, Singapore, 10–13 December 2007; p. P0686.
18. Islam, A.; Majumder, S.P. Effect of atmospheric turbulence on the BER performance of an optical CDMA FSO link with SIK receiver. *Optik* **2019**, *179*, 867–874. [CrossRef]
19. Kozawa, Y.; Habuchi, H. Theoretical Analysis of Optical Wireless CDMA with Modified Pseudo Orthogonal M-Sequence Sets. In Proceedings of the IEEE Global Telecommunications Conference (GLOBECOM), Honolulu, HI, USA, 30 November–4 December 2009; p. 11170402.
20. Ochiai, N.; Kushibiki, S.; Matsushima, T.; Teramachi, Y. Performance Analysis of Synchronous Optical CDMA System with EWO signaling. *IEICE Trans. Fundam.* **2003**, *J86-A*, 957–968. [CrossRef]
21. Hsieh, C.P.; Chang, C.Y.; Yang, G.C.; Kwong, W. A Bipolar-bipolar Code for Asynchronous Wavelength-time Optical CDMA. *IEEE Trans. Commun.* **2006**, *54*, 2572–2578.
22. Haas, H.; Yin, L.; Wang, Y.; Chen, C. What is LiFi? *J. Light. Technol.* **2016**, *34*, 1533–1544. [CrossRef]
23. Elgala, H.; Mesleh, R.; Haas, H. Indoor optical wireless communication: Potential and state-of-the-art. *IEEE Commun. Mag.* **2011**, *49*, 56–62. [CrossRef]
24. Oubei, H.M.; Shen, C.; Kammoun, A.; Zedini, E.; Park, K.H.; Sun, X.; Liu, G.; Kang, C.H.; Ng, T.K.; Alouini, M.S.; et al. Light based underwater wireless communications. *Jpn. J. Appl. Phys.* **2018**, *57*, 08PA06. [CrossRef]
25. Luo, P.; Ghassemlooy, Z.; Le Minh, H.; Bentley, E.; Burton, A.; Tang, X. Fundamental analysis of a car to car visible light communication system. In Proceedings of the 2014 9th International Symposium on Communication Systems, Networks & Digital Sign (CSNDSP), Manchester, UK, 23–25 July 2014; pp. 1011–1016.
26. Cove, T.; Thomas, J. *Elements of Information Theory*; John Wiley & Sons, Ltd.: Hoboken, NJ, USA, 2005; pp. 689–721.
27. Ansari, I.S.; AlQuwaiee, H.; Zedini, E.; Alouini, M.S. Information Theoretical Limits of Free-Space Optical Links. In *Optical Wireless Communications: An Emerging Technology*; Springer International Publishing: Berlin/Heidelberg, Germany, 2016; pp. 171–208.
28. Otsuki, T. Performance Analysis of Atmospheric Optical PPM CDMA Systems. *J. Light. Technol.* **2003**, *2*, 406–411. [CrossRef]
29. Azim, A.W.; Bazzi, A.; Fatima, M.; Shubair, R.; Chafii, M. Dual-Mode Time Domain Multiplexed Chirp Spread Spectrum. *IEEE Trans. Veh. Technol.* **2023**, *72*, 16086–16097. [CrossRef]
30. Hazu, R.; Kozawa, Y.; Habuchi, H. Visual Secret Sharing Based RGB Parallel Transmission System. In Proceedings of the 24th International Symposium On Wireless Personal Multimedia Communications (WPMC2021), Okayama, Japan, 14–16 December 2021; pp. 1–5.
31. Tokunaga, T.; Habuchi, H.; Kozawa, Y. New two-layered pseudo-noise code for optical-wireless code-shift keying/SCDMA. In Proceedings of the 2017 Eighth International Workshop on Signal Design and Its Applications in Communications (IWSDA), Sapporo, Japan, 24–28 September 2017; pp. 149–153.
32. Komuro, N.; Habuchi, H.; Tsuboi, T. Nonorthogonal CSK/CDMA with Received-Power Adaptive Access Control Scheme. *IEICE Trans. Fundam.* **2008**, *E91-A*, 2779–2786. [CrossRef]
33. Komuro, N.; Habuchi, H. Intensity Modulation Direct Detection Optical Wireless Communication with Nonorthogonal Code Shift Keying. In Proceedings of the IEEE Global Conference on Consumer Electronics (GCCE), Osaka, Japan, 15–18 October 2019; pp. 723–726.

34. Komuro, N.; Habuchi, H. Design and Analysis of Optical Wireless Code Shift Keying with Nonorthogonal Sequences. *OSA Contin.* **2021**, *4*, 1437–1451. [CrossRef]
35. Luong, D.; Thang, T.; Pham, A. Effect of Avalanche Photodiode and Thermal Noises on the Performance of Binary Phase-shift keying-subcarrier-intensity Modulation/free-space Optical Systems over Turbulence Channels. *IET Commun.* **2013**, *7*, 738–744. [CrossRef]

photonics

MDPI

Review

Hybrid Photonic Integrated Circuits for Wireless Transceivers

Tianwen Qian [1,*], Ben Schuler [1], Y. Durvasa Gupta [1], Milan Deumer [1], Efstathios Andrianopoulos [2], Nikolaos K. Lyras [3], Martin Kresse [1], Madeleine Weigel [1], Jakob Reck [1], Klara Mihov [1], Philipp Winklhofer [1], Csongor Keuer [1], Laurids von Emden [1], Marcel Amberg [1], Crispin Zawadzki [1], Moritz Kleinert [1], Simon Nellen [1], Davide de Felipe [1], Hercules Avramopoulos [2], Robert B. Kohlhaas [1], Norbert Keil [1] and Martin Schell [1,4]

1. Fraunhofer Institute for Telecommunications, Heinrich-Hertz-Institute, HHI, Einsteinufer 37, 10587 Berlin, Germany; ben.schuler@hhi.fraunhofer.de (B.S.); ydurvasa.gupta@hhi.fraunhofer.de (Y.D.G.); milan.deumer@hhi.fraunhfoer.de (M.D.); martin.kresse@hhi.fraunhofer.de (M.K.); madeleine.weigel@hhi.fraunhofer.de (M.W.); jakob.reck@hhi.fraunhofer.de (J.R.); klara.mihov@hhi.fraunhofer.de (K.M.); philipp.winklhofer@hhi.fraunhofer.de (P.W.); csongor.keuer@hhi.fraunhofer.de (C.K.); laurids.von.emden@hhi.fraunhofer.de (L.v.E.); marcel.amberg@hhi.fraunhofer.de (M.A.); crispin.zawadzki@hhi.fraunhofer.de (C.Z.); moritz.kleinert@hhi.fraunhofer.de (M.K.); simon.nellen@hhi.fraunhofer.de (S.N.); david.felipe@hhi.fraunhofer.de (D.d.F.); robert.kohlhaas@hhi.fraunhofer.de (R.B.K.); norbert.keil@hhi.fraunhofer.de (N.K.); martin.schell@hhi.fraunhofer.de (M.S.)
2. Photonics Communications Research Laboratory, National Technical University of Athens, 15572 Athens, Greece; efand@mail.ntua.gr (E.A.); hav@mail.ntua.gr (H.A.)
3. Optoelectronics Section, European Space Agency, Estec, Keplerlaan 1, 2201 AZ Noordwijk, The Netherlands; nikolaos.lyras@esa.int
4. Institute of Solid-State Physics, Technische Universität Berlin, Straße des 17. Juni 135, 10623 Berlin, Germany

* Correspondence: tianwen.qian@hhi.frauhofer.de or tianwen.qian@hhi.fraunhofer.de

Abstract: Recent advancements in hybrid photonic integrated circuits (PICs) for wireless communications are reviewed, with a focus on innovations developed at Fraunhofer HHI. This work leverages hybrid integration technology, which combines indium phosphide (InP) active elements, silicon nitride (Si_3N_4) low-loss waveguides, and high-efficient thermal-optical tunable polymers with micro-optical functions to achieve fully integrated wireless transceivers. Key contributions include (1) On-chip optical injection locking for generating phase-locked optical beat notes at 45 GHz, enabled by cascaded InP phase modulators and hybrid InP/polymer tunable lasers with a 3.8 GHz locking range. (2) Waveguide-integrated THz emitters and receivers, featuring photoconductive antennas (PCAs) with a 22× improved photoresponse compared to top-illuminated designs, alongside scalable 1 × 4 PIN-PD and PCA arrays for enhanced power and directivity. (3) Beam steering at 300 GHz using a polymer-based optical phased array (OPA) integrated with an InP antenna array, achieving continuous steering across 20° and a 10.6 dB increase in output power. (4) Demonstration of fully integrated hybrid wireless transceiver PICs combining InP, Si_3N_4, and polymer material platforms, validated through key component characterization, on-chip optical frequency comb generation, and coherent beat note generation at 45 GHz. These advancements result in compact form factors, reduced power consumption, and enhanced scalability, positioning PICs as an enabling technology for future high-speed wireless networks.

Keywords: hybrid integration; wireless transceivers; high-speed wireless communication; photonic integrated circuits

1. Introduction

The rapid growth of global data traffic, driven by emerging technologies such as 5G/6G networks, the Internet of Things (IoT), and artificial intelligence (AI), has created an unprecedented demand for high-speed, high-capacity wireless communication systems [1,2]. Traditional radio-frequency (RF) technologies, while widely deployed, face fundamental limitations in bandwidth, energy efficiency, and signal integrity, particularly as we approach the millimeter-wave (mmWave) terahertz (THz) regime [3–5]. In this context, photonic technologies offer significant advantages [6]. By leveraging the ultra-wide bandwidth capabilities of mature optical fiber communication technology, photonic systems enable the generation and modulation of signals in the optical domain, which can be efficiently converted into electrical signals up to THz range through photomixing [7–9].

Early demonstrations of photonic wireless systems relied on discrete components, such as fiber-coupled lasers, external Mach–Zehnder modulators (MZMs), and free-space optical setups. They achieved impressive data rates, for example, up to 100 Gbit/s in the 100 GHz band [10], but were limited by their large footprint, high power consumption, and susceptibility to alignment issues [3,5].

Recent advancements in photonic integration have enabled the development of chip-scale systems. Here, silicon photonics (SiPh) has gained significant attention due to its compatibility with CMOS processes, which enables low-cost and high-volume production of integrated modulators and detectors [11,12]. However, it faces challenges such as the lack of efficient on-chip light sources and the lack of volumes required for CMOS-style processes. This limitation has spurred the development of hybrid integration strategies, combining silicon with III-V materials like indium phosphide (InP) to incorporate high-performance lasers and amplifiers [13,14].

On the other hand, InP excels in monolithic integration of active components, such as lasers, amplifiers, and high-speed modulators, making them highly suitable for compact and efficient photonic transceivers [15,16]. However, InP is limited in passive component performance, and for that, low-loss materials like silicon nitride (Si_3N_4) [17,18] and polymers [19,20] are better suited. Si_3N_4 is renowned for its ultra-low optical loss [17]. Polymer material platforms, such as PolyBoard [19], offer low-loss waveguides and feature versatile micro-optical benches that can integrate bulk optic functionalities onto the chip. Its various refractive index contract waveguide system enables seamless co-integration with other materials. Additionally, lithium niobate ($LiNbO_3$) has gained attention for its exceptional electro-optic properties, enabling high-speed modulators with low power consumption [21–24]. Furthermore, plasmonics-enabled modulators have demonstrated ultra-high frequency responses, reaching the sub-terahertz range [25].

Fraunhofer HHI has been developing solutions that provide compact form factors, reduced power consumption, and enhanced scalability for such photonic integrated circuits (PICs), for example, using the PolyBoard platform to integrate active components such as lasers, modulators, and photodiodes, as well as nonlinear and nonreciprocal bulk optic functionalities through hybrid integration [26–29].

In this work, we present a new generation of PICs for wireless communications, involving the InP, Si_3N_4, and polymer material platforms. This integration scheme offers fully integrated PICs with building blocks such as tunable lasers, on-chip optical isolators, optical phased arrays (OPAs), modulators, semiconductor optical amplifiers (SOAs), positive-intrinsic-negative photodiodes (PIN-PDs), and waveguide-integrated photoconductive antennas (PCA). The generation of low-phase-noise wireless carriers through the on-chip frequency comb generation and injection locking yields a photonic engine for microwave signal generation. Moreover, the combination of an OPA and InP

PIN-PD-based antenna array yields a directional and reconfigurable antenna array for wireless communication.

2. Key Photonic Subsystems for Wireless Transceiver PICs

Hybrid integration takes the approach of using the advantage of each material platform. Table 1 lists the photonic functionalities offered on the material platforms that are going to be considered in this work for the implementation of the photonic wireless transceivers. The hybrid approach allows for good yield management even for complex photonic integrated circuits, as only known good dies from each platform are used for the hybrid integration. The key photonic subsystems that are going to be discussed are listed below:

- Optical injection-locked laser source for wireless communications
- Photonic THz emitter and receiver
- Beam steering with OPA and optical beam-forming network (OBFN)

Table 1. Qualitative comparison of functionalities offered on different photonic platforms. N/A = not available; + = possible; ++ = good; +++ = very good.

Platforms	Waveguide Loss	Routing Capability	High-Speed Modulation	Tunable Lasers	Optical Isolator	O/E Functionalities (e.g., THz Antenna)
InP	+	+	+++	++	N/A	+++
PolyBoard	++	+++	N/A	+++	+	N/A
Si_3N_4	+++	+++	N/A	+++	N/A	N/A

2.1. Optical Injection-Locked Laser Source for Wireless Communications

Figure 1 illustrates the schematic for generating phase-locked optical signals, a critical step in generating a low-phase-noise mmWave and THz signal by means of photomixing [9]. This process employs the optical injection locking technique [30,31], which ensures the phase coherence and stability in the generated signals. The optical signal from the narrow linewidth master laser passes through two cascaded phase modulators (PMs). These PMs are independently controlled by sinusoidal RF signals, enabling the generation of an optical frequency comb (OFC). The generated OFC is then used to injection-lock the two slave lasers, producing phase-locked optical signals. The optical isolator is incorporated to prevent feedback into the master laser. The required photonic building blocks for this subsystem are discussed in detail below.

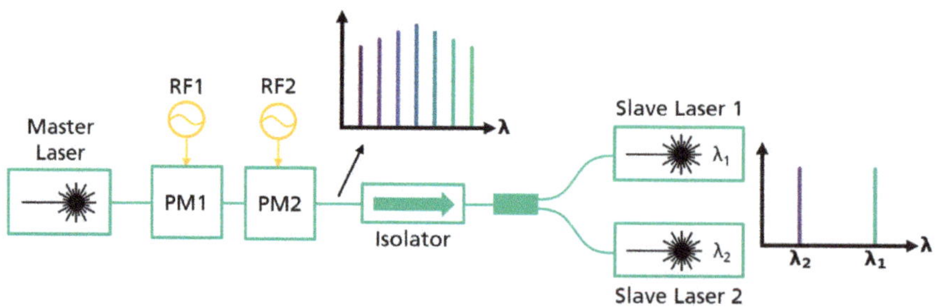

Figure 1. An optical injection locking system for generating phase-locked optical signals.

2.1.1. InP/Polymer Tunable Laser

Hybrid PIC-based external cavity diode lasers (ECDLs) have emerged as reliable, tunable, narrow linewidth laser sources. PICs with Vernier microring resonators (MRR) [32], distributed Bragg reflectors [33–35], Sagnac loop reflectors (SLRs) [36], and Mach–Zehnder interferometers [37] have been fabricated on silicon (Si) [37], silicon dioxide (SiO_2) [38], Si_3N_4 [32,34,35,39,40], polymers [33,41], and other materials [42,43].

Fraunhofer HHI's PolyBoard and InP foundry platforms offer hybrid InP/polymer tunable distributed Bragg reflector (DBR) lasers, operating in the telecom window with a center wavelength of approximately 1550 nm. These lasers offer a tuning range exceeding 20 nm, a narrow linewidth of 185 kHz, and optical power levels of up to 40 mW [41]. The cross-sectional schematic of the InP/polymer tunable laser is shown in Figure 2a. The DBR laser cavity consists of an InP active element serving as the gain medium and a polymer-based section incorporating a thermo-optically tunable phase section and Bragg grating. The left facet of the InP active section is high-reflection (HR) coated against air, while the right facet is anti-reflection (AR) coated against polymers.

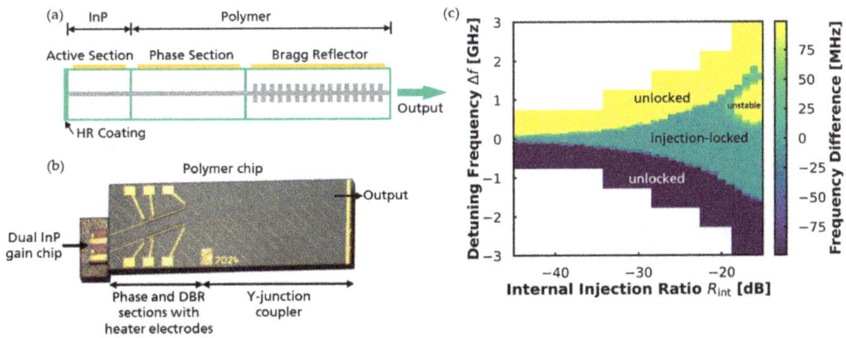

Figure 2. Hybrid InP/polymer tunable DBR laser: (**a**) cross-sectional schematic; (**b**) example micrograph of a dual hybrid InP/polymer tunable DBR laser, combined with a Y-junction coupler to the output; (**c**) spectral analysis for determining the injection locking range as a function of internal injection ratio R_{int} and detuning frequency Δf.

Compared to the other types of ECDLs, this design offers a significant advantage: a large mode-hop-free tuning range of up to 1.5 THz (12 nm) [44], while other types of ECDLs are still limited in the hundreds of GHz range [45,46]. This could enable seamless tuning of photonic-generated wireless carriers targeting mmWave and THz applications. Additionally, its compact footprint allows for easy scalability with arrays of InP gain chips; an example of a dual hybrid InP/polymer DBR laser is shown in Figure 2b. Furthermore, the PolyBoard platform supports additional hybrid integration features, such as on-chip optical isolators. Therefore, we chose this hybrid InP/polymer tunable DBR laser to explore the implementation of optical injection locking systems to generate phase-locked optical signals.

The characterization of hybrid DBR lasers in a master–slave configuration has been experimentally verified to achieve stable injection locking. By defining the detuning frequency Δf as the difference frequency between the master laser and slave laser, and the internal injection ratio $R_{int} = \frac{P_{injection}}{P_{slave}}$ as the ratio of the injected power to the slave laser power, the locking behavior is charted in Figure 2c. A locking range of 0.2 GHz to 3.8 GHz was observed. Under stable injection-locked conditions, the slave laser follows the frequency of the master laser, and the measured frequency difference is less than 20 MHz. The boundaries of the locking range are visible as sharp transitions. Outside the locking range, the frequency difference increases significantly. For a large internal injection ratio R_{int}, a small unstable region is observed within the locking range, corresponding to the undamped relaxation oscillations [47]. In this region, the frequency deviation increases abruptly, as the relaxation oscillations dominate the power distribution and alter the optical frequency.

2.1.2. InP Electro-Optic Modulator

There are several different ways of generating OFCs using integrated photonic approaches, e.g., Kerr soliton combs in high-Q microresonators [48,49], which offer ultra-low phase noise and broad bandwidth but require precise control of pump power and dispersion, mode-locked lasers [48,50], which provide excellent coherence and high repetition rates but have limited tunability due to fixed cavity designs and gain-switched laser combs [51], which are simple and cost-effective with moderate tunability but suffer from spectral flatness and phase noise limitations.

Here, we chose two cascaded InP-based electro-optic modulators, which are modulated with sinusoidal RF driving currents and thus generate an OFC [52–54]. This offers a highly flexible and frequency-tunable approach, making them particularly suitable for microwave photonics applications. E/O combs can be precisely controlled via microwave driving signals, enabling tailored comb spacing and phase coherence, which are critical for generating high-quality microwave signals.

The generic InP foundry integration platform at HHI provides high-bandwidth traveling-wave modulators based on the quantum-confined Stark effect (QCSE) [52,55]. Additionally, SOAs can be integrated directly after the phase modulator to enhance the power of each comb line. Figure 3a shows the InP chip with two cascaded phase modulators followed by a SOA, and an example of a generated OFC is shown in Figure 3b.

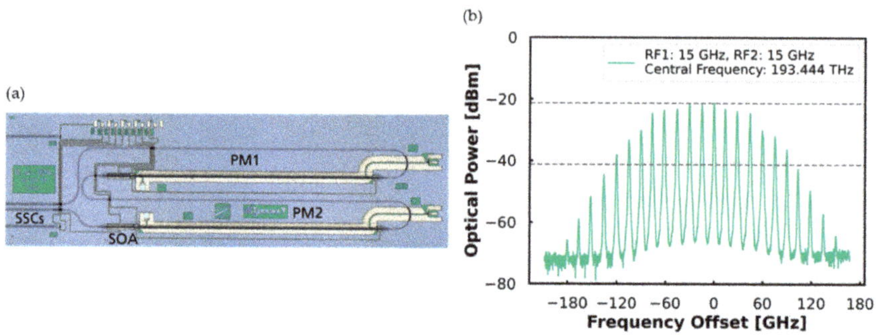

Figure 3. (**a**) InP chip with two cascaded PMs for generating OFCs; SOA is integrated after the second PM, and the coupling interfaces are facilitated by spot-size converters (SSCs); (**b**) example of generated OFC spectrum. The measured OFC would support a bandwidth of 225 GHz.

Based on the traveling wave electrode structure, high-speed MZMs have been optimized and offered as a standard building block on the InP foundry platform [15]. Figure 4 shows their typical DC and RF characteristics.

Figure 4a presents the extinction ratio (ER) map of a MZM with a 5 mm long active section for a TE-polarized optical input at 1550 nm. U and I represent the voltages applied across the arms of the electrodes. The half-wave voltage (V_π) is 1.4 V at a bias voltage of −10 V, corresponding to a $V\pi \cdot L$ value of 0.7 V·cm. The measured maximum ER is approximately 31 dB. Figure 4b illustrates the small-signal RF response of the MZMs with active lengths of 3 mm, 4 mm, and 5 mm, with corresponding 3 dB bandwidths of 42 GHz, 34 GHz, and 32 GHz, respectively. Figure 4c depicts the large-signal characterization of the MZM with a 5 mm-long active section for 64 Gbps NRZ, 80 Gbps NRZ, and 24 GBd PAM4.

The extinction ratios for these modulation formats are 9.3 dB, 8 dB, and 5.5 dB, respectively. For more details on the characterization, please refer to [15].

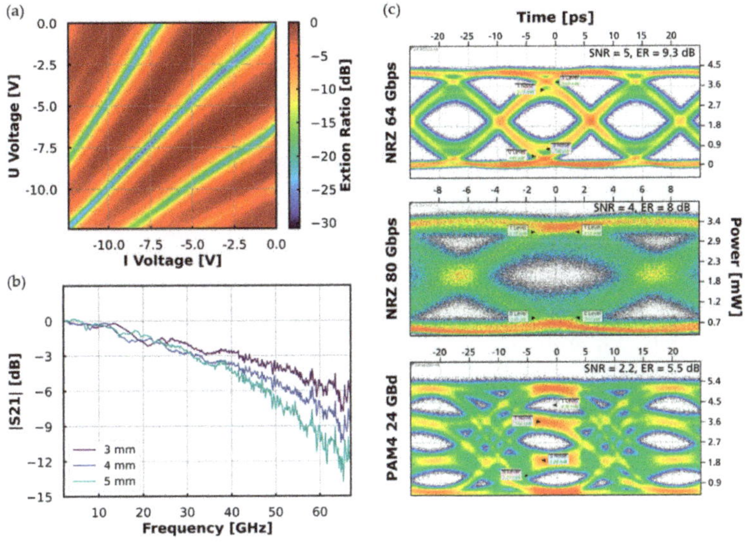

Figure 4. Characterization of an InP MZM: (**a**) DC ER map of a 5 mm long MZM [15]; (**b**) high-frequency response of InP MZM with different lengths [15]; (**c**) eye diagrams of a 5 mm long MZM for NRZ 64 Gbps, NRZ 80 Gbps, PAM4 24 GBd. [15].

2.1.3. Optical Isolators

Optical isolators are essential in the injection locking technique to prevent feedback from the slave lasers, ensuring stable operation of the system. The PolyBoard platform's micro-optical bench allows for the integration of bulk nonreciprocal optical elements directly into a photonic chip. The isolator incorporates a Faraday rotator sandwiched between a 0° polarization filter and a 45° polarization filter, followed by a 22.5° half-wave plate (HWP) to ensure TE polarization in the forward transmission direction. The optical coupling between the waveguide and free space is optimized through gradient-index (GRIN) lenses, which can be placed in the etched U-groove structures on the PolyBoard. Figure 5a shows a close-up picture of the integrated optical isolator. Over 30 dB isolation across the C-band with an excess loss below 1.5 dB can be achieved, as shown in Figure 5b [28,29,56].

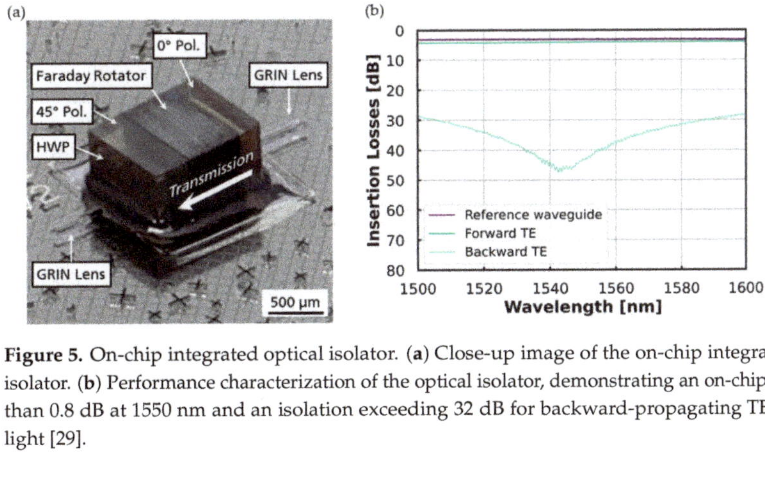

Figure 5. On-chip integrated optical isolator. (**a**) Close-up image of the on-chip integrated optical isolator. (**b**) Performance characterization of the optical isolator, demonstrating an on-chip loss of less than 0.8 dB at 1550 nm and an isolation exceeding 32 dB for backward-propagating TE-polarized light [29].

2.2. Photonic THz Emitter and Receiver

By superimposing the two phase-locked optical signals, a stable optical beat note is generated, which is then converted into a THz signal using a photomixer. On the receiver side, the beat note serves as a local oscillator to down-convert the incident THz signal to an intermediate frequency for electronic acquisition. For the transmitter (Tx), commonly PIN-PDs or uni-traveling-carrier (UTC) PDs are employed as photomixers [57–60], while photoconductive antennas (PCAs) are better suited for the receiver (Rx) [61–63].

For our transmitter, waveguide-integrated PIN-PDs, based on mature processes originally developed for fiber-optic communication systems, have demonstrated coherent links with data rates of up to 160 Gbit/s and well-defined radiation patterns [64,65]. For coherent communication, an InP in-phase/quadrature (I/Q) modulator imprints the data onto one of the optical signals before superposition, as shown in Figure 6. An I/Q modulator is typically constructed using two parallel MZMs. All these building blocks, i.e., lasers, MZMs, and PDs, are readily available on the InP platform.

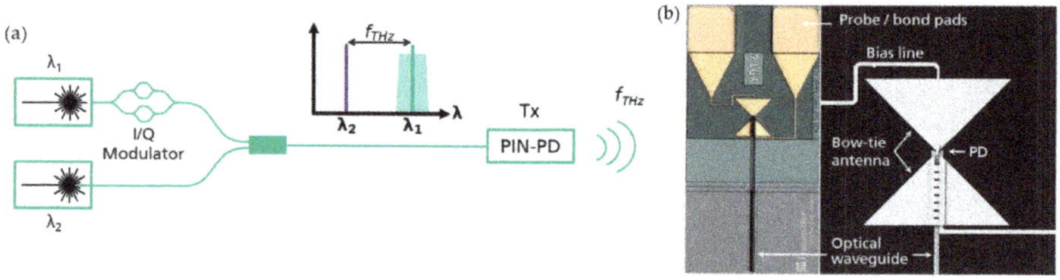

Figure 6. Photomixing using a waveguide-integrated PIN-PD and device micrographs. (**a**) Schematic of the photomixing scheme for an I/Q-modulated optical beat signal using a PIN-PD. (**b**) Micrograph and scanning electron micrograph of a waveguide-integrated PIN-PD chip with bow-tie antenna for THz radiation.

On the receiver side, as shown in Figure 7, the antenna receives the incident THz signal, inducing a voltage variation at frequency f_{THz} across the photoconductor located at the antenna's feed point. At the same time, the optical beat signal fed into the photoconductor acts as a local oscillator (LO) with the difference frequency of the two phase-locked laser signals f_{LO}, which modulates the conductance of the photoconductor. This oscillating conductance in the feeding point of the antenna leads to a down-conversion of the THz signal to an intermediate frequency (IF) $f_{IF} = |f_{THz} - f_{LO}|$ that can be measured with conventional electronics. Note that photonic integration requires waveguide-integrated PCAs and not traditional top-illuminated designs [66]. By integrating the PCA with optical waveguides, evanescent coupling enhances the light coupling efficiency into the photoconductive material. Unlike top-illuminated PCAs, which suffer from reflections at the air-semiconductor interface, waveguide-integrated PCAs minimize these losses by coupling light from below, resulting in improved optical absorption and higher output signal strength. Recent developments in waveguide-integrated PCAs have demonstrated up to 22-fold improvement in photoresponse compared to state-of-the-art top-illuminated designs [66], enabling flexible integration of THz receivers into photonic integrated circuits.

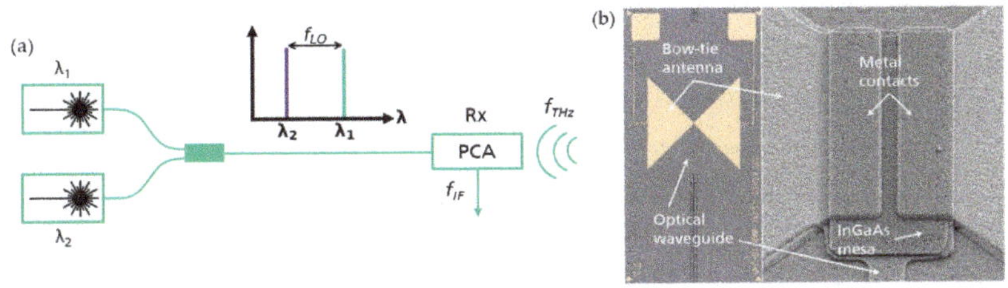

Figure 7. Photomixing using a waveguide-integrated PCA and device micrographs. (**a**) Schematic of the photomixing process for an optical beat signal using a PCA. (**b**) Micrograph and scanning electron micrograph of the novel waveguide-integrated PCA chip [67].

Scaling-Up with Antenna Arrays

To address the limitations of emitted power and antenna directivity [59,65,68–70] in mmWave and THz communication systems, arrays of antennas are implemented. The antenna pitch must be smaller than the targeting wavelength to achieve beam steering capability, necessitating waveguide-integrated designs to enable the required compact spacing, which discrete components cannot provide. This approach enables scalable and high-performance THz transmitters and receivers. A 1 × 4 THz emitter array based on PIN-PDs integrated with SOA [68] and a 1 × 4 THz receiver array based on PCAs are depicted in Figures 8 and 9, respectively. SSCs are included on the chip facet to facilitate the coupling towards standard fiber array or other chip platforms.

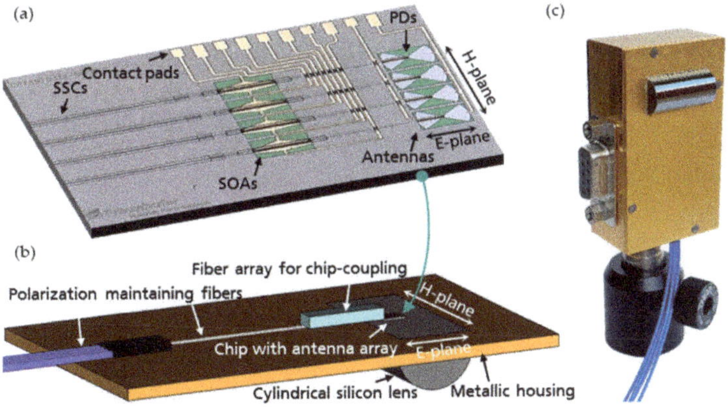

Figure 8. Fabricated 1 × 4 THz emitter array and its packaging. (**a**) Micrograph of the fabricated 1 × 4 THz emitter array based on PIN-PDs and integrated with SOAs [68]. (**b**) The CAD model of the packaged antenna array [68]. (**c**) The packaged module [68].

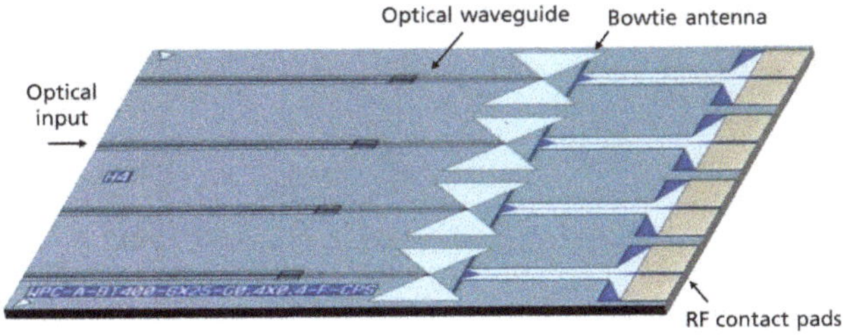

Figure 9. Micrograph of fabricated 1 × 4 THz receiver array based on waveguide integrated PCA.

2.3. Beam Steering with OPA and OBFN

Beam steering can be achieved by integrating OPAs with antenna arrays, where the phase of each individual antenna element is precisely controlled [71]. Figure 10 illustrates a schematic design of such an OPA, which utilizes two phase-locked optical signals as inputs. Power-splitting elements, such as multimode interferometers (MMIs), distribute the signals to the desired number of antennas. To maintain phase correlation between the two signals, the optical path lengths must be identical, necessitating low-loss waveguide routing. By introducing phase shifting in one of the optical signals, a wide steering angle can be achieved with tuning speeds in the MHz and GHz range [69,72,73].

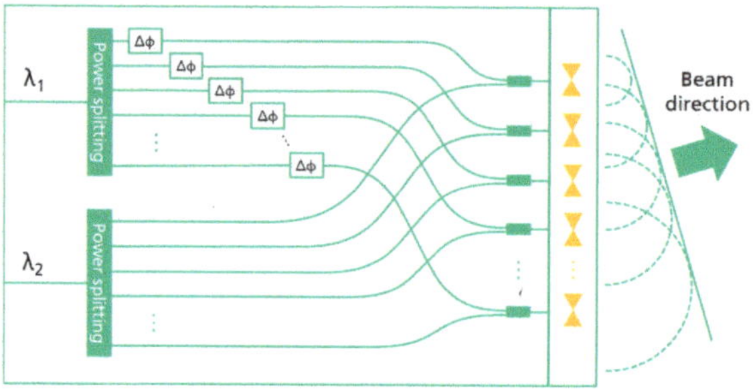

Figure 10. Illustration of the schematic design of OPA targeted for THz beam steering.

The PolyBoard platform offers waveguide propagation loss down to 0.8 dB/cm [19] and also includes building blocks, such as efficient thermo-optical phase shifters with low crosstalk. These phase shifters leverage the platform's high thermo-optical coefficient and low thermal conductivity, outperforming other photonic integration platforms [68].

Figure 11 shows a packaged 1 × 4 OPA array based on the PolyBoard platform. By controlling it in conjunction with the 1 × 4 THz emitter array shown in Figure 8, an increased output power of 10.6 dB, higher directivity, and continuous beam steering across 20° at 300 GHz have been successfully demonstrated in [68]. Note that the use of a silicon lens imposes a limitation on the steering angle due to total internal reflection within the silicon. To achieve wider steering angles, dielectric rod waveguide (DRW) antennas have been explored, demonstrating compatibility with the hybrid integration approach [74,75]. Alternative methods, such as metasurface-based beam steering, offer the potential for compact designs with wide-angle steering capabilities; however, many of these approaches remain largely theoretical [76,77]. For a larger antenna array, to overcome the beam squint effect [78–80], using an optical Blass matrix-based OBFN to introduce time delay into the signal is also possible with the platform solution [81–83].

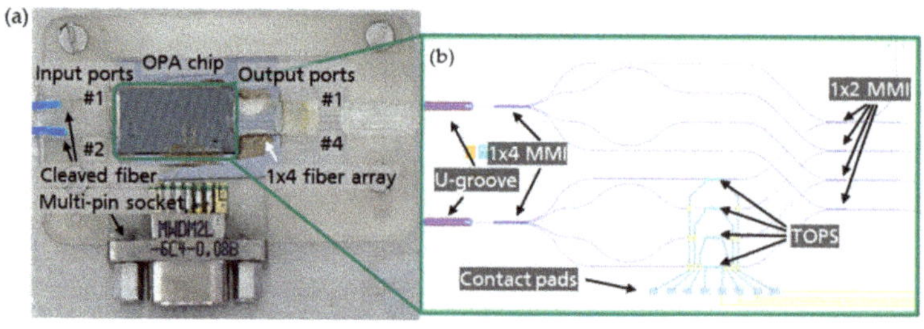

Figure 11. (**a**) Photograph of the assembled and packaged OPA chip. Wire bonding to a multi-pin plug allows stable connection and driving of the OPA in the beam steering testbed. The augmented layout (**b**) shows the polymer-based OPA chip with four thermo-optical phase shifters (TOPS) to set the phase of the beat signal at the respective output port [68].

3. Towards Full-Photonic Wireless Transmitter and Receiver PICs

3.1. Optical Interfacing Between PIC Technologies

Moreover, integrating the aforementioned photonic building blocks into the hybrid PIC requires efficient coupling between each platform. Numerical simulations were performed using ANSYS Lumerical MODE (version: 2024 R1), optimizing the waveguide width for each material platform to achieve optimal mode overlap and maximize mode-field diameters, thereby enhancing coupling tolerance. Table 2 provides an overview of the waveguide properties for the different material platforms used in the transceiver PICs. Parameters such as n_{eff} are the effective refractive index; $MFD_{x,y}$ are mode-filed diameters where the optical power has dropped to $1/e^2$ of the maximum intensity. The waveguide width varied in the simulation is listed, as well as the simulated coupling loss towards the PolyBoard. Figure 12 shows the schematic of the waveguide cross-section of (a) InP SSC, (b) PolyBoard, and (c) Si_3N_4 [17], and their corresponding simulation results.

Table 2. Waveguide properties of each material platform. N/A = not applicable.

Platforms	n_{eff} @1550 nm	$MFD_{x,y}$ [μm]	Waveguide Width [μm]	Simulated Coupling Loss Towards PolyBoard [dB]
InP [16]	3.2	(3.5, 4.4)–(10.0, 7.0)	[0.1–1.0]	1.3
PolyBoard	1.46–1.45	(4.3, 4.3)–(9.0, 7.8)	[0.8–3.2]	N/A
Si_3N_4 [17]	1.535–1.446	(1.5, 1.2)–(10.0, 10.0)	[0.8–3.7]	0.1

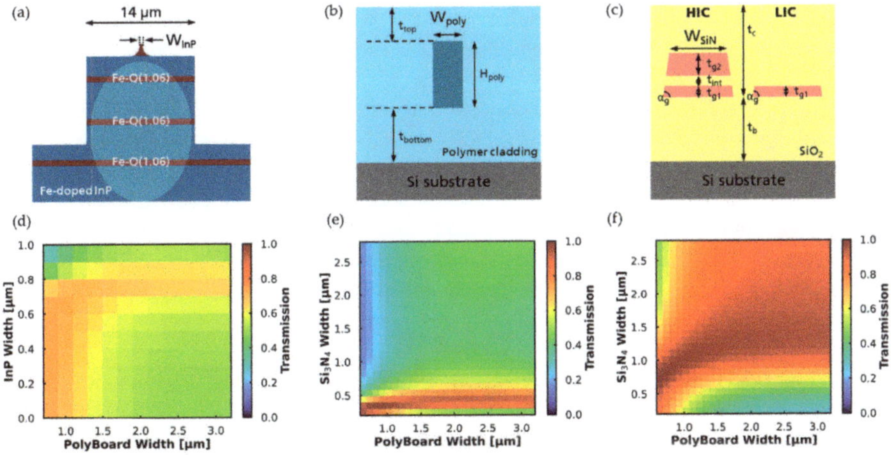

Figure 12. Cross-section of (**a**) InP SSC; (**b**) PolyBoard taper, with typical layer thicknesses for the polymer top cladding (t_{top}) and polymer bottom cladding (t_{bottom}) of 40 μm and 80 μm, respectively; (**c**) Si_3N_4 ADS waveguide for HIC (left) and LIC (right), with bottom stripe thickness t_{g1} = 75 nm, intermediate layer thickness t_{int} = 100 nm, top stripe thickness t_{g2} = 175 nm, and an etching angle of α_g = 82°. The typical layer thicknesses for the SiO_2 bottom cladding (t_b) and the top cladding (t_c) are 8 μm and 15 μm [17]. Simulation of power transmission between different waveguide interfaces: (**d**) InP and PolyBoard; (**e**) Si_3N_4 HIC and PolyBoard; (**f**) Si_3N_4 LIC and PolyBoard.

The SSC on the InP platform is implemented by inverse-tapering the InP Q(1.06) waveguide width, W_{InP}, enabling evanescent coupling to the substrate mode within the so-called dilute waveguide [16]. This waveguide is formed by three layers of Fe-doped passive InP Q(1.06) waveguides. The PolyBoard taper is achieved through an inverse taper by varying the waveguide width, W_{poly}. A refractive index contrast of Δn = 0.03

(n_{core} = 1.48, $n_{cladding}$ = 1.45) is used, while the waveguide height, H_{poly}, is fixed at 3.2 μm. For the Si_3N_4 platform, the asymmetric double-stripe (ADS) waveguide offers the natural transition between high-index-contract (HIC) and low-index-contract (LIC) waveguides, combining small bending radii with ultra-low-loss propagation [17]. In this case, the width of the Si_3N_4 waveguide, W_{SiN}, is varied to optimize coupling with the PolyBoard waveguide.

Figure 12d shows that the highest transmission between InP and the polymer occurs at a polymer waveguide width of 0.8 μm, corresponding to a coupling loss of 1.3 dB. Additionally, the transmission remains nearly constant for InP widths ranging from 0.1 μm to 0.65 μm, ensuring robustness against width variations caused by etching or lithography. An InP waveguide width of 0.25 μm is chosen, allowing for an etching tolerance of ±100 nm. For both Si_3N_4 HIC and LIC waveguides, optimal coupling is achieved with a 0.8 μm polymer waveguide, resulting in a coupling loss of 0.1 dB. Although simulations suggest additional viable options, we selected the 0.8 μm polymer waveguide for interfacing with InP and Si_3N_4 waveguides to maintain design simplicity.

3.2. *Hybrid Wireless Transmitter PIC*

With the above-mentioned photonic building blocks, the first prototypes of the hybrid wireless transmitter and receiver PIC have been demonstrated. Both PICs use PolyBoard as the motherboard for tunable lasers, waveguide routing with optical isolators, and interfacing with InP and Si_3N_4 chips. The other active photonic functions, such as the OFC generation, I/Q data modulation, and THz emitter and receivers, as well as the passive functions, such as OBFN based on Si_3N_4, are integrated on different chiplets. They are butt-coupled on the west and east sides of the PolyBoard based on the current assembly limitations. The active coupling method is used during the coupling process to spare the additional alignment loops for each interface and platform.

Figure 13 shows the assembled wireless transmitter PIC. Starting from the northwest side, four hybrid InP/Polymer tunable DBR lasers are integrated; one is implemented as the master laser, and the rest are implemented as slave lasers for generating phase-locked optical signals. An InP modulator chip is placed on the northeast side with two phase modulators for OFC generation, one I/Q modulator, and one MZM for generating modulated carriers. The phase-locked optical signals are then routed towards the OBFN chip on the southeast side with beam-forming functionality. Finally, the processed optical signals are received by an InP antenna chip with a SOA integrated wireless emitter array in a 4 × 4 configuration.

The assembled PIC is characterized by testing its coupling with various interfaces using a fiber array connected to the PolyBoard chip, with waveguide tap couplers (average insertion loss of 16.6 dB in TE polarization) enabling measurements. A fiber-to-chip coupling loss of 1.0 dB per facet is assumed during the characterization process. Figure 14a,b show the spectra of the hybrid tunable DBR lasers with varying heating power on the grating section, measured after passing through the InP modulator chip. Figure 14c shows the spectra of the slave laser measured directly through the fiber array. Additionally, Figure 14d validates the successful coupling of the InP antenna chip by comparing fiber-chip and chip-chip characterization before and after the assembly. The outermost SOAs are measured.

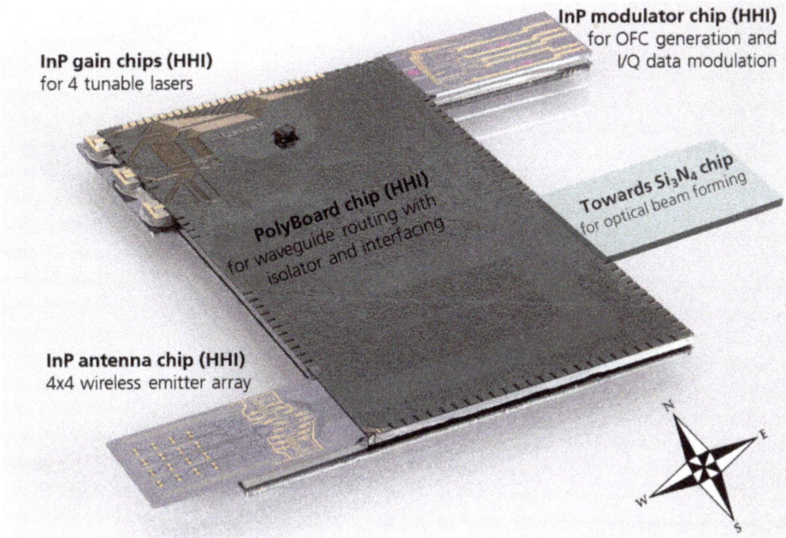

Figure 13. Hybrid wireless transmitter PIC integrating InP, PolyBoard, and Si_3N_4 platforms.

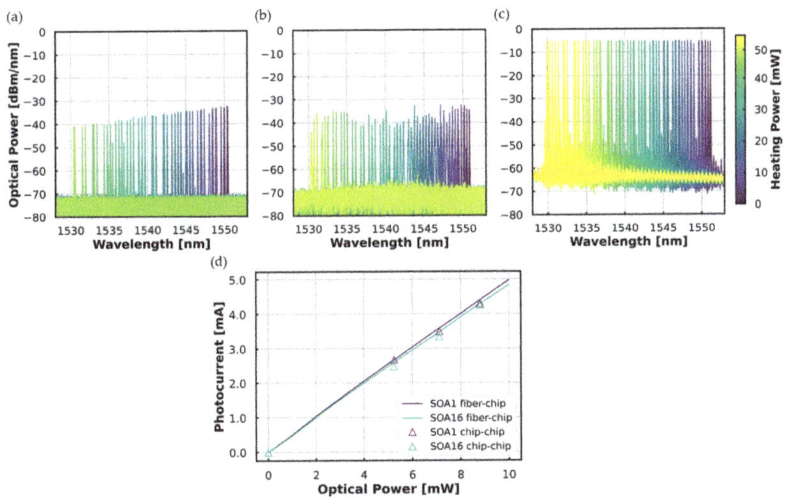

Figure 14. Characterization of the assembled THz transmitter PIC. By varying the power applied onto the grating heater, tuning spectra have been recorded: (**a**) the master laser going through the InP modulator PIC; (**b**) the slave laser going through the InP modulator PIC; (**c**) the slave laser measured directly using a fiber array. (**d**) Comparison of fiber-to-chip and chip-to-chip characterization based on the photocurrent generated in the outermost SOAs at varying optical power levels.

3.3. Hybrid Wireless Receiver PIC

The receiver PIC follows a configuration similar to that of the transmitter PIC. The InP modulator chip incorporates two cascaded phase modulators to generate the OFC. Instead of using an optical beamforming network, a phased array based on phase shifters and MMIs is integrated on the PolyBoard. This phased array is coupled to the single-channel and two-channel InP antenna chips based on waveguide integrated PCAs, which are located on the southeast side of the PolyBoard chip. The assembled wireless receiver PIC is shown in Figure 15.

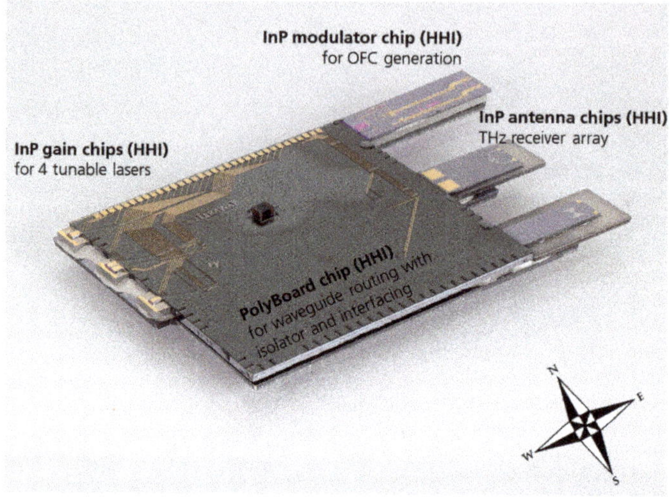

Figure 15. Hybrid wireless receiver PIC integrating InP and PolyBoard platforms.

The receiver PIC is characterized in a similar manner using a fiber array coupled to the PolyBoard chip. Figure 16a–c show the spectra of the master laser and two slave lasers with varying the heating power on the grating section, while Figure 16d displays the I-V characteristic of the coupled PCA chip. Additionally, the receiver PIC is further characterized to validate the on-chip optical injection locking [84]. As shown in Figure 16e, the OFC is generated by operating the master laser and applying 7.5 GHz RF signals to the cascaded phase modulators. The resulting OFC is measured through the fiber array accessed via the test ports. By locking the slave lasers to the third-order harmonics of the comb, a phase-locked optical beat note at 45 GHz is demonstrated in Figure 16f.

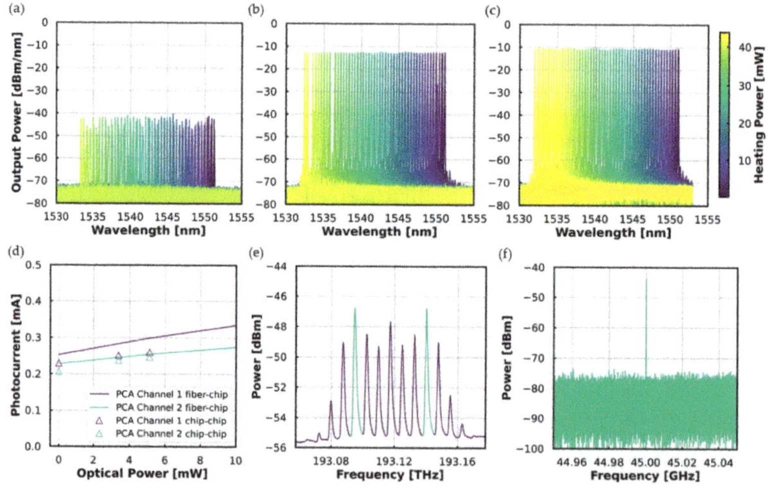

Figure 16. Characterization of the assembled THz receiver PIC. By varying the power applied onto the grating heater, tuning spectra have been recorded: (**a**) The master laser going through the InP modulator chip; (**b**) the slave laser 1; (**c**) the slave laser 2. (**d**) Comparison between fiber-chip and chip-chip characterization of the two-channel PCA at different optical power levels. (**e**) On-chip OFC with a spacing of 7.5 GHz is generated through the cascaded phase modulators and measured through the test ports. The comb lines marked in green are used for injection locking [84]. (**f**) Phase-locked optical beat note measured at 45 GHz [84].

4. Discussion

While the prototype of hybrid wireless transceiver PICs demonstrates significant advancements in realizing complex photonic systems through hybrid integration, several challenges remain, including power budget constraints, thermal management, packaging, and performance optimization of individual building blocks.

Taking the hybrid wireless receiver PIC as an example, the target optical power at the waveguide-integrated PCA exceeds 0 dBm. In an ideal scenario, the hybrid DBR laser is expected to deliver 13 dBm output power. Considering the passive components on the PolyBoard—including the phase shifter network (approximated as three 3-dB couplers, resulting in ~9 dB loss), waveguide propagation losses (~2 dB), and simulated coupling loss (~1.3 dB)—the power budget is already near its limit. This imposes strict requirements on coupling accuracy, the performance of individual chiplets, and overall thermal management, given that all individual components were characterized under controlled lab conditions. Additionally, the limited OFC bandwidth achieved in this work highlights packaging challenges, especially when photonic components require RF connectivity.

Future work should consider power budget optimizations with strategic trade-offs. One potential approach is to distribute complex photonic functionalities across subsystems and incorporate multiple SOAs at different stages to enhance the optical power delivered to antenna arrays, thereby increasing the effective radiated power. Furthermore, the hybrid integration approach can be expanded to include alternative material platforms, such as $LiNbO_3$, to achieve broader OFCs extending into the THz range, enhancing bandwidth and system flexibility. While scaling to larger transmitter and receiver antenna arrays is a natural progression, robust packaging solutions and effective thermal management strategies will be crucial to ensuring reliable operation and efficient heat dissipation.

5. Conclusions

This paper has reviewed the recent development of hybrid PICs for wireless communications at Fraunhofer HHI, focusing on the integration of key photonic subsystems: injection-locked laser sources, photonic THz emitters/receivers, and OPAs. We demonstrated cascaded PMs based on InP, generating OFCs with a 225 GHz bandwidth, and hybrid InP/polymer tunable lasers achieving a locking range of 3.8 GHz in a master–slave configuration. Waveguide-integrated PCAs on the InP platform showed a 22-fold improvement in photoresponse compared to traditional top-illuminated designs, while 1×4 arrays of PIN-PDs and PCAs were implemented for scalable wireless communication. The integration of polymer-based OPAs with 1×4 InP PIN-PD antenna arrays enabled continuous beam steering across 20° at 300 GHz, with an increased output power of 10.6 dB and improved directivity. Prototypes of hybrid wireless transceiver PICs were successfully assembled and experimentally demonstrated a phase-locked optical beat note at 45 GHz via on-chip OFC generation. These transceivers integrate tunable lasers, on-chip OFC generation, I/Q modulators, and THz emitter/receiver arrays, demonstrating the feasibility of fully integrated PICs for high-speed wireless communication.

In conclusion, the advancements presented in this work demonstrate the potential of hybrid PICs as a key enabler for next-generation wireless networks in the mmWave and THz regimes. By addressing power budget challenges through multi-stage SOA integration, leveraging wider OFCs, and scaling up antenna arrays, hybrid photonic integration technology is poised to play a pivotal role in shaping the future of high-speed wireless communication.

Author Contributions: Conceptualization, T.Q. and D.d.F.; methodology, T.Q.; software, T.Q. and B.S.; validation, T.Q., B.S. and D.d.F.; formal analysis, T.Q. and B.S.; investigation, T.Q. and B.S.; resources, T.Q. and B.S.; data curation, T.Q., B.S., Y.D.G., M.D., S.N., E.A. and N.K.L.; writing—original draft preparation, T.Q.; writing—review and editing, D.d.F., M.S., B.S., Y.D.G., M.D., S.N., M.K. (Martin Kresse), M.W., J.R., K.M., P.W., C.K., L.v.E. and M.K. (Moritz Kleinert); visualization, T.Q. and M.A.; supervision, M.S., N.K., R.B.K. and H.A.; project administration, T.Q., C.Z. and D.d.F.; funding acquisition, D.d.F., C.Z., N.K., R.B.K. and H.A. All authors have read and agreed to the published version of the manuscript.

Funding: This research was funded by the European Commission, through the EU Horizon 2020 project TERAWAY (Grant agreement ID: 871668).

Institutional Review Board Statement: Not applicable.

Informed Consent Statement: Not applicable.

Data Availability Statement: The data related to the paper are available from the corresponding authors upon reasonable request.

Acknowledgments: The authors acknowledge the EU Horizon 2020 project TERAWAY (Grant agreement ID: 871668), the EU Horizon Europe projects SPRINTER (Grant agreement ID: 101070581) and TERA6G (Grant agreement ID: 101096949), all funded by the European Commission, for their support of broader research efforts in this field. The numerical simulations in this work were performed using ANSYS Lumerical software under a commercial license authorized by CADFEM Germany GmbH. We acknowledge ANSYS, Inc. for providing access to their simulation tools under proper licensing agreements. We also sincerely thank Robert Grootjans (r.grootjans@lionix-int.com), Ilka Visscher (i.visscher@lionix-int.com), Ahmad W. Mohammad (a.w.m.mohammad@lionix-int.com), and Chris Roeloffzen (c.g.h.roeloffzen@lionix-int.com) from LioniX International BV for their valuable knowledge contributions on the Si_3N_4 chip used in this study.

Conflicts of Interest: The authors declare no conflicts of interest.

References

1. Leitenstorfer, A.; Moskalenko, A.S.; Kampfrath, T.; Kono, J.; Castro-Camus, E.; Peng, K.; Qureshi, N.; Turchinovich, D.; Tanaka, K.; Markelz, A.G.; et al. The 2023 terahertz science and technology roadmap. *J. Phys. D Appl. Phys.* **2023**, *56*, 223001. [CrossRef]
2. Song, H.-J.; Lee, N. Terahertz Communications: Challenges in the Next Decade. *IEEE Trans. Terahertz Sci. Technol.* **2022**, *12*, 105–117. [CrossRef]
3. Nagatsuma, T.; Ducournau, G.; Renaud, C.C. Advances in terahertz communications accelerated by photonics. *Nat. Photonics* **2016**, *10*, 371–379. [CrossRef]
4. Song, H.-J.; Nagatsuma, T. Present and Future of Terahertz Communications. *IEEE Trans. Terahertz Sci. Technol.* **2011**, *1*, 256–263. [CrossRef]
5. Nagatsuma, T.; Carpintero, G. Recent Progress and Future Prospect of Photonics-Enabled Terahertz Communications Research. *IEICE Trans. Electron.* **2015**, *98*, 1060–1070. [CrossRef]
6. Seeds, A.J.; Shams, H.; Fice, M.J.; Renaud, C.C. TeraHertz Photonics for Wireless Communications. *J. Lightwave Technol.* **2015**, *33*, 579–587. [CrossRef]
7. Degli-Eredi, I.; An, P.; Drasbæk, J.; Mohammadhosseini, H.; Nielsen, L.; Tønning, P.; Rommel, S.; Monroy, I.T.; Heck, M.J.R. Millimeter-wave generation using hybrid silicon photonics. *J. Opt.* **2021**, *23*, 43001. [CrossRef]
8. Safian, R.; Ghazi, G.; Mohammadian, N. Review of photomixing continuous-wave terahertz systems and current application trends in terahertz domain. *Opt. Eng.* **2019**, *58*, 1. [CrossRef]
9. Brown, E.R. THz Generation by Photomixing in Ultrafast Photoconductors. *Int. J. High Speed Electron. Syst.* **2003**, *13*, 497–545. [CrossRef]
10. Li, X.; Yu, J.; Zhang, J.; Dong, Z.; Li, F.; Chi, N. A 400G optical wireless integration delivery system. *Opt. Express* **2013**, *21*, 18812–18819. [CrossRef]
11. Siew, S.Y.; Li, B.; Gao, F.; Zheng, H.Y.; Zhang, W.; Guo, P.; Xie, S.W.; Song, A.; Dong, B.; Luo, L.W.; et al. Review of Silicon Photonics Technology and Platform Development. *J. Lightwave Technol.* **2021**, *39*, 4374–4389. [CrossRef]
12. Rahim, A.; Hermans, A.; Wohlfeil, B.; Petousi, D.; Kuyken, B.; van Thourhout, D.; Baets, R. Taking silicon photonics modulators to a higher performance level: State-of-the-art and a review of new technologies. *Adv. Photonics* **2021**, *3*, 24003. [CrossRef]

13. Besancon, C.; Néel, D.; Make, D.; Ramírez, J.M.; Cerulo, G.; Vaissiere, N.; Bitauld, D.; Pommereau, F.; Fournel, F.; Dupré, C.; et al. AlGaInAs Multi-Quantum Well Lasers on Silicon-on-Insulator Photonic Integrated Circuits Based on InP-Seed-Bonding and Epitaxial Regrowth. *Appl. Sci.* **2022**, *12*, 263. [CrossRef]
14. Xiang, C.; Jin, W.; Huang, D.; Tran, M.A.; Guo, J.; Wan, Y.; Xie, W.; Kurczveil, G.; Netherton, A.M.; Liang, D.; et al. High-Performance Silicon Photonics Using Heterogeneous Integration. *IEEE J. Sel. Top. Quantum Electron.* **2022**, *28*, 8200515. [CrossRef]
15. Gupta, Y.D.; Binet, G.; Diels, W.; Abdeen, O.; Gaertner, T.; Baier, M.; Schell, M. Implementation, Modelling and Verification of High-Speed Mach-Zehnder Phase Modulators in an Open Access InP Foundry Platform. *J. Lightwave Technol.* **2023**, *41*, 3498–3504. [CrossRef]
16. Soares, F.; Baier, M.; Gaertner, T.; Grote, N.; Moehrle, M.; Beckerwerth, T.; Runge, P.; Schell, M. InP-Based Foundry PICs for Optical Interconnects. *Appl. Sci.* **2019**, *9*, 1588. [CrossRef]
17. Roeloffzen, C.G.H.; Hoekman, M.; Klein, E.J.; Wevers, L.S.; Timens, R.B.; Marchenko, D.; Geskus, D.; Dekker, R.; Alippi, A.; Grootjans, R.; et al. Low-Loss Si3N4 TriPleX Optical Waveguides: Technology and Applications Overview. *IEEE J. Sel. Top. Quantum Electron.* **2018**, *24*, 4400321. [CrossRef]
18. Buzaverov, K.A.; Baburin, A.S.; Sergeev, E.V.; Avdeev, S.S.; Lotkov, E.S.; Bukatin, S.V.; Stepanov, I.A.; Kramarenko, A.B.; Amiraslanov, A.S.; Kushnev, D.V.; et al. Silicon Nitride Integrated Photonics from Visible to Mid-Infrared Spectra. *Laser Photonics Rev.* **2024**, *18*, 2400508. [CrossRef]
19. Kleinert, M.; Nuck, M.; Conradi, H.; De Felipe, D.; Kresse, M.; Brinker, W.; Zawadzki, C.; Keil, N.; Schell, M. A platform approach towards hybrid photonic integration and assembly for communications, sensing, and quantum technologies based on a polymer waveguide technology. In Proceedings of the 2019 IEEE CPMT Symposium Japan (ICSJ), Kyoto, Japan, 18–20 November 2019; pp. 25–30.
20. Dangel, R.; La Porta, A.; Jubin, D.; Horst, F.; Meier, N.; Seifried, M.; Offrein, B.J. Polymer Waveguides Enabling Scalable Low-Loss Adiabatic Optical Coupling for Silicon Photonics. *IEEE J. Sel. Top. Quantum Electron.* **2018**, *24*, 8200211. [CrossRef]
21. Chen, G.; Gao, Y.; Lin, H.-L.; Danner, A.J. Compact and Efficient Thin-Film Lithium Niobate Modulators. *Adv. Photonics Res.* **2023**, *4*, 2300229. [CrossRef]
22. Honardoost, A.; Abdelsalam, K.; Fathpour, S. Rejuvenating a Versatile Photonic Material: Thin-Film Lithium Niobate. *Laser Photonics Rev.* **2020**, *14*, 2000088. [CrossRef]
23. Zhu, D.; Shao, L.; Yu, M.; Cheng, R.; Desiatov, B.; Xin, C.J.; Hu, Y.; Holzgrafe, J.; Ghosh, S.; Shams-Ansari, A.; et al. Integrated photonics on thin-film lithium niobate. *Adv. Opt. Photonics* **2021**, *13*, 242. [CrossRef]
24. Sattari, H.; Prieto, I.; Zarebidaki, H.; Leo, J.; Choong, G.; Arefi, F.; Orvietani, M.; Della Torre, A.; Mettraux, A.; Dubois, F.; et al. Thin-film lithium niobate PICs: Advancements and potential applications in telecom and beyond. In *Integrated Photonics Platforms III*; SPIE: Strasbourg, France, 2024; pp. 86–89.
25. Burla, M.; Hoessbacher, C.; Heni, W.; Haffner, C.; Fedoryshyn, Y.; Werner, D.; Watanabe, T.; Massler, H.; Elder, D.L.; Dalton, L.R.; et al. 500 GHz plasmonic Mach-Zehnder modulator enabling sub-THz microwave photonics. *APL Photonics* **2019**, *4*, 056106. [CrossRef]
26. Andrianopoulos, E.; Tokas, K.; Felipe, D.d.; Theurer, M.; Weigel, M.; Pagano, A.; Kanta, K.; Kresse, M.; Megas, G.; Tsokos, C.; et al. Integrated 800 Gb/s O-band WDM optical transceiver enabled by hybrid InP-polymer photonic integration. *J. Opt. Commun. Netw.* **2024**, *16*, D44. [CrossRef]
27. Qian, T.; Deumer, M.; Gupta, Y.D.; Nellen, S.; Schuler, B.; Conradi, H.; Kresse, M.; Reck, J.; Mihov, K.; Kleinert, M.; et al. Hybrid Polymer THz Receiver PIC with Waveguide Integrated Photoconductive Antenna: Concept and 1st Characterization Results. In Proceedings of the Optical Fiber Communication Conference (OFC) 2022, San Diego, CA, USA, 6–10 March 2022; W3D.6, ISBN 978-1-55752-466-9.
28. Conradi, H.; Qian, T.; Kresse, M.; Reck, J.; de Felipe, D.; Kleinert, M.; Weigel, M.; Zawadzki, C.; Keil, N.; Schell, M. Tunable DBR Laser with Integrated Optical Isolator. In Proceedings of the Optical Fiber Communications Conference (OFC) 2021, San Francisco, CA, USA, 6–10 June 2021; pp. 1–3, ISBN 978-1-943580-86-6.
29. Kleinert, M.; de Felipe, D.; Conradi, H.; Kresse, M.; Jehle, L.; Weigel, M.; Qian, T.; Mihov, K.; Reck, J.; Zawadzki, C.; et al. Hybrid Polymer Integration for Communications, Sensing and Quantum Technologies from the Visible to the Infrared. In Proceedings of the 2021 European Conference on Optical Communication (ECOC), Bordeaux, France, 13–16 September 2021; pp. 1–4, ISBN 978-1-6654-3868-1.
30. Yao, J. Microwave Photonic Systems. *J. Lightwave Technol.* **2022**, *40*, 6595–6607. [CrossRef]
31. Yao, J. Microwave Photonics. *J. Lightwave Technol.* **2009**, *27*, 314–335. [CrossRef]
32. Fan, Y.; Oldenbeuving, R.M.; Roeloffzen, C.G.; Hoekman, M.; Geskus, D.; Heideman, R.G.; Boller, K.-J. 290 Hz Intrinsic Linewidth from an Integrated Optical Chip-based Widely Tunable InP-Si3N4 Hybrid Laser. In Proceedings of the Conference on Lasers and Electro-Optics, Washington, DC, USA, 25–29 June 2017. JTh5C.9.

33. De Felipe, D.; Zhang, Z.; Brinker, W.; Kleinert, M.; Novo, A.M.; Zawadzki, C.; Moehrle, M.; Keil, N. Polymer-Based External Cavity Lasers: Tuning Efficiency, Reliability, and Polarization Diversity. *IEEE Photonics Technol. Lett.* **2014**, *26*, 1391–1394. [CrossRef]
34. Xiang, C.; Morton, P.A.; Bowers, J.E. Ultra-narrow linewidth laser based on a semiconductor gain chip and extended Si_3N_4 Bragg grating. *Opt. Lett.* **2019**, *44*, 3825–3828. [CrossRef]
35. Kharas, D.; Plant, J.J.; Loh, W.; Swint, R.B.; Bramhavar, S.; Heidelberger, C.; Yegnanarayanan, S.; Juodawlkis, P.W. High-Power (>300 mW) On-Chip Laser With Passively Aligned Silicon-Nitride Waveguide DBR Cavity. *IEEE Photonics J.* **2020**, *12*, 1504612. [CrossRef]
36. Zhang, Y.; Yang, S.; Guan, H.; Lim, A.E.-J.; Lo, G.-Q.; Magill, P.; Baehr-Jones, T.; Hochberg, M. Sagnac loop mirror and micro-ring based laser cavity for silicon-on-insulator. *Opt. Express* **2014**, *22*, 17872–17879. [CrossRef]
37. Tang, R.; Kita, T.; Yamada, H. Narrow-spectral-linewidth silicon photonic wavelength-tunable laser with highly asymmetric Mach-Zehnder interferometer. *Opt. Lett.* **2015**, *40*, 1504–1507. [CrossRef] [PubMed]
38. Debregeas, H.; Ferrari, C.; Cappuzzo, M.A.; Klemens, F.; Keller, R.; Pardo, F.; Bolle, C.; Xie, C.; Earnshaw, M.P. 2kHz Linewidth C-Band Tunable Laser by Hybrid Integration of Reflective SOA and SiO_2 PLC External Cavity. In Proceedings of the 2014 International Semiconductor Laser Conference, Palma de Mallorca, Spain, 7–10 September 2014; pp. 50–51.
39. Maier, P.; Chen, Y.; Xu, Y.; Bao, Y.; Blaicher, M.; Geskus, D.; Dekker, R.; Liu, J.; Dietrich, P.-I.; Peng, H.; et al. Sub-kHz-Linewidth External-Cavity Laser (ECL) With Si_3N_4 Resonator Used as a Tunable Pump for a Kerr Frequency Comb. *J. Lightwave Technol.* **2023**, *41*, 3479–3490. [CrossRef]
40. Xiang, C.; Guo, J.; Jin, W.; Wu, L.; Peters, J.; Xie, W.; Chang, L.; Shen, B.; Wang, H.; Yang, Q.-F.; et al. High-performance lasers for fully integrated silicon nitride photonics. *Nat. Commun.* **2021**, *12*, 6650. [CrossRef] [PubMed]
41. De Felipe, D.; Happach, M.; Nellen, S.; Brinker, W.; Kleinert, M.; Zawadzki, C.; Möhrle, M.; Keil, N.; Göbel, T.; Petermann, K.; et al. Hybrid polymer/InP dual DBR laser for 1.5 μm continuous-wave terahertz systems. In Proceedings of the Terahertz, RF, Millimeter, and Submillimeter-Wave Technology and Applications IX SPIE OPTO, San Francisco, CA, USA, 25 February 2016; Sadwick, L.P., Yang, T., Eds.; SPIE: San Francisco, CA, USA, 2016; p. 974719.
42. Wang, M.; Fang, Z.; Zhang, H.; Lin, J.; Zhou, J.; Huang, T.; Zhu, Y.; Li, C.; Yu, S.; Fu, B.; et al. Recent Progresses on Hybrid Lithium Niobate External Cavity Semiconductor Lasers. *Materials* **2024**, *17*, 4453. [CrossRef]
43. Zhang, L.; Li, X.; Luo, W.; Shi, J.; Sun, K.; Qiu, M.; Zheng, Z.; Kong, H.; Zhou, J.; Zhang, C.; et al. Review of 1.55 μm Waveband Integrated External Cavity Tunable Diode Lasers. *Photonics* **2023**, *10*, 1287. [CrossRef]
44. Qian, T.; Liebermann, P.; Kresse, M.; Mihov, K.; Schell, M. A hybrid photonic integrated signal source with >1.5 THz continuous tunability and <0.25 GHz accuracy for mmW/THz applications. In Proceedings of the 24th European Conference on Integrated Optics, Enschede, The Netherlands, 19–21 April 2023.
45. Pintus, P.; Guo, J.; Jin, W.; Tran, M.A.; Peters, J.; Xiang, C.; Liang, J.; Ohanian, O.J.; Bowers, J.E. 225 GHz Mode-Hop-Free Tuning with a Narrow Linewidth Integrated InP/Si Laser. In *CLEO: Science and Innovations*; Optica Publishing Group: Washington, DC, USA, 2022; ISBN 978-1-957171-05-0.
46. van Rees, A.; Fan, Y.; Geskus, D.; Klein, E.J.; Oldenbeuving, R.M.; van der Slot, P.J.M.; Boller, K.-J. Ring resonator enhanced mode-hop-free wavelength tuning of an integrated extended-cavity laser. *Opt. Express* **2020**, *28*, 5669–5683. [CrossRef]
47. Lang, R. Injection locking properties of a semiconductor laser. *IEEE J. Quantum Electron.* **1982**, *18*, 976–983. [CrossRef]
48. Okawachi, Y.; Kim, B.Y.; Lipson, M.; Gaeta, A.L. Chip-scale frequency combs for data communications in computing systems. *Optica* **2023**, *10*, 977. [CrossRef]
49. Heffernan, B.M.; Kawamoto, Y.; Maekawa, K.; Greenberg, J.; Amin, R.; Hori, T.; Tanigawa, T.; Nagatsuma, T.; Rolland, A. 60 Gbps real-time wireless communications at 300 GHz carrier using a Kerr microcomb-based source. *APL Photonics* **2023**, *8*, 066106. [CrossRef]
50. Chang, L.; Liu, S.; Bowers, J.E. Integrated optical frequency comb technologies. *Nat. Photonics* **2022**, *16*, 95–108. [CrossRef]
51. Weng, W.; Kaszubowska-Anandarajah, A.; He, J.; Lakshmijayasimha, P.D.; Lucas, E.; Liu, J.; Anandarajah, P.M.; Kippenberg, T.J. Gain-switched semiconductor laser driven soliton microcombs. *Nat. Commun.* **2021**, *12*, 1425. [CrossRef]
52. Sinatkas, G.; Christopoulos, T.; Tsilipakos, O.; Kriezis, E.E. Electro-optic modulation in integrated photonics. *J. Appl. Phys.* **2021**, *130*, 010901. [CrossRef]
53. Yu, J.; Li, K.; Chen, Y.; Zhao, L.; Huang, Y.; Li, Y.; Ma, J.; Shan, F. Terahertz-Wave Generation Based on Optical Frequency Comb and Single Mach-Zehnder Modulator. *IEEE Photonics J.* **2020**, *12*, 7900808. [CrossRef]
54. Wang, X.; Wang, W.; Zhao, A.; Zhu, D.; Pan, S. Stable OFC generator based on cascaded phase modulators. In Proceedings of the Eleventh International Conference on Information Optics and Photonics (CIOP 2019), Xi'an, China, 6–9 August 2019; Wang, H., Ed.; SPIE: Bellingham, WA, USA, 2019; p. 56, ISBN 9781510631731.
55. Weng, W.; Kaszubowska-Anandarajah, A.; Liu, J.; Anandarajah, P.M.; Kippenberg, T.J. Frequency division using a soliton-injected semiconductor gain-switched frequency comb. *Sci. Adv.* **2020**, *6*, eaba2807. [CrossRef] [PubMed]

56. Conradi, H.; Felipe, D.d.; Kleinert, M.; Nuck, M.; Zawadzki, C.; Scheu, A.; Keil, N.; Schell, M. High Isolation Optical Isolator: A new Building Block for PolyBoard Photonic Integrated Circuits. In Proceedings of the 2018 European Conference on Optical Communication (ECOC), Rome, Italy, 23–27 September 2018; pp. 1–3, ISBN 978-1-5386-4862-9.
57. Nagatsuma, T.; Ito, H.; Ishibashi, T. High-power RF photodiodes and their applications. *Laser Photonics Rev.* **2009**, *3*, 123–137. [CrossRef]
58. Nellen, S.; Ishibashi, T.; Deninger, A.; Kohlhaas, R.B.; Liebermeister, L.; Schell, M.; Globisch, B. Experimental Comparison of UTC- and PIN-Photodiodes for Continuous-Wave Terahertz Generation. *J. Infrared Millimeter Terahertz Waves* **2020**, *41*, 343–354. [CrossRef]
59. Deumer, M.; Nellen, S.; Lauck, S.; Keyvaninia, S.; Berrios, S.; Kieper, M.; Schell, M.; Kohlhaas, R.B. Ultra-Wideband PIN-PD THz Emitter with > 5.5 THz Bandwidth. *J. Infrared Millimeter Terahertz Waves* **2024**, *45*, 831–840. [CrossRef]
60. Seddon, J.P.; Natrella, M.; Lin, X.; Graham, C.; Renaud, C.C.; Seeds, A.J. Photodiodes for Terahertz Applications. *IEEE J. Sel. Top. Quantum Electron.* **2022**, *28*, 3801612. [CrossRef]
61. Deumer, M.; Breuer, S.; Berrios, S.; Keyvaninia, S.; Schwanke, G.; Schwenson, L.; Lauck, S.; Liebermeister, L.; Nellen, S.; Schell, M.; et al. Continuous wave THz receivers with rhodium-doped InGaAs enabling 132 dB dynamic range. *Opt. Express* **2024**, *32*, 29855–29867. [CrossRef]
62. Deumer, M.; Breuer, S.; Kohlhaas, R.; Nellen, S.; Liebermeister, L.; Lauck, S.; Schell, M.; Globisch, B. Continuous wave terahertz receivers with 4.5 THz bandwidth and 112 dB dynamic range. *Opt. Express* **2021**, *29*, 41819. [CrossRef]
63. Fernandez Olvera, A.; Roggenbuck, A.; Dutzi, K.; Vieweg, N.; Lu, H.; Gossard, A.; Preu, S. International System of Units (SI) Traceable Noise-Equivalent Power and Responsivity Characterization of Continuous Wave ErAs:InGaAs Photoconductive Terahertz Detectors. *Photonics* **2019**, *6*, 15. [CrossRef]
64. Nellen, S.; Lauck, S.; Peytavit, E.; Szriftgiser, P.; Schell, M.; Ducournau, G.; Globisch, B. Coherent Wireless Link at 300 GHz With 160 Gbit/s Enabled by a Photonic Transmitter. *J. Lightwave Technol.* **2022**, *40*, 4178–4185. [CrossRef]
65. Nellen, S.; Lauck, S.; Schwanke, G.; Deumer, M.; Kohlhaas, R.B.; Liebermeister, L.; Schell, M.; Globisch, B. Radiation pattern of planar optoelectronic antennas for broadband continuous-wave terahertz emission. *Opt. Express* **2021**, *29*, 8244–8257. [CrossRef] [PubMed]
66. Deumer, M.; Nellen, S.; Berrios, S.; Breuer, S.; Keyvaninia, S.; Liebermeister, L.; Schell, M.; Kohlhaas, R.B. Advancing terahertz photomixers through the integration of photoconductive antennas with optical waveguides. *APL Photonics* **2025**, *10*, 036105. [CrossRef]
67. Deumer, M.; Nellen, S.; Breuer, S.; Kohlhaas, R.B.; Schwenson, L.; Wenzel, K.; Liebermeister, L.; Schell, M.; Globisch, B. Waveguide-integrated photoconductive THz receivers. In Proceedings of the 2022 47th International Conference on Infrared, Millimeter and Terahertz Waves (IRMMW-THz), Delft, The Netherlands, 28 August–2 September 2022; pp. 1–2.
68. Nellen, S.; Qian, T.; Schwanke, G.; Lauck, S.; Felipe, D.d.; Kleinert, M.; Deumer, M.; Liebermeister, L.; Baier, M.; Globisch, B.; et al. Photonic-enabled beam steering at 300 GHz using a photodiode-based antenna array and a polymer-based optical phased array. *Opt. Express* **2022**, *30*, 44701–44716. [CrossRef]
69. Heck, M.J. Highly integrated optical phased arrays: Photonic integrated circuits for optical beam shaping and beam steering. *Nanophotonics* **2017**, *6*, 93–107. [CrossRef]
70. Che, M.; Matsuo, Y.; Kanaya, H.; Ito, H.; Ishibashi, T.; Kato, K. Optoelectronic THz-Wave Beam Steering by Arrayed Photomixers With Integrated Antennas. *IEEE Photonics Technol. Lett.* **2020**, *32*, 979–982. [CrossRef]
71. Uchendu, I.; Kelly, J.R. Survey of Beam Steering Techniques Available for Millimeter Wave Applications. *PIER B* **2016**, *68*, 35–54. [CrossRef]
72. Guo, Y.; Guo, Y.; Li, C.; Zhang, H.; Zhou, X.; Zhang, L. Integrated Optical Phased Arrays for Beam Forming and Steering. *Appl. Sci.* **2021**, *11*, 4017. [CrossRef]
73. Zhang, H.; Zhang, Z.; Lv, J.; Peng, C.; Hu, W. Fast beam steering enabled by a chip-scale optical phased array with 8×8 elements. *Opt. Commun.* **2020**, *461*, 125267. [CrossRef]
74. Lu, P.; Haddad, T.; Tebart, J.; Roeloffzen, C.; Stöhr, A. Photonic Integrated Circuit for Optical Phase Control of 1×4 Terahertz Phased Arrays. *Photonics* **2022**, *9*, 902. [CrossRef]
75. Carpintero, G.; Headland, D.; Schwanke, G.; Deumer, M.; Nellen, S.; Lauck, S.; Liebermeister, L.; Spanidou, K.; Ali, M.; Rivera, A.; et al. Photonic-Enabled Terahertz Phase Arrays Using Dielectric Rod Waveguides for 6G Wireless Communications. In Proceedings of the 2024 IEEE/MTT-S International Microwave Symposium—IMS 2024, Washington, DC, USA, 16–21 June 2024; pp. 74–77.
76. Fu, X.; Yang, F.; Liu, C.; Wu, X.; Cui, T.J. Terahertz Beam Steering Technologies: From Phased Arrays to Field-Programmable Metasurfaces. *Adv. Opt. Mater.* **2020**, *8*, 1900628. [CrossRef]
77. Shao, Z.; Shen, R.; Xia, W.; Ghasempour, Y.; Sengupta, K.; Rangan, S. A Hybrid Antenna-Metasurface Architecture for mmWave and THz Massive MIMO. In Proceedings of the 2023 57th Asilomar Conference on Signals, Systems, and Computers, Pacific Grove, CA, USA, 29 October–1 November 2023; pp. 1610–1616, ISBN 979-8-3503-2574-4.

78. Afeef, L.; Arslan, H. Beam Squint Effect in Multi-Beam mmWave Massive MIMO Systems. In Proceedings of the 2022 IEEE 96th Vehicular Technology Conference (VTC2022-Fall), London, UK, 26–29 September 2022; pp. 1–5, ISBN 978-1-6654-5468-1.
79. Wang, B.; Gao, F.; Xing, C.; An, J.; Li, G.Y. Wideband Beamforming for Hybrid Phased Array Terahertz Systems. In Proceedings of the ICC 2021—IEEE International Conference on Communications, Montreal, QC, Canada, 14–23 June 2021; pp. 1–6, ISBN 978-1-7281-7122-7.
80. Wang, B.; Jian, M.; Gao, F.; Li, G.Y.; Lin, H. Beam Squint and Channel Estimation for Wideband mmWave Massive MIMO-OFDM Systems. *IEEE Trans. Signal Process.* **2019**, *67*, 5893–5908. [CrossRef]
81. Camponeschi, F.; Gemmato, V.; Scotti, F.; Rinaldi, L.; Mohammad, A.W.; Roeloffzen, C.G.; van Dijk, P.W.; Ghelfi, P. Multibeam Beamforming Demonstration of a Hybrid Integrated Photonic Module for a Synthetic Aperture Radar Receiver. *J. Lightwave Technol.* **2024**, *42*, 7604–7611. [CrossRef]
82. Reza, M.; Mohammad, A.; Serafino, G.; Roeloffzen, C.; van Dijk, P.; Mohammadhosseini, H.; Abbasi, A.; Desoete, B.; Ghelfi, P. Design of an Integrated-Photonics RF Beamformer for Multi-Beam Satellite Synthetic Aperture Radar. In Proceedings of the 2020 International Topical Meeting on Microwave Photonics (MWP), Matsue, Japan, 24–26 November 2020; pp. 87–90.
83. Tsokos, C.; Andrianopoulos, E.; Raptakis, A.; Lyras, N.; Gounaridis, L.; Groumas, P.; Timens, R.B.; Visscher, I.; Grootjans, R.; Wefers, L.S.; et al. True Time Delay Optical Beamforming Network Based on Hybrid Inp-Silicon Nitride Integration. *J. Lightwave Technol.* **2021**, *39*, 5845–5854. [CrossRef]
84. Andrianopoulos, E.; Lyras, N.K.; Qian, T.; Deumer, M.; Megas, G.; Schwanke, G.; Gupta, Y.D.; Groumas, P.; Tegegne, Z.G.; Schuler, B.; et al. Demonstration of On-Chip Optical Frequency Comb Generation and Optical Injection Locking. In Proceedings of the Optical Fiber Communication Conference (OFC) 2024, San Diego, CA, USA, 24–28 March 2024.

MDPI AG
Grosspeteranlage 5
4052 Basel
Switzerland
Tel.: +41 61 683 77 34

Photonics Editorial Office
E-mail: photonics@mdpi.com
www.mdpi.com/journal/photonics

www.ingramcontent.com/pod-product-compliance
Lightning Source LLC
Chambersburg PA
CBHW061826210926
51850CB00006B/818